A Course in Dynamic Meteorology

Foreword

I have gone through the manuscript of the book entitled "A Course in Dynamic Meteorology" by Mr. Pandharinath very carefully. It is an excellent work and I would be delighted to see it in print.

The weather plays a significant role in almost every human activity and its importance is being increasingly realised in our country which is so dependent on the monsoons. What is not often understood is that the science that governs atmospheric processes is extremely complex. Numerical weather modelling involves extensive use of equations and mathematical techniques. Dynamic Meteorology is that fundamental branch of atmospheric science which enables us to quantify atmospheric motion and make accurate predictions 5 to 10 days ahead.

Atmospheric science, meteorology, oceanography, and environmental science are being taught at several Indian Universities and an increasing number of students are opting for such courses at the postgraduate level. There are also many institutions in the country which impart formal training to those who intend to take up a career in meteorology. Unfortunately, there is no text book available that deals with dynamic meteorology for the tropical regions and particularly in the Indian context.

Mr. Pandharinath, who is an experienced meteorologist, has done great service by writing book. He has developed the subject systematically and covered all important aspects. I am confident that this book will meet a longstanding demand of both teachers and students, and it will be welcomed addition to many libraries.

I strongly recommend the publication of Mr. Pandharinath's book.

R.R. Kelkar

Director General of Meteorology (Retd.),
Indian Meteorological Dept.,
Pune.

12th September 2004

Preface

The book is designed as an introduction to Dynamic Meteorology for under graduate and postgraduate level students, who are familiar with Calculus and Thermodynamic.

Meteorology is a specialised subject and encompass fundamentals of Physics, Mathematics, and host of other subjects. Though there are a few books in India which deal with the Science of Weather but hardly there is any book dealing with Dynamic Meteorology.

The Sun is the source of energy for all weather systems and life on Earth. The various estimates in incident solar energy, terrestrial radiation, atmospheric stratification and various atmospheric thermodynamic processes are described.

The subject has been treated lucidly but there are several approximations and hence proofs are not rigorous. The physical processes involved in the atmosphere are treated with the fundamental equations of motion, hydrostatic equation, ideal gas equation, thermodynamic equations, equation of continuity or mass conservation, convergence, divergence, circulation, and vorticity.

In the end, general circulation of the atmosphere is described as zonal and meridional averages.

A number of books were referred to collect relevant material but mostly reference was made of WMO technical notes/publications.

The author is greatly indebted to Dr. R.R.Kelkar, Director General of Meteorology (Retd.), Indian Meteorological Department for his encouragement in writing this book. Further, the author expresses his indebtedness to Shri Nikhil Shah, Proprietor of BS Publications, without whose patronage the book would not have been published.

Any suggestions for the improvement of the book and corrections will be welcomed and thankfully acknowledged.

Author

Published by :

BSP BS Publications

A unit of **BSP Books Pvt., Ltd.**

4-4-309/316, Giriraj Lane, Sultan Bazar,
Hyderabad - 500 095
Phone : 040 - 23445605, 23445688
e-mail : info@bspbooks.net

ISBN : 978-93-52300-00-6 (HB)

A Course in Dynamic Meteorology

Navale Pandharinath
M.Sc. (Maths), M.Sc. (Statistics)

Director (Retd.),
Indian Meteorological Deparment,
Hyderabad.

BSP **BS Publications**
A unit of **BSP Books Pvt., Ltd.**

4-4-309/316, Giriraj Lane, Sultan Bazar,
Hyderabad - 500 095
Phone : 040 - 23445605, 23445688

Contents

CHAPTER 8

The Operator ∇ (del)

CHAPTER 9

The Continuity Equation

CHAPTER 10

Mathematical Equations of Motion

CHAPTER 1

The Sun

1.1 Introduction

Sun has been revered from the dawn of civilization as the prime source of light, warmth and life itself. The orientation of temples, churches were made to face east, the cardinal direction of the sun rise as a mark of worship. Some of the important names of the sun god are: Surya (Indian), Mithras (Iranian), Helios (Greek), Sol (Roman), Shamosh (Assyrian), Ormuzd (Persian) and Tezcatlipoca (Mexican).

The sun is a normal size yellow star, located on one of the spiral arms of the Milky way galaxy at a distance of about 3×10^4 light years from the centre of the galaxy (Light year is an astronomical unit of distance traveled by light in one year).

The bright luminous region of the visible sun, visible to the naked eye, is called photosphere. It consists of very hot gases under high pressure. Telescopic observations of the photosphere indicate mottled surface with fine grains. In addition to this it frequently shows dark (cool) regions, called sun spots and bright (hot) regions called faculae. If we obscure the photosphere (like in the total solar eclipse) three additional layers are observed outside the photosphere. They are called (i) the reversing layer (ii) the chromosphere (iii) the corona. These are shown in the Fig. 1.1 and described below.

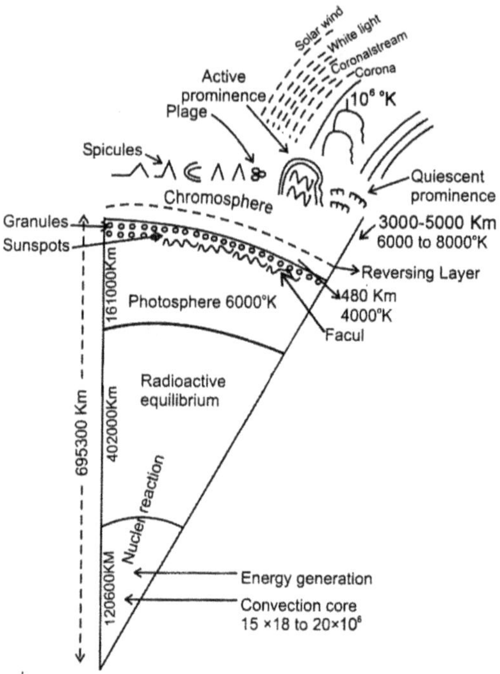

Fig. 1.1 Schematic representation of strcuture and activities of the sun with the exaggerated photosphere, chromosphere.

1.2 Photosphere

It is defined as the layer of the Sun that produces radiations of the visible part of the solar spectrum. However the energy is produced in the deeper central region of the Sun's sphere. Photosphere consists of hot gases under high pressures. The surface of the photosphere will be constantly changing without any regularity or rhythm. It appears as mottled or granular which is due to the boiling motion of the gases. The dimensions of the granules vary from 200 to 1200 Km with a life span of about 10 minutes, after which it changes its shape with the formation of another granule. Granules are not seen by naked eye. At any instant the whole surface of the Sun contains about a million granules. In addition to granules, the photosphere shows bright (hot) regions called faculae and dark (cold) regions called Sunspots, which are regions of strong magnetic fields.

1.2.1 Chromosphere / Sun's Atmosphere

During a total solar eclipse a reddish ring of light is observed around the Sun's photosphere which is called chromosphere or colour sphere of the Sun. The depth of the chromosphere varies from about 3000 to 5000 Km above the photosphere. It consists of hydrogen and helium gases at comparatively low pressure but has temperature about 20,000 °K. Fibrilles and Spicules are observed in the chromosphere. Extended structures approaching granules are fibrilles which are observed around the center of the solar disc, while spicules are observed at the edge of the disc (In Latin spicule means 'edge'). The density of the gases decrease continuously with height from the center. Temperature rises in the chromosphere slowly from 4,400 °K. Between the photosphere (temperature about 6,000 °K) and chromosphere (of temperature 1.5 to 2.0 millions °K) there is a cooler gas (temperature about 4,000 °K). This is called reversing layer. This reversing layer is responsible for Fraunhofer absorption lines. Above this reversing layer temperature rises.

Individual spicules have life period of 3 to 5 minutes, during which they elongate to a height of 12,000 km above the base of chromosphere with velocity 25 to 30 km/s. Their diameter is about 750 km.

1.2.2 Corona

Above the chromosphere lay corona of the sun, which spreads far out into space. It has a silvery white colour. The temperature of the corona is about 1,000,000 °K (1.5 to 2.0 million °K). During the total solar eclipse corona shows spectacular light of hollo. Solar winds originate in the corona of speed 100 Km/s or more. The plasma particles in corona attain such velocities which exceed the escape velocity of the Sun's atmosphere. This is called Solar wind. Solar wind depends on the solar activity. Associated with solar wind, solar mass, about 4×10^8 Kg/s escape from the Sun.

1.2.3 Solar Activity

If the frequency of observable features of the Sun (such as Sunspots, solar wind, solar flares etc.) are more than normal then the Sun is said to be 'active' or 'disturbed', but when they are absent the Sun is said to be quiet. There seems to be a solar activity cycle of about 11 years, which is based on the observations of Sunspots. Sunspots begin to appear at latitude 35° on both hemispheres of the Sun. This is Sunspot minima phase. As the cycle advances the number of spots increase both in size and number and spread to solar equator. This is Sunspot maxima phase. Subsequently these spots decrease gradually. Again a new cycle begins when Sunspots reappear at latitude 35°.

The average duration of time between two successive Sunspot minima (or maxima) is about 11 years. As said earlier Sunspots are regions of cooler temperature and hence they are regions of lower energy. In addition to the Sunspots some localized features are observed where the energy emission is more than normal. These are called active regions.

A Facula appears before Sunspots and persists even after the visible appearance of the Sunspot. Faculae are also noticed at high solar latitudes in the absence of Sunspots.

Associated with the Sunspot groups in the photosphere, increased solar activity is also observed in chromosphere due to large bright regions called plages. Some times clouds of bright gas observed projecting outwards from chromosphere which are called prominences. In the neighbourhood of active regions prominences change rapidly in size, appearance and brightness. The sun's thermal radiation remains constant and is not effected by solar activity, but affects short wave radiation (wavelength less than 100 nm) (1 nm = 10^{-9} m). The phenomena of solar activity is still not clearly understood.

Depending on solar activity the intensity of solar wind and the size of corona changes. Sunspots are always accompanied by faculae around them. The magnetic fields of Sunspots observed to exceed a thousand times as compared to the magnetic field out side the Sunspots. The polarity of the magnetic field in Sunspot areas change in 22 years, which is twice the period of Sunspot cycle. There is no comprehensive theory of solar activity. During 70 years between 1645 to 1715, no Sunspots were observed and before 1645 there does not appear 11 year Sunspot cycle. These facts were brought the light by E. Maunder, the superintendent Greenwich observatory, in 1890. The Sunspot minima or complete absence of it during the above 70 years is named Maunder Minima.

1.2.4 Auroras

Auroral phenomena develop due to solar wind splashes which enter into the upper atmosphere of the earth.

1.2.5 Cosmic Rays

The solar activity influences the intensity of cosmic rays (high energy particles of the galactic origin) that are falling on the earth. Cosmic rays propagate along the galactic magnetic field and are deviated by the earth magnetic field. During quiet Sun the solar wind is weaker, as a consequence the solar magnetosphere shrinks and thus allows the cosmic rays to penetrate to the earth.

The Sun's thermal radiation remains constant and is not effected by the solar activity. However solar activity affects short wave radiation (λ-less than 100 nm). The Phenomena of solar activity is still not clearly understood.

The effect of solar gravity on the earth's biosphere is insignificant compared to the gravity of the earth itself. ($g_{\odot} << g_{\otimes}$ on earth biosphere).

1.2.6 Solar Flares

Solar flares are the most spectacular activity of the chromosphere. It has a short period of life. It consists of intense eruption and brightness in its neighborhood for a few minutes. The brightness gradually subsides within few hours and returns to original state. The flares are the result of gigantic nuclear explosions that occur inside the Sun. There may be a few large flares in a year but there maybe hundred (100) odd small flares daily. The large flares emit electromagnetic radiation along with UV, visible, IR, X-rays, radio waves, and high energy charged particles of plasma. These particles may have velocities as much as 10^5 km/s. The plasma particles as said earlier consists of 91% protons and 9% ionised helium. These plasma particles on reaching the earth interacts with atmosphere and earth's magnetic field. It causes ionospheric storm (High frequency radio blackout) and magnetic disturbances. The solar wind associated with these flares have velocities about 300 to 800 km/s.

1.2.7 The Heterogeneous Rotation of the Sun

The Sun itself rotates in forward direction (coinciding with planets revolution) about an axis which makes an angle of about 7° 15' with the normal to the ecliptic. The north pole of the Sun is seen on earth between 7 June to 7 December and the south pole of the Sun is seen during 8 December to 6 June. Unlike the earth (which has its all particles same angular velocity) the gaseous sphere of the Sun rotates a full turn at its equator in 25 days and at its poles in 35 days. This heterogeneous rotation is called differential. The angular velocity of Sun's rotation also changes with depth. However the connection of the Sun, its magnetic field and differential rotation are all related and interact with one another.

1.2.8 Characteristics of the Sun's Radiation

In very hot materials (negatively charged particles) electrons are in continuous motion at high speeds. These high speed particles collide with one another and cause vibrations. This gives rise the electromagnetic waves. Very high speed particles vibrate at high speeds produce short waves. Slower particles collision produce long waves. In the Sun the gases which are at high temperature (6,000 °K) produce electrically charged particles with different speeds. As a consequence of their collisions electromagnetic waves of different wavelengths are emitted by the Sun.

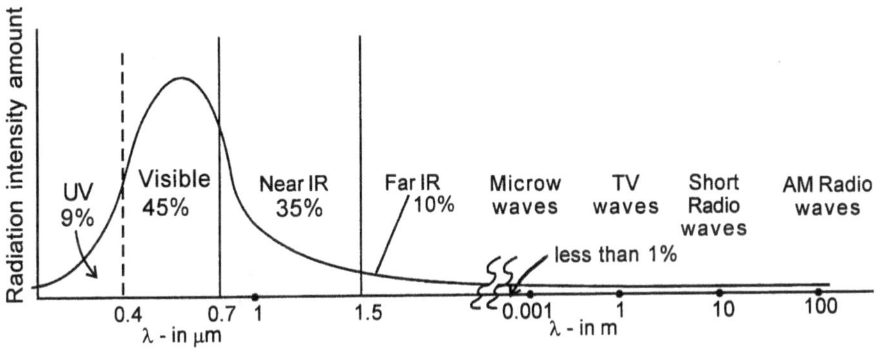

Fig. 1.2 Sun's electromagnetic spectrum.

These waves constitute the electromagnetic spectrum of the Sun, which extends from gamma rays ($\lambda = 10^{-13}$ m) to radio waves ($\lambda = 10^3$ m). This range is 10^{16}m. However about 99% of solar radiation lies between 0.15 – 4.0 μm. Of this

1. about 9% lies in UV-radiation (0.15 μm to 0.38 μm). This causes photo chemical effects, bleaching, Sunburn etc.
2. 44% lies in visible light [0.38 μm to 0.7 μm : violet to red].
3. 46% lies in IR (0.7 μm to 2.3 μm). This causes radiant heat with some chemical effects.

All these radiations travel with speed 3×10^8 m/s in vacuum. The Sun emits energy at an average surface temperature of 6,000 °K while the earth emits at its average surface temperature of 288 °K (15 °C).

According to Wien's Law :

$$\lambda_{max} = \frac{\text{Constant}}{T}$$

Where constant $= 2897\ \mu mK \simeq 3000\ \mu mK$

For the Sun : $\lambda_{max} = \dfrac{3000\mu mK}{6000K} = 0.5\ \mu m$

i.e., the Sun emits maximum amount of radiant energy at wavelength of 0.5 μm.

For the earth :

$$\lambda_{max} = \frac{3000\mu mK}{300K} = 10\ \mu m \text{ (taking } 288\ ^oK \simeq 300\ ^oK)$$

i.e., the earth emits maximum amount of radiation energy at wavelength of 10 μm.

Because of these facts the terrestrial radiation is called long wave radiation while the solar radiation is called short wave radiation.

The mean earth-Sun distance is $a_\oplus$ = IAU $\simeq 1.5 \times 10^{11}$ m. The angular diameter of the Sun as observed from the earth is 9.3×10^{-3} radians, i.e., slightly more than half a degree.

$$\frac{\text{Arc}}{\text{Radius}} = \text{Radian}$$

$$\text{Arc} = \text{Radius} \times \text{Radian}$$

Sun's diameter = earth-Sun distance × Angular diameter of the Sun

$$= 1.5 \times 10^{11} \times 9.3 \times 10^{-3} = 13.95 \times 10^8 \text{ m}$$

The radius of the Sun $\simeq 6.97 \times 10^8$ m.

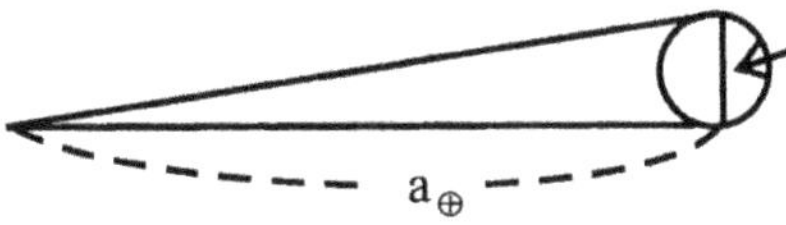

Fig. 1.3 Sun's diameter

1.3 Gravitational Attraction of the Sun

We know the mass of the Sun

$$m_\odot = 1.99 \times 10^{30} \text{ kg}$$
$$R_\odot \simeq 6.96 \times 10^{8} \text{ m.}$$

From the formula

$$g_\odot = \frac{G m_\odot}{R_\odot^2}$$

where $G = 6.67 \times 10^{-11}$ Nm^2/kg

We have $g_\odot \simeq 274$ m/s^2;

or $g_\odot \simeq 30\ g_\oplus$;

where $g_\oplus = 9.81$ m/s^2 (earth's gravitational attraction)

1.3.1 Density of the Sun ($\rho_\odot$)

$$\rho_\odot = \frac{m_\odot}{V_\odot} = \frac{3}{4}\frac{m_\odot}{\pi R_\odot^3}$$

$$= \frac{3}{4} \times \frac{1.99 \times 10^{30}}{\left[6.96 \times 10^{8}\right]^3} \times \frac{7}{22}$$

$$\simeq 1.41 \times 10^{3} \text{ Kg/m}^3$$

$$= 1.41 \times \text{density of water}$$

Note : The density of the earth = 5.5×10^3 Kg/m^3

Definition

The solar constant $S_\odot$: The amount of solar radiation (energy) incident on unit area in unit time on a surface held at right angles to the solar beam at the outer boundary of the atmosphere.

$$S_\odot = 1.36 \times 10^{3} \text{ w/m}^2 = 1360 \text{ Joules/m}^2.$$
$$= 1.359 \times 10^{3} \text{ w/m}^2$$

Unit area = 1 m^2

W = watts

Unit time = 1 second.

The solar constant varies $\pm$ 2% due to variation in the solar output and varies about 3.5% due to changes in the earth-Sun distance.

Solar luminosity :

$$L_{\odot} = 4\pi a_{\otimes}^{2} \times S_{\odot}, \quad \text{where } a_{\otimes} = \text{Earth-Sun distance}$$

$$4\pi \times (1.5 \times 10^{11})^2 \times 1.36 \times 10^3$$

$$\simeq 3.83 \times 10^{26} \text{ W}$$

Sun's brightness S :

$$S = \frac{L_{\odot}}{4\pi R_{\odot}^{2}} = \frac{3.83 \times 10^{26}}{4\pi.\left[6.96 \times 10^{8}\right]^{2}} \ 1$$

$$= \simeq 6.29 \times 10^{7} \text{w/m}^2$$

[The mass of the Sun consists of about 80% Hydrogen, 19% Helium and 1% of other elements. The Sun is a burning gaseous globe with radius about 7×10^5 Km and mass about 2.0×10^{30} Kg, with average temperature, photosphere about 6000 °K]

1.4 Solar Energy

The fusion furnace of the Sun converts about 4 million tons of hydrogen into helium and radiates an amount of energy 3.8×10^{23} kw per second as a byproduct. The solar energy incident on 1 m^2 area held normal to the Suns rays at the outer boundary of the atmosphere in one second is 1.38×10^3 w/m^2. It is virtually constant, and is called solar constant (accepted by IGY).

A house with peak load requirement of 2.5 kw would need only a solar collector of 3.6 m^2 with 100% efficiency. The total solar energy received at the earth in one day is equivalent to 35 lakh nuclear explosions or 10 thousand hurricanes or 100 million thunderstorms or 100 billion Tornadoes. If this energy were stored it will satisfy to world's needs of industries & domestic use for about 100 years.

It is estimated that the Solar Energy Received Per Year (SERPY) is 1.5×10^{15} MWh. 1 SERPY is sufficient for all the needs of world for a period of 28000 years. On a clear sunny day, 1 acre of land in a day receives solar energy which is equivalent to about 16 bbl (bbl = billion barrels) of crude oil. The total solar energy intercepted by the earth

$$= 2.55 \times 10^{18} \text{ cal min}^{-1}$$

$$= 3.67 \times 10^{21} \text{ cal day}^{-1}$$

Insolation (incoming solar radiation) as it passes through the atmosphere part of it is scattered, part of it is absorbed and the rest is transmitted.

According to one estimate, the solar radiation received at the outer boundary of the atmosphere is about 1.74×10^{14} kw. Of this about (30%) 5.2×10^{13} kw is reflected back to space as short wave radiation, 8.2×10^{13} kw (about 47%) is absorbed by the land, ocean and atmospheric system, while the remaining 4.0×10^{13} kw (about 23%) is utilised in driving the hydrologic cycle through evaporation, convection and precipitation. About 3.7×10^{11} kw ($\simeq$ 0.21%) of the incident solar energy is utilized (input into Biosphere) for driving winds, waves, convection and sea currents and another 4.0×10^{9} kw (0.023% of the incident solar energy) is used up in photosynthesis process on earth. A green canopy of plant kingdom consumes solar energy about 0.01 ly/min or 7 watts/m^2.

According to Sellers, the disposition of solar radiation in the earth-atmosphere system is given by the mathematical relation.

$$Q_s = C_r + A_r + C_a + A_a + (Q + q)(1 - \alpha) + (Q + q)\,\alpha \qquad .....(1.1)$$

where Q_s = The amount of solar radiation incident at the top of the atmosphere = 263 kly/yr.

Q = The direct beam solar radiation incident on horizontal surface at the ground = 82 kly/yr

q = Diffused solar radiation incident on horizontal surface at the ground = 58 kly/yr

α = Albedo of the surface $\simeq$ 0.1143

C_r = Solar radiation reflected and scattered back to the space by clouds

A_r = Solar radiation reflected and scattered back to the space by aerosol (air molecules, dust and water vapour) of the atmosphere.

C_a = Solar radiation absorbed by the clouds

A_a = Solar radiation absorbed by the atmospheric aerosols.

Solar energy per unit area expressed in langley (ly) or kilolangley (kly). 1 ly = 1 cal/cm^2, 1 kly = 1000 ly. It is estimated that Sun radiates each minute about 56×10^{26} cal of energy in all.

The solar constant

$$S_\odot = \frac{56 \times 10^{26} \text{ Cal/min}}{4\pi\left(a_\oplus^2\right)} \qquad [1 \text{ cal/cm}^2 \text{ min} = 0.6975 \text{ Kw/m}^2]$$

$$= 697.5 \text{ w/m}^2$$

$$= \frac{56 \times 10^{26}}{4\pi\left(1.5 \times 10^{13} \text{ cm}\right)^2} \simeq 2.0 \text{ ly/min} = 1359 \text{ w/m}^2$$

The total solar energy intercepted by the earth in unit time

$$= \pi R_{\oplus}^2 \; S_{\odot}$$
$$= \pi\,[6.40 \times 10^8 \text{ cm}]^2\, S_{\odot}$$
$$\simeq 2.55 \times 10^{18} \text{ Cal/min}$$
$$\simeq 3.67 \times 10^{21} \text{ Cal/day}$$

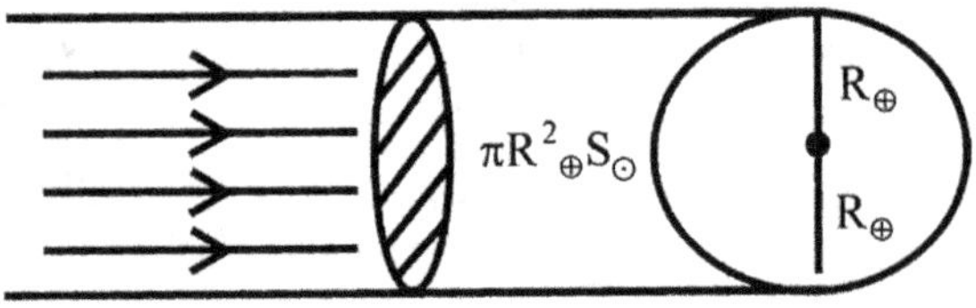

Fig. 1.4 Total solar energy intercepted by the earth.

The amount of solar energy received on unit area in unit time at the top of the atmosphere is

$$Q_s = \frac{\pi R_{\oplus}^2 S_{\odot}}{4\pi R_{\oplus}^2} = \frac{S_{\odot}}{4} = 0.5 \text{ ly/min}$$

$$= \frac{0.5 \times 60 \times 24 \times 365.25}{1000}$$

$$= 263 \text{ kly/year}$$

$$= 339.75 \text{w m}^{-2}$$

$$Q_s = \frac{S_{\odot}}{4} = \frac{1359}{4}$$

$$Q_s = \frac{1359}{4} = 339.75 \text{ w/m}^2$$

$$= 263 \text{ ly/year}$$

The average albedo of the atmosphere = 32%

Therefore the insolation incident at the surface of the earth 68% of 339.75 w/m^2 $\simeq$ 231 w/m^2

The efficiency of the atmosphere in conversion of insolation is only 1 to 2%.

The annual average solar energy received in northern hemisphere for the layer 1000 to 75 hPa is given below

$$\text{PE} = 567.5 \times 10^6 \text{ J m}^{-2}$$

$$\text{Internal energy} = 1674.8 \times 10^6 \text{ J m}^{-2}$$

$$\text{Total PE} = 2242.3 \times 10^6 \text{ J m}^{-2}$$

$$\text{and KE} = 1153.4 \times 10^3 \text{ J m}^{-2}$$

an average amount of KE is

$$= \frac{1153.4 \times 10^3}{2242.3 \times 10^6} = \frac{1}{2000} \text{ of total PE}$$

On an average

1. 78 kly/year (or 30%) of incident solar radiation ($C_r + A_r$) is reflected and scattered back to the space by cloud and atmospheric constituents. Cr $\simeq$ 62 kly/year (24% of Q_s) and $A_r \simeq$ 16 kly/year (6% of Q_s)
2. Absorbed by clouds & atmosphere 45 kly/year (C_a = 8 kly/year or 3% of Q_s, A_a = 37 kly/year or 14% of Q_s)
3. 16 kly/year $(Q + q)\alpha$ is reflected by ground
4. 124 kly/year $(Q + q)(1 - \alpha)$ is absorbed by ground (land and sea).

i.e.,

$$Q_s = C_r + A_r + C_a + A_a + (Q + q)(1 - \alpha) + (Q + q)\alpha$$

viz $$263 = 62 + 16 + 8 + 37 + 124 + 16$$

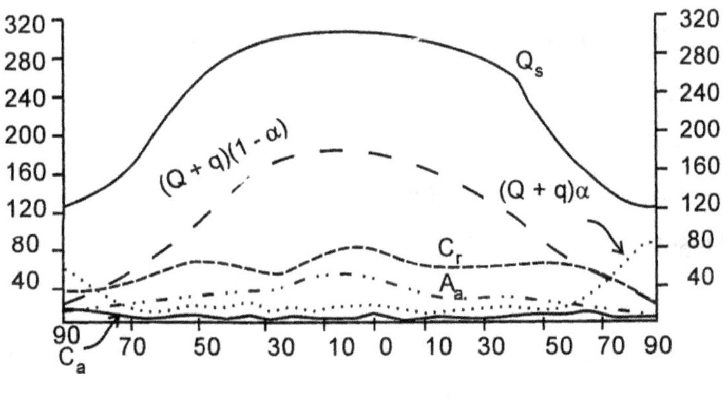

Fig. 1.5 Average insolation balance sheet (not to any scale).

During an average year the insolation balance sheet	kly/year
Q_s = Solar energy incident on the top of the atmosphere	263
C_r = reflected by clouds (24% of Q_s)	63
A_r = reflected by atmospheric aerosols (6 % of Qs)	15
∴ Total reflected back to space by the atmosphere (C_r + A_r) (30% of Q_s)	78
(Q + q) α reflected from the earth's surface	16
∴ Total reflected by the earth-atmospheric system	94
C_a = Absorbed by clouds (3% of Q_s)	8
A_a = Absorbed by atmospheric aerosols and water vapour (14% of Q_s)	37
(C_a + A_a)	45
(Q + q) (1– α) = absorbed at the earth's surface	124
Total solar radiation absorbed by the earth-atmosphere system	169

According another estimate the solar radiation and terrestrial heat balance is as follows.

1. *At the earths surface*

Heat gain

4.7×10^{13} kW (27%)	by direct radiation
2.8×10^{13} kW (or 16%)	by diffuse radiation
6.96×10^{12} kW (or 4%)	by turbulent transfer
Total 8.2×10^{13} kW (or 47%)	

Heat loss

4.18×10^{13} kW or (24%)	by radiation
4.0×10^{13} kW (or 23%)	by conduction (evaporation)
Total 8.2×10^{13} kW (or 47%)	

2. *At the Atmosphere*

Heat gain

2.5×10^{13} kW or (15%)	by absorption of insolation
2.7×10^{13} kW (or 16%)	by absorption of terrestrial radiation
4.0×10^{13} kW (or 23%)	by conduction
Total 9.2×10^{13} kW (or 53%)	

Heat loss

8.5×10^{13} kW (about 50%)	by radiation
6.96×10^{12} kW (about 4%)	by turbulent transfer
Total 9.2×10^{13} kW (about 53%)	

According to another estimate the disposal of insolation (100%) is as follows,

Where E_r, E_a insolation reflected, absorbed by the earths surface other notation same as given above (Insolation 100%)

$$E_r + C_r + A_r = 7\% + 23\% + 6\% = 36\%$$

$$E_a + C_a + A_a = 47\% + 3\% + 14\% = 64\%$$

$$C_r + C_a = 23\% + 3\% = 26\%$$

$$A_r + A_a = 6\% + 14\% = 20\%$$

$$E_r + E_a = 7\% + 47\% = 54\%$$

1. At the Atmosphere :

$$C_r + A_r = 23\% + 6\% = 29\%$$

$$C_a + A_a = 3\% + 14\% = 17\%$$

$$\text{Total } C_r + A_r + C_a + A_a = 46\% \qquad \text{.....(i)}$$

2. At the surface of the earth :

$$E_r + E_a = 7\% + 47\% = 54\% \qquad \text{.....(ii)}$$

3. Insolation for the atmosphere (45 kly-year)

Outgoing long wave radiation for the atmosphere 117 kly/year. Net radiation balance for the atmosphere 45-117 = –72 kly/yr. .....(iii)

4. Insolation for the earth's surface (124 k ly/year)

Outgoing long wave radiation energy for the earth's surface 52 kly/yr. Net radiation balance for the earth's surface 124 – 52 = 72 kly/yr. .....(iv)

5. For the earths-atmosphere system balance

(iii) + (iv) = –72 + 72 = 0. This result shows that earth-atmosphere system radiates the same amount of energy as it is receiving. .....(v)

The approximate average annual latitudinal distribution of solar radiation in the earth atmosphere system is shown in the Fig. 1.5.

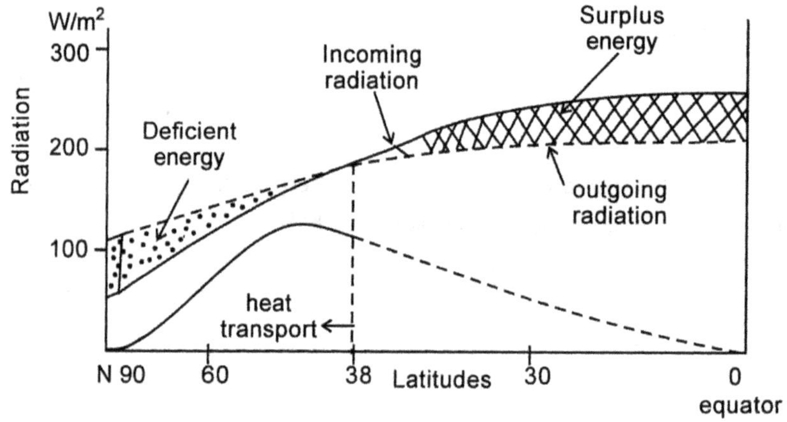

Fig. 1.5 Mean radiation balance northern hemisphere, solid line incoming solar radiation, dotted line outgoing earth radiation.

It is estimated that the earth would absorb 61% insolation if the whole sky were clear, but it would absorb only 28% of insolation if the sky were completely over cast. On an average the mean cloud coverage of the earth is slightly more than 50% and under this condition the earth would absorb only 43% the insolation. Assuming this average condition, the distribution insolation is given below.

Reflection and back scattering	35%	$\simeq$ 92 kly/yr]
Absorption by ozone	2%	$\simeq$ 5 kly/yr]
Absorption by clouds, water vapour, dust etc	20%	$\simeq$ 53 kly/yr]
Absorption by the earth's surface	43%	$\simeq$ 113 kly/yr]

The earth atmosphere together absorbs 65% of solar radiation which warms the earth and atmosphere.

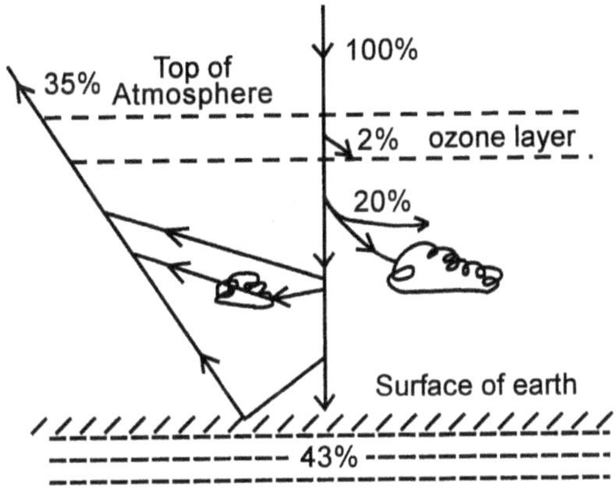

Fig. 1.6 Disposition of insolation.

Questions

1. Describe the physical features of the sun and photosphere including its composition.
2. Describe the chromosphere and corona of the sun. What is reversing layer and its importance?
3. Write short note on : (i) Solar activity, how it effects cosmic rays (ii) solarflares (iii) Heterogeneous rotation of the sun.
4. Describe the electromagnetic spectrum of the sun's radiation.
5. Using Wien's law, find the wavelengths of the sun and the earth which emit maximum amount of radiant energy. Assume the surface temperature of the sun 6000 °K and the earth 300 °K.
6. Find the radius of the sun, assuming mean earth-sun distance as 1.5×10^8 Km and angular diameter of the sun as 9.3×10^{-3} radiance as observed from the earth.

7. Find the density and gravitational attraction of the sun taking the mass of the sun as 1.99×10^{30} Kg and radius 6.96×10^{5} Km, universal gravitation constant $G = 6.67 \times 10^{-11}$ N m^2/kg^2.

8. Define solar constant, langley, kilo-langley. Given the estimated radiant energy of the sun as 56×10^{26} cal per minute and the earth-sun distance as 1.5×10^{8} Km, find value of solar constant in ly/minute, kly/year.

9. Write the mathematical expression for the disposition of solar radiation in the earth-atmosphere system as given by Seller. State the various terms used in the expression. Also find the amount of solar energy incident at the top of the atmosphere treating solar constant as 2.0 ly/minute.

CHAPTER 2

Measurements of Solar Radiation

A simple Sunshine recorder (glass ball) burns card during Sunshine and thus provides the number of Sunshine hours in a day, from which monthly, seasonal and annual averages are worked out. A variety of sophisticated instruments that are used for measuring radiation are given below, as classified by WMO.

Radiometer (or Actinometer) is a generic term used for any instrument that measures radiation energy. The basic types of radiometers are :

1. Pyrheliometer
2. Pyranometer
3. Pyregeometer
4. Pyrradiometer
5. Net pyrradiometer

Pyrheliometer is used for measuring intensity of direct insolation (in coming solar radiation).

Pyranometer is used for measurement of global radiation (i.e., combined intensity of direct insolation and diffused sky radiation).

Pyregeometer is used for measurement of hemispherical long wave radiation of the earth together with reflected atmospheric radiation flux.

Pyrradiometer (or total hemispherical radiometer) is used for measurement of hemispherical long wave radiation together with global radiation.

Net Pyrradiometer is used for measurement of net all-wave radiation.

Terrestrial Radiation : Long wave radiation emitted by the planet earth and its atmosphere.

Total Radiation : Sum of solar and terrestrial radiation.

$$\text{Albedo of the earth : } = \frac{\text{Radiation reflected by the earth's surface}}{\text{Radiation incident on the earth's surface}},$$

Albedo of a surface :

$$= \frac{\text{Amount of global radiation from the Sun and sky reflected by the surface}}{\text{Amount of global radiation from the Sun and sky incident on the surface}}$$

(In Latin 'albus' meaning 'white', thus Albedo stands for a kind of a degree of whiteness.)

Insolation : Intensity of direct solar radiation on a horizontal surface or simply incoming solar radiation.

Attenuation of Solar Radiation : It is the loss of solar energy caused by scattering by air molecules or by selective absorption of certain molecules (such as water vapour, ozone, CO_2 etc.,).

Diffuse or Sky Radiation : Scattered and reflected hemispherical insolation except the solid angle subtended by the Sun's disc, **or**

The solar radiation which is received indirectly at the surface of the earth after being scattered or diffusely reflected by the atmosphere is called Sky radiation.

Direct Solar Radiation : Insolation from the solid angle of the Sun's disc on a surface perpendicular to the axis of the cone, including scattered and un-reflected solar radiation in that cone.

Global Solar Radiation : It's the downward direct and diffused solar radiation incident on a horizontal surface, **or**

Total radiation reaching the earth's surface i.e., (direct + sky) radiation. Insolation at the outer boundary of the earth's atmosphere is given by

$$I_o = S_\odot \sin \alpha \qquad .....(2.1)$$

Where $S_\odot$ = Solar constant, α = Angle of radiation (Solar beam) with the horizontal and is given by the relation.

$$\sin \alpha = \sin \delta \sin \phi + \cos \delta \cos \phi \cos \tau \qquad .....(2.2)$$

Where δ = Angle of declination,

ϕ = local latitude

τ = Sun's hour angle

Insolation at the surface of the earth, supposing there is no atmosphere, is a function of

1. Solar output
2. Distance from the Sun
3. Altitude of the Sun, and
4. The length of the day

The solar energy output is effected by the Suns heterogenous rotation on its axis, Sunspots cycle, Sun flares, Solar winds etc. It is effected by the variation of earth-Sun distance, it varies with the altitude of the Sun, maximum when the altitude is about 90°, and the number of hours of sun-shine in day (which varies with the season). A small part of the insolation that reaches the earth (in the form of short wave radiation) is absorbed and converted into long wave radiant energy (that is heat). Heat thus gained provides the fuel for all the weather and climatic processes. This conversion takes place both in horizontal and vertical directions and causes variety of temperature conditions.

Pyrometers : Radiation laws (Stefan-Boltzman law, Wiens Law) are used to estimate the temperature of hot bodies. These estimates are approximates because the laws used are strictly applicable to black-bodies and very rarely hot bodies behave like black-bodies.

Instruments to measure high temperatures, using emitted radiation are called pyrometers. These instruments measure very high temperatures (above 900 °K).

A radiation pyrometer is shown in the Fig. 2.1. Through the aperture marked XX radiation from the hot body enters. This is converged by a concave mirror C. The focused beam is made to pass through a limiting diaphragm D before it is incident on the thermocouple T. The E.M.F of the thermocouple is recorded on a milli-voltmeter connected to the terminals of the thermocouple T.

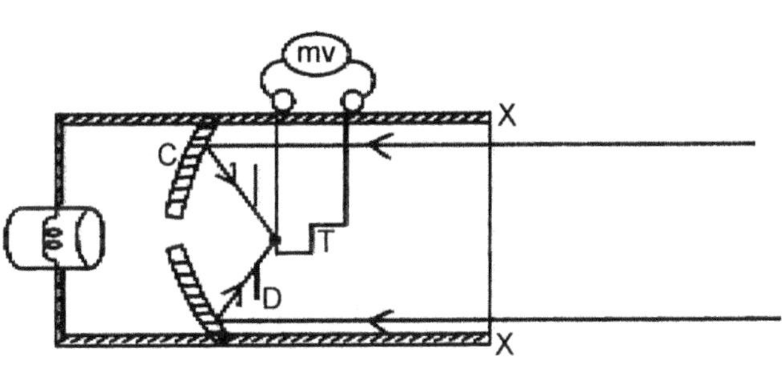

Fig. 2.1 Pyrometer.

The E.M.F developed is proportional to the temperature difference between the hot and cold junctions. Let T and T_o be the temperatures of the hot body and the receiving cold junction.

Then $\qquad$ E.M.F α $(T^4 - T_0^4)$

or $\qquad$ E.M.F α T^4, as $T >>> T_o$

2.1 Dual Nature of Light

Wave theory of light very convincingly answers many optical phenomena, but fails to answer black body radiation, photoelectric effect and X-ray absorption. The latter is satisfactorily answered by particle aspect of light. The fact is that radiant energy is quantized. The dual nature of light is best explained by atomic theory. The wave-particle duality of matter is substantiated by experiments. A wave is specified by a frequency υ, wavelength λ, phase velocity u, amplitude A and intensity I. They are not all independent.

$$u = \lambda\upsilon \qquad .....(2.3)$$

A particle (or corpuscle) on the other hand, is specified by a mass m, velocity v, momentum p and energy E. The connection between the wave and particle nature of light was given by Planck. The relation between the energy of the photon E and the frequency υ is given by

$$E = h\upsilon \qquad .....(2.4)$$

where h is Planck's constant. Compton gave the relation between wavelength of moving particle and momentum of the matter as

$$p = \frac{h}{\lambda} \qquad .....(2.5)$$

de Broglie gave

$$\lambda = \frac{h}{mv} \qquad .....(2.6)$$

and $\qquad E = mc^2 \qquad .....(2.7)$

[$E = mc^2$ Einstein's equation relativistic energy expression]

It follows $\qquad u = \lambda\upsilon = \dfrac{h}{mv} \times \dfrac{E}{h}$

relativistic $\qquad u = \dfrac{mc^2}{mv} = \dfrac{c^2}{v}$

Here $\qquad v$ (= speed) $<<< C, \quad u >$ speed of light.

2.2 Explanation of Some Terms

All things irrespective of their size emit radiation. The Sun, the earth, the stars, this book, trees, your body, air are all radiating a wide range of electromagnetic waves depending on their temperature. Radiation ceases when the body acquires a temperature zero degrees K (absolute).

Solar Radiation : the radiation emitted by the Sun.

Terrestrial Radiation : the radiation emitted by the planet earth including its atmosphere.

Total Radiation : Sum of solar and terrestrial radiation.

Radiation : Emission or transfer of energy in the form of electromagnetic waves or particles.

Radiation emitted by a body depends on :

1. The nature of its surface,
2. Its temperature, and
3. Its surface area.

A dull or black surface emits more radiation than a silver polished surface.

Radiation emitted is directly proportional to area of its surface. A body at higher temperature emits more radiation than at a lower temperature.

A black surface absorbs more radiation falling on it compared to a bright polished surface, which reflects more. Different surfaces absorb radiant energy falling on them differently.

The absorptive power of a surface (a_λ) for a certain temperature and wavelength is given by

$$a_\lambda = \frac{\text{The amount of radiation absorbed by the surface}}{\text{The amount of radiation incident on its surface}}$$

Black Body : A hypothetical body which absorbs all the radiation falling on it at any temperature without reflecting or transmitting any part of it is called a black body.

For a black body the absorptive power is unity.

A Gray Body : A body which at a certain given temperature emits a fixed proportion of black body radiation at that given temperature in all wavelengths.

Note : A white body which scatters the visible radiation falling on it may act as a black body for a different wavelength of radiation. For example, snow acts as black body for wavelengths greater than 1.5 μm.

Emissive Power or Spectral Density : Radiant energy distributed over a range of wavelengths. When radiation at a particular wavelength (λ), is considered, i.e., in the infinitesimal wavelength interval λ to $\lambda + d\lambda$, it is called spectral density or monochromatic.

Prevost's Law of Exchanges : If two bodies at the same temperature absorb different amounts of energy then their thermal radiation at this temperature should also be different.

2.3 The Emissive Properties of Black Bodies

The emissive properties of a black body are given by a number of laws which are given below.

(i) *Planck's Law* : The spectral density of the radiance of a black body ($L_{\lambda B}$) is a function of the temperature (T) and wavelength (λ).

$$L_{\lambda B} = f(T, \lambda) = \frac{C_1\, l^{-5}}{\left[e^{\frac{C_2}{\lambda T - 1}}\right]}$$

where C_1 = first radiation constant = 3.74×10^{-16} J m^2 s^{-1}

C_2 = second radiation constant = 1.439×10^{-2} m °K

The spectral density of the radiance of a black body is independent of direction. This implies Lamberts cosine law.

Lambert's Cosine Law : The flux per unit solid angle emitted in any direction from a unit radiating surface is proportional to the cosine of the angle between the direction of radiation and normal to the surface.

(ii) *Stefan-Boltzman Law* : Integrated black body radiance (L_B) over the entire spectrum is obtained by integrating both sides of Planck's equation w.r.t λ

Thus we have

$$L_B = \int_0^\infty \frac{C_1 \lambda^{-5}}{\left[e^{\frac{C_2}{\lambda T - 1}}\right]} d\lambda$$

We obtain $L_B = \dfrac{\sigma T^4}{\pi}$ (This is Stefan-Boltzman Law)

Where σ = Stefan-Boltzmamn constant = 5.67×10^{-8} J m^{-2} °K^{-1} s^{-1}

L_B is independent of direction.

This law states that the total emittance of a black body is proportional to the fourth power of its absolute temperature. This law is generally used for estimating the temperature of hot bodies.

(iii) ***Wien's Displacement Law*** : For a given temperature the wavelength (λ_{max}) of maximum radiance is given by

$$\lambda_{max}\ T = \text{const.}$$

where constant $= 0.2898 \times 10^{-2}$ m $°K^{-1}$.

In other words the frequency ($\upsilon = \frac{1}{\lambda}$) corresponding to maximum emissive power of a black body is proportional to its absolute temperature

$$\lambda_{max} = \frac{\text{const}}{T} \qquad \left(\upsilon\ \alpha\ T \text{ or } \frac{1}{\lambda}\ \alpha\ T\right)$$

This relation shows that the maximum energy of emission shifts progressively to shorter wavelengths as the temperature rises. This law is used for determining the temperature of the stars and the Sun.

(iv) ***Kirchhoff's Law*** : The ratio of the total radiant emittance of a gray body to its absorptance is equal to the total radiant remittance of a black body at the same temperature.

$$\frac{e_\lambda}{a_\lambda} = e_b$$

Emissivity ε_λ : For a particular temperature and wavelength

$$\varepsilon_\lambda = \frac{L_\lambda}{L_{\lambda B}}$$

where

L_λ = Radiance of a radiating object

$L_{\lambda B}$ = Black body radiance.

In case of black body $\varepsilon_\lambda = 1$

If ε_λ = constant, but less than unity, for all wave lengths, the object is called gray body.

Comparison of total energy of various phenomena, treating the total solar energy intercepted by the earth 3.67×10^{21} cal/day or 2.55×10^{18} cal/min as unity

Solar energy received per day	1
Monsoon circulation	10^{-2}
Strong earth quake energy- (Richter scale 8)	10^{-2}
Average depression	10^{-3}
Average hurricane	10^{-4}
Kinetic energy of general circulation of atmosphere	10^{-5}
Average magnetic storm	10^{-7}
Average thunderstorm	10^{-8}
Nagasaki atomic bomb explosion (August 1945)	10^{-8}
Average earth quake (Richter scale 6)	10^{-8}
10 mm of rain	10^{-8}
Average tornado	10^{-11}
Average lightning stroke	10^{-13}
Average dust devil	10^{-15}

Questions

1. What is radiation? What are the basic types of radiometers and what they measure?
2. Define the albedo of the earth and the albedo of a surface. How are they used in satellite based remote sensing.
3. Write briefly on: (i) the dual nature of light, (ii) black body and gray body, (iii) Planck's law, (iv) Stefan-Boltzman law, (v) Wien's law and (vi) Kirchhoffs law.

CHAPTER 3

Infrared Radiation (IR)

For all practical purposes insolation consists in the electromagnetic radiation (about 99%) in the range 0.15 to 4.0 μm with peak energy around 0.5 μm in the visible range, while the terrestrial radiation (about 99%) in the range of 4 to 80 μm with peak energy around 10 μm, which is in invisible range. Insolation consists mostly in shortwaves while terrestrial radiation in infrared region. Terrestrial radiation is also called ground radiation and some times nocturnal radiation (In Latin 'terra' meaning 'Earth').

The global value of IR is about 258 kly/year which is slightly less than the incoming solar radiation of 263 kly/year. The earths surface with its average temperature of 288 ^{O}K ($\approx$ 300 °K) is assumed to emit and absorb energy as a gray body in infrared region.

Atmosphere is virtually transparent to insolation (short wave radiation) but it readily absorbs terrestrial (IR) radiation. Atmosphere is both selectively absorbs and emits radiation, consequently called selective absorber. The selective absorbers are water vapour (absorbing range 1 to 8 μm and greater than 12 μm), Carbon-dioxide (absorbs at 4 μm and between 13 to 17 μm), Ozone (absorbs UV-radiation between 0.2 to 0.3 μm and IR radiation at 9.6 μm). Molecular oxygen (absorbs UV-radiation below wavelength of 0.2 μm). However practically no radiation is absorbed by these gases and water vapour between 8 to 11 μm range. i.e., this range is transparent to both long wave and short wave radiations. This wave lengths range (8 to 11 μm) is called *atmospheric window.*

[Radiation is also subjected to scattering as it passes through a medium but without loss of energy. In meteorology it is called *diffuse radiation.*]

It may be noted here that substances which are not efficient absorbers of short wave radiation (or solar radiation) may become most efficient absorbers and emitters of long wave (or IR) radiation from the earth.

A Green House : In cold countries glass houses are built to protect the green plants from severe cold. A glass house is transparent to solar radiation i.e., solar radiation reaches the ground through glass. However a glass house is opaque to the long wave (terrestrial) radiation i.e., it will not allow the long wave radiation to escape from the house and thus keeps it warm.

The atmospheric gases CO_2, water vapour , ozone in the lowest troposphere allow the solar radiation to pass through it, however they do not allow the terrestrial long wave radiation to escape to the space as in a green house. Hence the term green house effect is applied to this atmospheric phenomena. The gases CO_2, O_3, atomic oxygen, methane N_2O, chloro-fluorocarbons which do not allow the terrestrial radiation to escape to the space are called green house gases.

In the absence of green house effect the surface of the earth would have been cooler by 30 to 40 °C.

Aerosols : Solid particles (lithometeors) and liquid droplets (hydrometeors) suspended in the air are called aerosols. Typical constituents of aerosols are silicate, salts (NaCl, $Mg\,SO_4$, $NaNO_3$, NH_4Cl, NH_4NO_3), acids (H_2SO_4, $H\,NO_3$), metal oxides, organic combustion products, volcanic dust, cosmic dust. Aerosols are present in the atmosphere in enormous number whose effective diameters range from 0.01 μm to 40 μm. The concentration of aerosols vary several orders of magnitude both in time and space. They play vital role in fog/cloud formation and in radiative processes.

Aerosols are mainly formed by dispersion, combustion and photochemical processes.

They are divided into three categories depending on their diameters.

1. Aitken nuclei–diameter less than 0.2 μm
2. Large nuclei–diameter between 0.2 to 2.0 μm
3. Giant nuclei–diameter more than 2.0 μm

Aitken nuclei are present in abundance in the atmosphere as compared to large and giant nuclei.

Scatter of electromagnetic wave radiation depends on the wavelength (λ) and particle (molecule) diameter (d).

1. **Rayleigh Scattering ($d < \lambda$)**

   The radiation wave (or light) scattering that occurs in aerosols when the particle diameters are smaller than wavelength (λ) of interacting radiation is called Rayleigh scattering (R-scatter). R-scatter $\propto 1/\lambda^4$, i.e., when λ is small R-scatter is stronger or more effective.

   In visible range λ - varies 0.4 μm (for blue end) to 0.8 μm (for red end). The R-scatter is about sixteen times larger at blue end than at red end.

   Blue Colour of the Sky : Atmospheric particles/aerosols scatters more blue light, as their wave lengths are smaller (λ_b = 0.49 μm) in comparison to larger wavelengths of orange (λ_o = 0.647 μm) or red (λ_r = 0.71 μm). Since the atmosphere contains abundance of Aitken nuclei and because R.scatter of blue light is much more than others, as a result the sky appears blue.

   Red Colouration of Sky : At the time of Sunrise or Sunset, the Sun's rays travel longer in the atmosphere as compared to noon. At Sunrise or Sunset there is both excessive scatter as well as absorption of shorter wavelength but longer wavelengths of orange, red light are less absorbed, less scattered. This causes red colouration of the sky during sun rise and sun set.

   Note : Different colouration of the sky will be discussed separately along with Auroras and Rainbows (optical phenomena).

2. **Mie scattering ($d \simeq \lambda$)**

   The scattering of light/radiation that occurs in aerosols (medium of particles) when their diameters are roughly equal to the wavelengths (λ) of the interacting radiation is called Mie Scattering (M-scatter). M-scatter affects longer wavelengths prominently than R-scatter. M-scatter is prominent during dust storms, volcanic eruptions, haze and cloudy weather.

3. **Non selective scatter (N S scatter : $d >>> \lambda$)**

   The scattering of light/radiation that occurs in aerosol (medium of particles) when their diameters are greater than the wavelengths of the interacting radiation is called N. S. Scatter. The diameters of water droplets vary from 5 μm to 100 μm These droplets scatter equally all visible and

IR wavelengths i.e., the scatter is non-selective w.r.t λ. Under N. S. scatter conditions all wavelengths of visible light spectrum (blue, green and red lights) scatter equally without any preference of wavelengths and result into white light. Because of this mist, fog, clouds generally appear white.

In general the atmospheric particles vary in their diameters (d) 10^{-4} μm to 10^3 μm. As a result, depending on the presence of dominant particles [$d < \lambda$; $d \simeq \lambda$ and $d > \lambda$], Rayleigh, Mie or N. S. Scattering occurs in the atmosphere.

3.1 Radiative Balance and Horizontal Transport of Heat

Past available records show that the average temperature of the earth as a whole is about 15 °C and is virtually constant. This implies that the amount of heat energy that is received by the earth (mainly as short wave radiation) is equal to the amount of heat lost by the earth by way of long wave radiation. Overall the earth-atmosphere system radiates nearly the same amount of energy as it is receiving.

However the latitudinal distribution of incoming radiation to the earth and the outgoing radiation from the earth are different which is shown in the Fig. 3.1. (it is not to the scale). The figure shows that near the equator more energy is absorbed than emitted by the earth. The earth-atmosphere system thus shows surplus energy between latitudes 0 to 35° (area hatched with cross lines). Contrary to this more energy is emitted to space than it is absorbed between latitudes 35 to 90°, indicating deficient energy in these latitudes (area marked with dots). The earth as a whole neither getting heated up nor becoming colder. This signifies that the excess of heat energy absorbed in low latitudes being transported to higher latitudes or else the regions in low latitudes would have much higher temperature and the region between mid and higher latitudes would be cooled down to much lower temperature. This transfer of heat from low to higher latitudes takes place through the action of wind and ocean currents. Ocean currents carry about 30% required heat transport from equator to pole wards.

Note : Calculations indicate that, in the absence of CO_2 and water vapour the average equilibrium temperature of the earth would have been –20 °C as against the observed average temperature of 15 °C.

3.2 Liquids

The force with which like molecules attract each other is called cohesion. If we are to convert liquid into vapour, we have to provide extra energy to separate two molecules. The energy supplied must be at least equal to the intermolecular interaction energy. The attraction of unlike molecules is called adhesion.

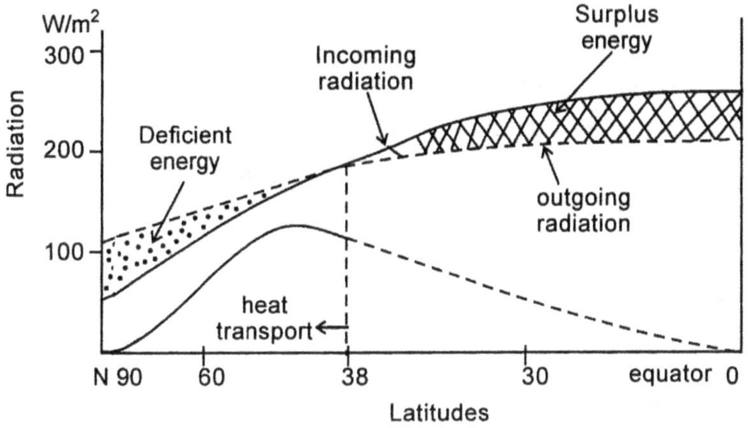

Fig. 3.1 Mean radiation balance northern hemisphere. Solid line incoming solar radiation, dotted line outgoing earth radiation.

The phenomena by virtue of which the free surface of liquid behaves like an elastic membrane under tension trying to contract in order to have minimum surface area is termed surface tension.

$$\text{Surface tension} = \sigma = \frac{\text{Force}}{\text{length}} = \frac{F}{l}$$

Its unit is Newton per meter (N m^{-1}).

The phenomena of rise or fall of a liquid in a very narrow tube (capillary tube) is called capillarity.

In case of water the force of adhesion between water and glass is greater than the force of cohesion between the molecules of water. By virtue of this water rises up till the downward weight of the water in the capillary tube above the outer surface water in the vessel balances the upward force due to surface tension along the line of contact of water and glass in the capillary tube Fig. 3.2.

In case of mercury the force of cohesion between mercury molecules is greater than the force of adhesion between mercury molecules and glass. Consequently mercury depressed in the tube Fig. 3.3. Smaller the radius of the capillary tube the greater is the rise or fall of liquid in the tube.

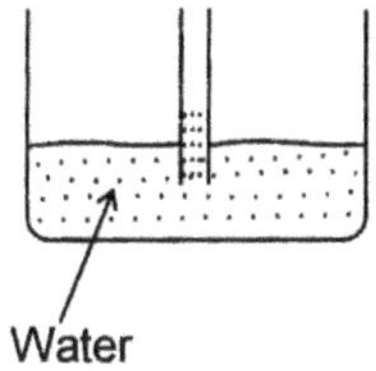

Fig. 3.2 Water in the Capillary tube

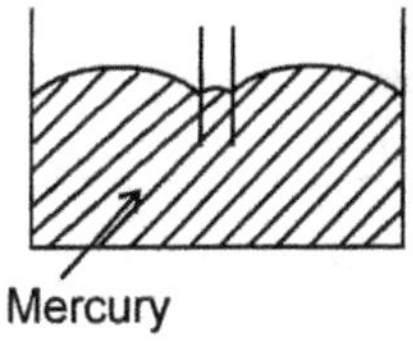

Fig. 3.3 Mercury in the Capillarity tube

Streamline Flow : The flow of liquid is said to be streamlined or orderly if the particles of the liquid move along fixed paths known as streamlines. The velocity of the particles moving one after the other at any point on the streamline remains same in magnitude and direction at that point.

Total energy of a liquid : The total energy at any point in a flowing liquid or fluid contains (i) Pressure energy (ii) Potential energy and (iii) Kinetic energy.

Pressure energy per unit mass $= \frac{p}{\rho}$,

potential energy per unit mass $= gh$, and

kinetic energy per unit mass $= \frac{1}{2}v^2$.

Total energy per unit mass of flowing liquid $= \frac{p}{\rho} + gh + \frac{v^2}{2}$

where p = pressure ρ = density g = gravity
h = height v= velocity

Questions

1. Describe the salient features of radiation wavelengths of sun and the earth. Why are they called short wave and long wave radiations?
2. What is atmospheric window? Why is it called atmosphere a selective absorber?
3. What is greenhouse effect? Write the main atmospheric green house gases. What would have been the temperature of the earth in the absence of green house gases?
4. Write briefly on : (i) Aerosols (ii) Rayleigh scattering, (iii) Blue colouration of the sky, (iv) Red colouration of the sky, (v) Mie scattering and (vi) Non-selective scatter.
5. Write briefly the radiative balance of heat maintained on earth.
6. Find the volume of one mole of an ideal gas at STP, given R = 8.314 J $mol^{-1}\ K^{-1}$.
7. Find the volume of one mole of an ideal gas at p = 1 atm, T = 273 °K and R = 0.08207 atm. $mol^{-1}\ K^{-1}$. (use pv = nRT).

CHAPTER 4

Atmosphere

Air, which we call, is a mechanical mixture of gases. The various gases, solid and liquid particles in the air, that envelope the earth are bound to it by the gravitational attraction. This envelope of gases, solid and liquid particles is called atmosphere. The atmosphere extends above the surface of the earth to great heights. There is no sharp boundary between the atmosphere and the outer space. However, conventionally, the height of the atmosphere is taken as 1000 km. More than 50% of the mass of the atmosphere lies below an altitude of 5.5 km and 98% of the mass lies below an altitude of 30 km (Above 700 km of altitude a vary thin atmosphere of Hydrogen and Helium atoms exists). The lowest one kilometer, which is called planetary boundary layer, contains 10% of the mass of the atmosphere. All biological and human activities are confined to this planetary boundary level baring aircraft flights. Infact the vertical dimension of the atmosphere is very shallow compared to the size of the earth, it is as thin as the skin on an apple. The mass of the atmosphere is about 5.6×10^{18} kg (mass of the earth is about 6×10^{24} Kg), while the mass of oceans (Hydrosphere) water is about 1.4×10^{21} kg. This shows that the mass of the oceans is more than 250 times the mass of the atmosphere. The density of the dry air at the surface is 1.225 Kg/m^3 and at an altitude of 5.5 km this drops to 0.66 Kg/m^3 and at 30 km altitude it is about 0.013 Kg/m^3.

Geological evidence suggests that the age of the earth is about 4.5×10^9 years (4.5 billion years) and life began on earth between 0.6 to 1.0 billion years after earth has formed (i.e., life began on earth some where 4 to 3.5 billions years ago). It is estimated that in each second about one Kg of hydrogen from the upper atmosphere escapes into outer space. The mass of oxygen in the atmosphere is 10^{18} kg, that is one fifth of the mass of the atmosphere. The plant kingdom over the globe produces oxygen about 3×10^6 kg (3 million kg) per second which is consumed by the living beings on the whole earth. About 12 kg of cosmic dust per second falls from space on the earth.

4.1 Weather and Climate

The physical state of atmosphere at any given location rarely exhibits steady state even during short intervals of time. This physical state of the atmosphere constitutes weather at a location when considered over a short period of time. Weather is broadly described by the parameters temperature, pressure, relative humidity, wind direction and speed, cloud, precipitation. Climate on the other hand is the average weather over a place including variability and extremities. For this purpose the average has to be taken for a period of not less than 30 years. Both weather and climate are inseparable parts of meteorology. Settlements and civilizations have risen and fallen with the favourable and unfavourable climatic conditions. Weather knows no political boundaries and it favours or frowns equally on all nations. Weather and climate have all pervading influence on man.

4.2 Composition of the Atmosphere

The composition of the atmospheric gases, are practically in the same proportion (that is homogeneous) up to an altitude of 80 to 90 Km with the exception of water vapour, ozone and carbon-dioxide which are variable. The composition of dry air at sea level is given in Table 4.1.

Non-gaseous constituents, (called aerosols) such as dust, smoke, salt particles from sea spray, water particles are all variable.

All living organisms are composed of Carbon, Nitrogen, Oxygen, Hydrogen which are also the basic chemical elements of water and air shells of earth. A large part of living matter contains in green plants which entrap solar energy and construct complex compounds by photosynthesis. The main sources of plant feeding are CO_2 and Water. Plants use about 2% of the incident solar radiation for photosynthesis process. If we assume incident solar energy as 0.5 langleys per min near the earth, then 0.01 ly/min or 7 w/m^2 is consumed by plants.

The plant kingdom provides about 10^{17} kg of biomass annually and an equal quantity of oxygen. An average size of tree supplies about 3500 kg of oxygen per year which is sufficient for three people (because of this the plant kingdom is called the green lungs of the earth). A man requires about 3.13 kg of oxygen or 15 kg of air daily.

Table 4.1 Composition of the atmospheric gases.

Constituent	**Symbol**	**% by volume**	**Total weight × 10^{17} Kg**
Nitrogen	N_2	78	38.65
Oxygen	O_2	21	11.84
Argon	Ar	0.93	0.66
Carbon-dioxide	CO_2 variable	$\simeq$0.03	$\simeq$0.023
Neon	Ne	180×10^{-5}	0.00064
Helium	He	52×10^{-5}	0.00004
Methane	CH_4	15×10^{-5}	0.00004
Krepton	Kr	10×10^{-5}	–
Hydrogen	H	5×10^{-5}	–
Xenon	Xe	0.8×10^{-5}	–
Ozone	O_3	variable	–

At any location water vapour varies from nearly absent to about 4 % by volume (or about 3% by weight) of the atmosphere. However by continuous interchange in the hydrologic cycle, in a year the water vapour produces precipitation about 40 times greater in volume. It is practically absent above altitude of 10 to 12 km. Water vapour is fed to the atmosphere by evaporation of water from surface water bodies or by transpiration of plant kingdom and transported upwards by turbulence or connective currents. This will be discussed in detail in hydrologic cycle. The volume of the water vapour in the atmosphere is about 14×10^{12} m^3 or 0.001 percent of the volume of the earth's hydrosphere. Changes in the moisture concentration below an altitude of 6 km creates all weathers.

The total mass of the CO_2 in the atmosphere is about 0.023×10^{17} Kg while the mass of CO_2 in ocean water is about 1.4×10^{17} Kg which is more than 60 times the amount of CO_2 in the atmosphere. Sea water plays a peculiar role in respect of dissolved natural gases of N_2, O_2, CO_2 and H_2S. These gases as said earlier are closely related to living matter on the land and Sea. CO_2 enters the atmosphere by human and animal breathing, decay and burning of material containing carbon, and volcanic activity. More than 90% of the earth's CO_2 is dissolved in the sea waters. The solubility of CO_2 in sea water varies with temperature, so it enters or leaves the ocean waters with changes in temperature. Because of this the concentration of CO_2 in the atmosphere, particularly near the surface, changes. A large part of the atmospheric CO_2 is removed by the plant kingdom as mentioned above.

Ozone concentration in the atmosphere varies with altitude, latitude and time of the day. In general ozone is found in the atmosphere in variable small quantities up to an altitude of 50 Km. It is mainly found in the stratosphere with maximum concentration between altitudes 18–20 Km in a globe encircling ozone layer. The total ozone column above the earth shields the earth's surface from the lethal UVB (Ultra Violet Radiation).

UVB induces skin cancers, damages eye and suppress immune systems in human beings. UVB affects the productivity of aquatic and terrestrial ecosystem. One percent decline of ozone in atmosphere results in an estimated 3% rise in the potential incidence of skin cancer in humans.

Ozone molecule (O_3) consists of three atoms of oxygen. In the upper stratosphere ozone is formed by the absorption of UV-radiation by oxygen at wavelength 0.18 μm. This splits oxygen molecule (O_2) into two atomic oxygen atoms.

$$O_2 + \text{UVB} \rightarrow O + O$$

The atomic oxygen atom (O) reacts with an oxygen molecule (O_2) to form an ozone molecule.

$$\dot{O} \text{ (atomic)} + O_2 \rightarrow O_3 + \text{energy (exothermic reaction)}$$

Ozone molecules then tend to sink in the atmosphere to altitudes between 15–25 km. Ozone thus formed absorbs all wave lengths of UV-radiation below 0.3 μm. Even in these altitudes the concentration of ozone is still very small compared to oxygen (one ozone molecule to 10^6 normal oxygen molecules). The ozone molecules from the lower stratosphere entering into troposphere gradually decompose into oxygen and thus keep maintain the balance. Ozone is present near the surface of the earth in highly varying concentrations as part of photochemical smogs. Small amounts of ozone is formed near the ground by electrical discharges.

The composition of the atmosphere above 80 Km altitude is not homogeneous and the gases are less mixed. In these altitudes most of the gases are found in ion-atomic state which is caused by solar UV and X-radiation (If a neutral atom loses an electron it becomes positive ion, but if it gains an electron it becomes negative ion. This process is called ionization). Ionization also occurs below 80 Km altitude but in a lesser degree. Here the free electrons and ions mix together and neutralize to form molecules. In higher altitudes (above 100 Km) protons (ionized hydrogen atoms) and free electrons dominate but the atmosphere is very thin which gradually merges with the interplanetary gas.

4.3 Interplanetary Gas

The space between the planets is not empty but contains very low density material and hot gases mainly in the form of electrons and protons. This is called interplanetary gas. The planets and the earth in their orbital motions around the sun pass through this inter-planetary gas. The upper most earth's atmosphere at very high altitudes (above 1000 km) merges gradually with this interplanetary gas.

4.4 Vertical Divisions of the Atmosphere Based on Temperature

In 1962 World Meteorological Organization (WMO) decided to divide the atmosphere into four regions based on temperature change with altitude. These are :

(i) The Troposphere (ii) The Stratosphere

(iii) The Mesosphere (iv) The Thermosphere

These are shown in the Fig. 4.1.

Note : Altitude refers to height above mean sea level (amsl)

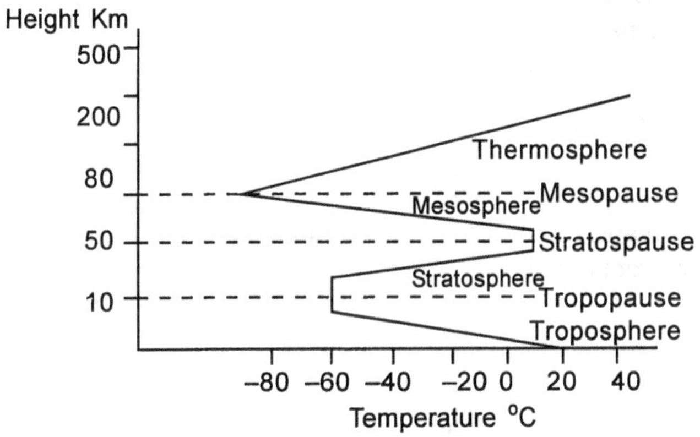

Fig. 4.1 Vertical divisions of atmosphere based on temperature.

4.5 Troposphere

The lowest layer of the atmosphere adjacent to the earth's surface is called troposphere. Tropos in Greek meaning 'turn'. The temperature turns (increases) above the troposphere and the sun turns above tropics during solstices. The altitude of troposphere in the equatorial region is about 18 Km, which drops to 8 Km in the polar regions [Decrease of temperature with increasing altitude is called lapse rate. Layers where temperature increases with increasing altitudes are called inversion layers].

In troposphere, in general, temperature falls with increasing height. On some occasions, particularly in winter, inversion layers are found near the surface of the earth. There may be found some shallow inversion layers in troposphere. The average lapse rate in troposphere is 6.5 °C/km (3.6° F/1000 ft). Troposphere contains about 75% of the mass of the atmosphere and virtually all the water vapour and aerosols. All weather systems and associated cloud systems are practically confined to troposphere only.

The top of the troposphere is called tropopause where temperature begins to rise or isothermal. Pause meaning break, indicating break in two regions of the vertical atmosphere. Tropopause is defied as follows. The lowest level at which the lapse rate decreases to less than or equal to 2 °C/km at least for a layer of 2 Km layer and above does not exceed 2 °C/km. The altitude of tropopause varies with latitude. The height of the tropical tropopause is about 18 Km (lower temperature about –80 °C) while that of polar tropopause is about 8km (constant temperature about –50 °C). Tropopause is not a continuous layer. In middle latitudes two tropopause are found, one with tropical characteristics and another with extra tropical characteristics, in between these two there is sub-tropical jet (around lat 30°). Double tropopause with less marked characteristics are observed with polar jet in between them (around lat 60°).

The altitudes of tropopause change from day to day and also in association with the movement of synoptic systems. The characteristics of tropopause change with time and place. There is sharp rise in temperature above the tropical tropopause, but in middle latitude tropopause there is slight fall of temperature. Troposphere is mainly warmed up by the underlying earth surface, consequently temperature falls with increasing altitude. The colder, denser air in the upper part of the troposphere sinks by gravity along with traces of ozone and forces the warm air near the surface upward. Thus sets in convection currents, clouding and heat transfer.

4.6 Stratosphere

The second stratum that begins, above the tropopause is called stratosphere. It extends above tropopause to an altitude 50 Km. There is an isothermal layer above tropopause to about 20 Km altitude, above which temperature rises gradually to about 32 Km altitude and then rises rapidly. The upper layer of the stratosphere has temperature equal to that of earth's surface temperature. The importance of this region lies in the presence of ozone and sudden warming in polar area, quasi biennial wind oscillation in equatorial area. As mentioned above the concentration of ozone lies at about 25 to 50 altitude, which absorbs the harmful solar radiation between wavelengths 220 nm to 290 nm. There is no ozone above stratosphere. Ozone absorbs about 2% of the insolation and as a result of which stratospheric temperature rises to more than 0 °C (but less than thermospheric temperature). There is not much convection in the stratosphere because it has cold temperature at its bottom and warm temperature at the top.

The small layer above the stratosphere, where temperature ceases to rise (isothermal) and that it separates the stratosphere with third region (of mesosphere) where temperature again falls, is called *stratopause*.

4.7 The Mesosphere

The region above the stratopause is called Mesosphere and it extends up to an altitude of 80 Km. In Greek 'Meso' means "medium". This region is characterized by fall of temperature above stratopause (> 0 °C) to about –95 °C near 80 Km altitude. This region presents noctilucent clouds in higher latitudes during summer. The top of the mesosphere is called mesopause. The three regions troposphere, stratosphere and mesosphere together is called homosphere. As stated earlier the composition of the atmosphere is nearly constant, except ozone, water vapour and carbon-dioxide in homosphere. The mean structure of temperature in the mesosphere, the temperature upto 100 Km altitude is shown in the Fig. 4.2. Both in the troposphere and the mesosphere the temperature in general decreases with altitude but in the stratosphere the temperature increases with altitude baring in winter in high latitudes. The stratospheric latitudinal temperature distribution is accounted by the radiative heating of ozone. The temperature is maximum in summer polar region and minimum in winter polar region due to solar radiative flux at the solstices. At an altitude above 60 km the temperature increases from summer pole to winter pole.

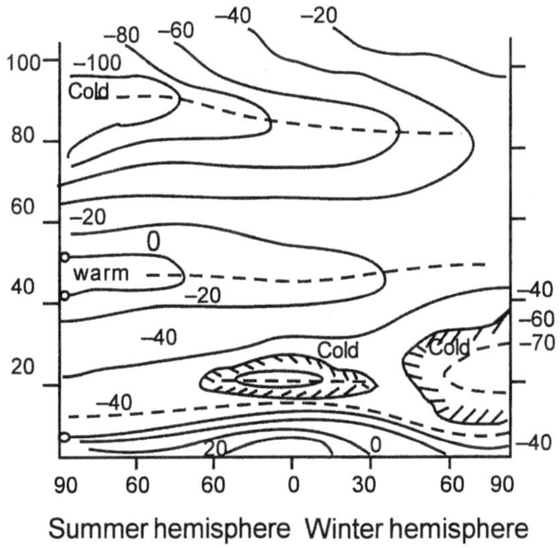

Fig. 4.2 Mean meridional cross section of temperature in °C. Dotted lines indicate tropopause, stratopause and mesopause.

The mean zonal (East-west direction) wind profiles indicate approximately thermal wind balance with temperature profile. Easterly Jet in summer hemisphere and westerly Jet in winter hemisphere with wind maxima occurring at about 60 Km altitude. In winter there is high latitude westerly Jet in lower stratosphere, which is called polar night Jet. This is related to stratospheric warming. In the equatorial stratosphere there exists quasi-biennial oscillations which will be the discussed separately.

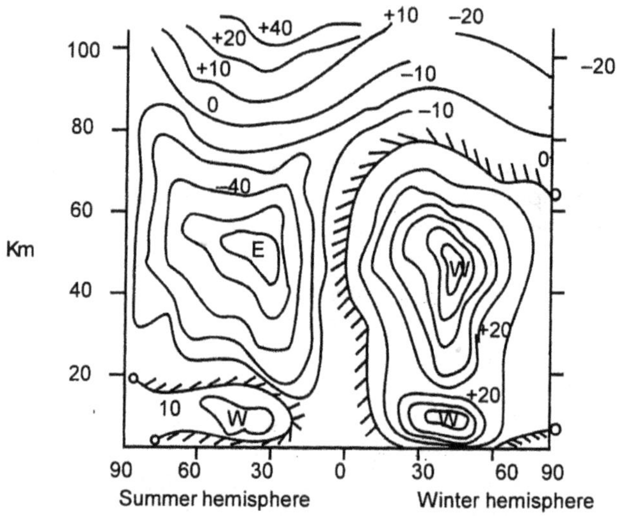

Fig. 4.3 Mean zonal wind in mps at the time of solstices.

4.8 The Thermosphere

The region above mesopause is called thermosphere or hetrosphere. Here the composition of the atmosphere is not homogeneous. The main gases stratify by their molecular weights. The lowest layer predominates with nitrogen and oxygen molecules [90 to 150 km], then oxygen (150–500 km) and at the top (500–1000 Km) hydrogen and then inter-planetary gas. The temperature of the thermosphere rises to 1000 to 1200 °K at 400 km but it has gas of very thin density 3×10^{-12} Kg/m^3 and pressure 10^{-8} mm of mercury. In this region gases are found mostly in atomic state due to photo-dissociation by the short wave radiation of the sun and they are not in the same proportion as in homosphere.

There is temperature rise above mesopause. Oxygen atoms are seen in the lower thermosphere which gradually increases with altitude. At about 130–150 Km altitude about 70% of the oxygen is found in atomic state due to the action of UV-radiation of the sun. At further up there are only hydrogen, helium atoms.

4.8.1 The Ionosphere

The upper atmosphere which is characterized by the presence of dense ions and free electrons is called ionosphere. These ions cause reflection of radio waves. The ionosphere is sub divided into D-region (50–90 km), E-region (90–140 km) and F-region (F_1 and F_2) (140–250 km, 250–500 km).

4.8.2 The D-Region

This occurs between 50 to 90 km (lies mainly in mesosphere). It reflects low-frequency radio waves, but absorbs medium and high frequency waves. This region is entirely dependent on solar radiation and disappears during nights. The D-region is well-developed during solar flares. Because of this, during solar flares complete break down in medium and high frequency radio communication takes place. This is called sudden ionospheric disturbances (SID).

4.8.3 The E-Region

This occurs between altitudes 90 to 140 km. This region strongly reflects medium and high frequency radio waves. E-region begins to weaken after sunset, but does not disappear completely. However during polar nights it disappears completely. The lower part of the E-region is marked by the recombination of ions and electrons. Under special conditions and irregular times a "sporadic E-layer" occurs. On such occasions the motion of ions and electrons can be detected by their effects on radio transmission.

4.8.4 The F-Region

This is further sub-divided into F_1 and F_2 regions extends above 140 km.

(a) *F_1-Region* : It is found only during day time when the sun is fairly high. When the sun is low and at night it merges with the higher F_2-region. F_1-region is important in the propagation of medium and high frequency radio waves.

(b) *F_2-Region* : This region is important in long-distance radio communication. Depending on location and conditions the ionization density reaches a peak at altitudes of 250–500 km. It gradually diminishes above 500 km but it has no upper limit. The upper F_2-region contains protons (ionized atomic hydrogen) and electrons. In fact this region is similar to inter planetary gas.

4.8.5 The Dynamo Region

In the region of 80–130 km altitude ions, atoms and molecules of neutral gases move together. The mean velocity of electrons are different from that of ions. Atmospheric tidal oscillations occur in this region with the creation of ionospheric currents. These ionospheric currents are called "dynamo currents", which are caused by the motion of the conducting atmosphere across the earth's magnetic field. This region is called the "dynamo region" because these currents are similar to the currents generated by a moving conductors in a dynamo. These electric currents are responsible for the regular variations of earth's magnetic field.

The regular variation of earth's magnetic field follows a 24 hr daily pattern. They are related to the solar or lunar day, and over a period of 24 hrs they pass through a regular cycle of increases and decreases. These are believed to be the effect of electric currents of tidal oscillations at altitudes 80 to 130 km referred above. Irregular variations may also occur suddenly in short intervals of time few seconds to days. These irregular variations are caused by the particles ejected from the sun. During solar winds magnetic storms occur particularly when these winds approach the outer atmosphere. As stated in solar flares, during intense solar activity charged solar particles enter the upper atmosphere and cause electrical phenomena.

This electrical phenomena in turn cause fluctuations in direction and strength of the earth's magnetic field. When the irregular variations are prominent magnetic storm results. During the period of magnetic storms communication system will be jammed or blackout over the entire world. Frequent magnetic storms would occur during the period of maximum sunspot activity, which has about eleven years cycle. The solar particles would be trapped and controlled by the earth's magnetic field. These particles become concentrated in two zones of radiation which surround the earth above the magnetic equator. These bands are called Van Allen radiation belts.

4.9 Exosphere

As stated earlier, the ionosphere consists of ions, electrons and neutral particles, extends outward till it merges with the inter planetary gas. It may be noted that even at about 1200 km altitude the number of neutral particles are more or less same as that of electrons. However at about 500-600 km altitude the density of neutral particles is so small that collision between them is very rare. The mean free path is very large, as a result some particles escape from the earth's gravitational attraction. Because of this the region above 500–600 km is named exosphere. It is estimated that each second 1 kg of hydrogen escapes from the earths atmosphere to outer space.

In contrast to the neutral particles, the electrically charged particle movements are controlled by the earths magnetic field. The collision between them still occurs at levels above 500–600 km. As a result the electrically charged particles do not escape from the earth's atmosphere.

Cosmic rays coming from the space undergo deviation in the earth's magnetic field and hence intensity depends on latitude. In the vicinity of the equator the deviation is manifested strongly, and the particles subjected to large deviation do not enter the atmosphere (this is latitude effect). The positively charged particles of the cosmic rays deviate towards east and the negatively charged particles towards west (this is called east-west effect). Intensity also depends on the longitude (this is called longitude effect).

4.10 Van Allen Radiation Belts

Radiation belts have been found at the outer most atmosphere. There are two delimited (assigning of boundaries) regions with high intensity of cosmic ionizing radiation. These belts being formed due to trapping of charged particles by the earths magnetic field. The inner belt extends from about 600 km to 6000 km while the outer belt 20000 to 60000 km from the earth. Inner radiation belt consists of high energy protons while the outer radiation belt formed by electrons of solar origin. Radiation belts are characteristic of all celestial bodies which have magnetic field. The moon has no magnetic field of its own and hence it has no radiation belts around it.

Questions

1. What is atmosphere? How is the mass distributed in the atmosphere with respect to height (vertical)?
2. Distinguish between weather and climate. What are its effects on civilization?
3. What is homogenous atmosphere? Write its composition.
4. What are the green lungs of the earth? What is the man's daily requirement of air? What is photosynthesis and the daily output of oxygen by a normal tree?

5. What do you know about the variation of water vapour at any location? How is that water vapour produces precipitation about 40 times greater in volume in a year?

6. Write briefly on the ozone formation in the atmosphere and ozone hole. What would happen to the life on earth with the depletion of ozone in the atmosphere?

7. Write briefly on the WMO-vertical divisions of the atmosphere based on temperature.

8. Write the principal features of troposphere, stratosphere and mesosphere.

9. Write briefly on (i) The thermosphere (ii) The ionosphere, (iii) The dynamo region, (iv) Exosphere (v) Van Allen radiation belts.

CHAPTER 5

Evolution of the Earth's Atmosphere

The atmosphere surrounding the earth with white clouds has gradually evolved with the seas, lakes, the green vegetation and brown soils of the land. The surface of the earth which is now teeming with life would have been impossible without the atmosphere, and the atmosphere would not be the same without plants and life on earth. The uniqueness of our living planet earth was realized only after the voyages of astronauts and the photographs taken by the NASA Apollo mission. It was described in the solar system that the atmosphere of the planets Mars and Venus consists of carbon-dioxide (CO_2) with little Nitrogen (N_2), while the atmosphere of the earth consists of mainly (N_2) and oxygen (O_2). The crucial parameters that determine the nature of the atmosphere depends on the size of the planets and its distance from the sun. Suppose the earth were 6% closer to the sun, which would increase solar radiation and temperature of the earth would be about 700 °K. This would not have allowed volcanic steam to condense to form oceans. The earth's atmosphere would have CO_2 and there would not have any life on earth. It seems earth alone has oceans of liquid water on its surface, which is probably essential for development of life.

About 4500 million years ago when earth was formed there was no life and the atmosphere was without oxygen. All terrestrial planets in their early history of formation had only Hydrogen (H) and Helium (He) in their atmosphere that acquired from solar nebula. Then these planets were completely devoid of atmospheres like the moon today. Subsequently new gases accumulated which were exhumed from their interior through volcanic activity. In this process earth acquired in its atmosphere largely water vapour, CO_2, N_2 and a number of other minor gas constituents. The temperature of the earth at that time probably was 260 °K. However water vapour, CO_2 are efficient green house gases, trapped IR-radiation and increased the temperature of the earth surface to about 300 ° K. In the beginning the atmospheric pressure was about 6 hPa (mb = millibars, hPa = hectopascales). Under these conditions atmospheric water vapour condensed to form oceans on the surface,and the atmosphere probably contained N_2, H, oxides of carbon, with traces of water vapour, ammonia and methane gases. This probably took place some 600 to 1000 million years after the earth was formed.

Geological evidence suggests that the first forms of life began at this time was bacteria and have been called "our ultimate grand parents". As the bacteria evolved into cyanobacteria forms, blue-green in colour, they used sunshine and produced oxygen and organic materials. As these bacteria colonized the planet earth and influenced the composition of atmosphere. As time passed for another 1500 million years or about 2000 million years ago, green plants evolved and the photosynthesis process added oxygen to the atmosphere. This stage favoured for the development of animal forms which breathed in oxygen and breathed out carbondioxide. The biological activity together with plant photosynthesis further changed the atmosphere to the present day composition. This seems to be the overall considered view of scientists regarding evolution of the atmosphere during its long history of 4 billion years.

5.1 Field Variables

Atmosphere is a continuous medium. The physical quantities that specify the atmosphere such as pressure, temperature, density, wind velocity etc. are continuous functions of space and time. They are called field variables or simply fields. Field variables are differentiable. A field is a part of space where every point of the space has a value for any measurable quantity such as pressure, temperature, humidity, specific volume etc,. Thus if we have measurements of pressure, temperature etc; in the atmosphere (space), the plot of these (pressure, temperature erc) provide us pressure field, temperature field etc. The plot of equal pressure, equal temperature lines etc, provide us isobars, isotherms etc; which enables the meteorologist to study these scalar fields to draw some conclusions of weather. The typical scales of atmospheric motions are given in the Table 5.1.

Table 5.1

Type of Motion	Horizontal scale 'm'
Molecular mean free path	10^{-7}
Small turbulent eddies	$10^{-2} - 10^{-1}$
Tides	$10^{-2} - 1$
Dust devils	1–10
Gusts of wind	$10 - 10^{2}$
Tornadoes	10^{2}
Thunderstorms	10^{3}
Warm/cold front, squall lines	$10^{4} - 10^{5}$
Hurricanes	10^{5}
Cyclones	10^{6}
Planetary waves/Rossby waves	10^{7}
Atmospheric tides and meanzonal wind	10^{8}

Question

1. Write briefly the considered scientific view of the evolution of the earth's atmosphere. Why is it called bacteria as our ultimate grand parents?

CHAPTER 6

Physical Variables

The principal state variables are pressure, temperature and the specific (or molar) volume. We know that

$$\rho = \text{density} = \frac{\text{mass}}{\text{volume}} = \frac{M}{V}$$

$$\alpha = \text{specific volume} = \frac{\text{volume}}{\text{mass}} = \frac{V}{M}$$

i.e.,

$$\alpha = \frac{1}{\rho}$$

$$p = \text{pressure} = \frac{\text{normal force}}{\text{area}} = \frac{\Delta F}{\Delta A}$$

6.1 Definition

An ideal (or perfect) gas is one in which there are no forces of molecular interaction.

Ideal gases obey the following four laws.

1. **Boyle's Law (Isothermal process)** : Temperature remaining constant, the volume (v) of a mass of gas is inversely proportional to the pressure (p)

$$V \alpha \frac{1}{p} \quad \text{when T is constant.}$$

2. Charle's Law (Isobaric process) : Pressure remaining constant, the volume of a given mass of gas is directly proportional to the absolute temperature (T)

$$V \alpha T \quad \text{when p is constant}$$

3. Gay-lussac's Law (Isochoric process) : Volume remaining constant, the pressure of a given mass of gas is directly proportional to its absolute temperature.

$$p \alpha T \quad \text{when V is constant}$$

4. Avogadro's Law : Equal volumes of different gases at the same temperature and pressure contain equal number of molecules

OR

Moles of different gases occupy the same volume at the same temperature and pressure

$$V \alpha N, \text{ where } V = \text{volume of gas, } N = \text{Number of gas molecules}$$

Mole : One mole (1 mol) of any substance is a quantity of matter expressed in grams (or kg) and numerically equal to molecular weight.

OR

A gram molecule or mole (kilogram molecule or k mole) is the amount of substance, expressed in grams (kilograms) equal to its molecular weight.

The number of atoms in a gram-atom (for all substances) = N_A [this number is called Avogadro number]

Number of molecules in a gram molecule = The number of atoms in a gram atom = N_A.

For all substances :

$$N_A = 6.023 \times 10^{23} \text{ mole}^{-1}$$
$$= 6.023 \times 10^{26} \text{ kmole}^{-1}$$

A mole thus may be defined as the mass of a substance which contains 6.023×10^{23} structural units (atoms, molecules or ions).

SI unit of pressure is one Pascal (1 Pa), equal to one Newton per square meter (1 N m^{-2})

1 bar = 10^5 Pa; 1 mb (millibar) = 10^2 Pa = hPa (hecto pascals); 1Pa = 1Nm^{-2}

6.2 Molecular Mass μ

It is the mass per mole. It is convenient to measure the amount of gas in terms of moles 'n' instead of the mass μ. Thus $M = n\mu$. Numerically μ is molecular weight. One mole of an ideal gas at STP occupies 22,414 cm^3 or 22.414 *l*.

All ideal gases at STP occupy 22.414 l

STP = Standard Temperature (273 °K), pressure (1 atm)

1 atm = one atmosphere = 1013×10^2 N m^{-2} or 1013×10^2 N/m^2

$= 1.013 \times 10^5$ Pa

$$n = \frac{M}{\mu}$$

where μ = molecular mass; M = mass

n = no of moles

R = 8.31×10^3 J k–$mole^{-1}$ K^{-1}

= 0.0821 litre. atm. $mole^{-1}$ K^{-1}

= 0.848 Kg f. m. mol^{-1} K^{-1}

= 8.31×10^7. erg. mol^{-1} K^{-1}

= 1.987 cal mol^{-1} K^{-1}

Ideal Gas Equation :

$$pV = nRT \qquad \text{.....(6.1)}$$

$$pV = \frac{M}{\mu} RT$$

$$p\frac{V}{M} = \frac{1}{\mu} RT \quad \text{or} \quad p\alpha = R'T$$

The general equation of state for an ideal gas is given by

$$pV = nRT \quad \text{or } p\alpha = R'T$$

where $$R' = \frac{R}{\mu}$$

p = pressure

v = volume

n = number of moles

R = Universal gas constant

T = Temperature in absolute degrees (K)

μ = molecular mass or molecular wt

Note : R = constant and it is same for all gases at sufficiently high temperature and low pressure.

R is numerically equal to work done by one mole of ideal gas when it is heated isobarically by one degree.

(i) In SI (mks) units (where pressure p is 1Pa or 1 N m^{-2}, unit of V is 1 m^3) the numerical value of R is given by

$$R = \frac{pV}{nT}$$

$$R = 8.314\ (N\ m^{-2})\ .\ m^3\ .\ mol^{-1}\ .\ K^{-1}$$

$$= 8.314\ J\ .\ mol^{-1}\ .\ K^{-1}$$

(ii) In cgs units (where pressure p is 1 dyne cm^{-2}, unit of volume 1 cm^3)

$$R = \frac{pV}{nT}$$

$$p = 76 \times 13.6 \times 980 \text{ dynes } cm^{-2}$$

$$V = 22400cc;$$

$$T = 273\ ^\circ K$$

$$n = 1 \text{ mole}$$

$$1\ J = 10^7 \text{ erg}$$

$$R = 8.314 \times 10^7\ (dyn \times cm^{-2}) \times cm^3\ .\ mol^{-1}\ .\ K^{-1}$$

$$= 8.314 \times 10^7\ erg \times .\ mol^{-1}\ K^{-1}$$

$$= 8.314\ J\ mol^{-1}\ K^{-1}$$

(iii) When volume is expressed in liters (l) pressure in atmospheres (atm) and temperature in kelvins, the value of universal constant R is given by

$$R = \frac{pV}{nT}$$

$$p = 1 \text{ atm}$$

$$V = 22.4 \text{ lit}$$

$$T = 273^\circ K$$

$$n = 1 \text{ mole}$$

$$R = 0.08207\ atm\ .\ mol^{-1}\ .\ K^{-1}$$

For a fixed mass of an ideal gas the product nR is constant. hence $\frac{pV}{T}$ is constant.

For different temperatures

$$T_1, T_2, T_3 \ldots \text{etc.}$$

corresponding pressures

p_1, p_2, p_3 ... etc.

and volumes V_1, V_2, V_3 ... etc.

We have $$\frac{p_1 V_1}{T_1} = \frac{p_2 V_2}{T_2} = \frac{p_3 V_3}{T_3} = \ldots = \text{constant.}$$

From Avogadro's law, the volume of 1 kg molecular weight of an ideal gas occupies volume 22.4 m^3 at 273 °K and 1 atmospheric pressure

$$1 \text{ atm} = 1013.25 \times 10^2 \text{ Nm}^{-2}$$
$$V = 22.4 \text{ m}^3$$
$$n = 1$$
$$T = 273$$

Substituting these in pV = nRT we get the value of R.

$$R = \frac{1013.25 \times 10^2 (\text{Nm}^{-2}).22.4\text{m}^3}{1\text{mol} \times 273\text{K}}$$

$$= 8.314 \times 10^3 (\text{Nm}^{-2}) \text{ mol}^{-1} . \text{ m}^3 . \text{ K}^{-1}$$
$$= 8.314 \times 10^3 \text{ J.k mole}^{-1} . \text{ K}^{-1}$$

Ex : The volume of one mole of an ideal gas at STP.

Given $p = 1 \text{ atm} = 1013.25 \times 10^2 \text{ Pa};$

$$T = 273 \text{ °K}$$
$$n = 1 \text{ mol}$$
$$R = 8.314 \text{ J mol}^{-1} \text{ K}^{-1}$$

$$\therefore \quad V = \frac{nRT}{p} = \frac{(1\text{mole})(8.314. \text{ J mol}^{-1} . \text{K}^{-1})(273\text{K})}{1013.25 \times 10^2 \text{ Pa}}$$

$$= 0.0224 \text{ m}^3$$
$$= 22.4 \text{ l}$$

$$1\text{m}^3 = 100 \times 100 \times 100 \text{ cm}^3 \qquad \text{liter} = 1000 \text{ cm}^3$$
$$= 1000 \times 1000 \text{ cm}^3$$
$$= 1000 \ l$$

Ex : A tank is attached to an air compressor which has a mass of air 115 gm, temperature 303° K, gauge pressure 4×10^5 Pa. What is the volume of air in the compressor ? Also find the volume at STP.

Given the molecular mass of dry air is 28.8g mol^{-1};

R = gas constant = 8.314 J mol^{-1} K^{-1}

Sol :

(i) p = pressure = gauge pressure + atmospheric pressure.

$= 4 \times 10^5$ Pa + 1013.25×10^2 Pa $\simeq 5.01 \times 10^5$ Pa

or 5.01×10^5 Nm^{-2}

$M = n\mu$; given M = 115 g; μ = 28.8g mol^{-1}

$\therefore$ $115 = n \times 28.8$ gives n = 3.99 mol

Ideal gas equation

$$pV = nRT$$

$$V = \frac{nRT}{p} = \frac{(3.99 \text{ mol})(8.314 \text{ Nm}^{-2})(303\text{K})}{(5.01 \times 10^5 \text{ Nm}^{-2})}$$

$$= \frac{100513.8}{501 \times 10^3} \text{ m}^3 = 0.02 \text{ m}^3 = 0.02 \times 1000 \ l$$

Volume of air in compressor = 20 *l*

(ii) At STP.

$p = 1013.25 \times 10^2$ N m^{-2}

$T = 273^\circ K$

$R = 8.314$ N m^{-2} mol^{-1} K^{-1}

$$V = \frac{nRT}{p} = \frac{3.99 \text{ mol} \times 8.314 \text{ J mol}^{-2} \text{ K}^{-1} \times 273\text{k}}{1013.25 \times 10^2 \text{ Nm}^{-2}}$$

$= 89.4 \times 10^{-3}$ m^3

$\simeq 89.4$ *l*

($\because$ 1 $m^3 = 10^3$)

Ex : Find the formula for variation of pressure with altitude in the atmosphere. Assume the atmosphere to be isothermal. What would be the pressure under these conditions over Mt. Everest whose altitude given as 8882 m and T = 273 °K.

Sol :

(i) From the hydrostatic equation (which will be dealt separately) we have

$$dp = -g\rho dz \quad .....(6.2)$$

where g = gravity, z is altitude

we know pv = nRT .....(6.3)

where

$$n = \frac{M}{\mu} \quad .....(6.4)$$

From eq. (6.3) and eq. (6.4)

$$pV = \frac{M}{\mu} RT$$

or
$$p = \frac{M}{V} \cdot \frac{RT}{\mu}$$

$$p = \rho \cdot \frac{RT}{\mu}$$

$$\therefore \quad \rho = \frac{p\mu}{RT}$$

We know density varies with pressure.

Substituting this value of ρ in hydrostatic equation we have

$$\therefore \; dp = -g \times \frac{p\mu}{RT} dz$$

$$\frac{dp}{p} = \frac{-g\mu}{RT} dz \qquad .....(6.5)$$

Let p_1, p_2 be the pressure at altitudes z_1, z_2

Integrating eq. (6.5) in proper limits

$$\int_{p_1}^{p_2} \frac{dp}{p} = \frac{-g\,\mu}{RT} \int_{z_1}^{z_2} dz$$

or
$$\ln \frac{p_2}{p_1} = -\frac{g\,\mu}{RT} (z_2 - z_1) \qquad .....(6.6)$$

Let p_0 be the pressure at the sea level (z = 0) and p be the pressure at altitude z.

From eq. (6.6) we have

$$\ln \frac{p}{p_0} = -\frac{g\mu}{RT}(z-0)$$

or
$$p = p_0 \, \bar{e}\left(\frac{g\mu}{RT}\right) . z \qquad .....(6.7)$$

(ii) p_0 = 1 atm = 1013.25×10^2 Nm^{-2}, T = 273, g = 9.8 ms^{-2}, z = 8882 m, $\mu = 28.8 \times 10^{-3}$ Kg mol^{-1}

$$\therefore \quad \frac{g\mu}{RT} z = \frac{9.8\,\text{m}\,\text{s}^{-2}\left(28.8\times10^{-3}\,\text{kg}\,\text{mol}^{-1}\right)(8882\,\text{m})}{8.34\,\text{J.mol}^{-1}\,\text{K}^{-1}\,(273\,\text{K})}$$

$$= 1.10$$

From eq. (6.7) we have

$$p = (1013.25 \times 10^2\ Nm^{-2})e^{-1.10}$$
$$= (1.01325 \times 10^5\ Pa)e^{-1.10}$$
$$= 0.336 \times 10^5\ Pa$$
$$p = 0.333\ atm\ .\ [\because\ 1\ atm = 1.01325 \times 10^5\ Pa].$$

6.3 Molecular Weight

The molecular weight of a substance may be defined as the number of times its one molecule is heavier than $\frac{1}{12}$th of the weight of one C^{12} atom.

Usually both atomic weight and molecular weight of a substance are expressed in grams and called gram atomic weight, gram molecular weight respectively. Gram atomic wt. of an element is defined as the quantity of the element in grams numerically equal to the atomic weight of the element. Similarly gram molecular wt. of a substance is defined as the quantity of a substance in grams numerically equal to the molecular wt. of the substance.

6.3.1 Mole

A mole is defined as the amount of a substance that contains the same number of atoms or molecules or ions (chemical units) as the number of atoms in exactly 12 grams of C^{12}.

6.3.2 Avogadro Number

The number of atoms in 12 grams of C^{12} is called Avogadro number, it is denoted by N_A. Its value has been determined as $N_A = 6.0225 \times 10^{23}$

$\therefore$ A mole of any substance is the mass of the substance that contains 6.0225×10^{23} molecules. For all substances number of molecules in a gram molecule = the number of atoms in a gram atom = N_A.

6.3.3 Dalton's Law of Partial Pressures

The dry air of atmosphere is a mixture of ideal gases. Dalton's law states that the total pressure of a mixture of ideal gases is equal to the sum of their partial pressures and each gas occupies the total volume.

Using this law we have for each gas

$$p_i V = n_i R_i T,\ i = 1, 2, 3, \ldots\ N.$$

Summing up we have

$$V \sum_{i=1}^{N} p_i = T \sum_{i=1}^{N} n_i R_i$$

p_i = partial pressure of i th constituent

n_i = number of moles of i th constituent

R_i = gas constant of i th constituent

Let

$$n = \sum ni, \quad R = \frac{1}{n} \sum n_i R_i, \quad p = \sum pi$$

The above equation becomes

$$Vp = nT \, . \, R$$

i.e. $PV = nRT.$

This is the gas equation for mixture of gases, where gas constant

$$R = \frac{1}{n} \sum n_i R_i,$$

determined by knowing the composition the atmosphere and molecular weight of each constituent

$$R_d = R \text{ for dry air} = 287 \text{J kg}^{-1} \text{K}^{-1}.$$

Using this constant, the mean molecular weight of dry air is found

$$n_d = \frac{R}{R_d} = \frac{8314}{287} \frac{\text{Jkmol}^{-1} \text{K}^{-1}}{\text{Jkg}^{-1} \text{K}} = 28.97$$

(n_d = no.moles of dry air)

Note : (i) For dry air

$C_p = 1004 \text{ J kg}^{-1} \text{K}^{-1}$ $C_v = 717 \text{ J kg}^{-1} \text{K}^{-1}$

$R = 287 \text{ J kg}^{-1} \text{K}^{-1}$ $g = 9.8 \text{ ms}^{-2}$

(ii) The average partial pressure, at sea level, of Nitrogen is about 760 hPa or 570 mm Hg column, Oxygen is about 240 hPa or 180 mm of Hg column and water vapour 10 hPa or 7.5 mm of Hg column.

6.4 Phase Change

Phase change is a transition of a substance from one phase to another. Water can exist in three states, namely solid, liquid or gas. As temperature decreases and pressure increases an ideal gas can change from the gas phase to the liquid phase, or the solid phase. A substance can exist in either solid, liquid or gas phase, or in two phases simultaneously or in all three phases along the triple line. At any point on the solid-liquid, solid-vapour or liquid-vapour surfaces, two phases can exist in equilibrium and along the triple line all three phases can coexist. A vapour, at the pressure and temperature, at which it can exist in equilibrium with liquid is termed saturated vapour and the liquid is called saturated liquid.

The phase diagram of p-T projection is given below Fig. 6.1. Keeping pressure constant, if a substance is heated, then the points on the dotted line L_1 shows different phases as it passes. Melting point & boiling points are shown p_1, p_2.

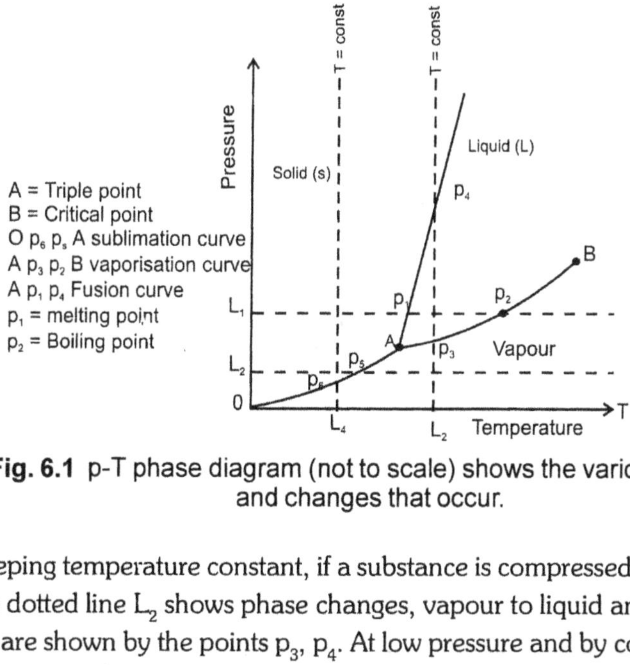

Fig. 6.1 p-T phase diagram (not to scale) shows the various phases and changes that occur.

Keeping temperature constant, if a substance is compressed, then the points on the dotted line L_2 shows phase changes, vapour to liquid and liquid to solid which are shown by the points p_3, p_4. At low pressure and by constant-pressure heating material transforms solid to vapour state p_5, the dotted line L_3 represents this process which is called sublimation. e.g., solid carbon-dioxide under goes sublimation (**Note**: there is no liquid phase). At low temperature, below triple point, if a substance is compressed by increasing pressure, vapour directly changes to solid phase p_6. This process is shown by the dotted line L_4.

6.4.1 Triple Point

For any substance there is one pressure and temperature at which all three phases namely solid, liquid and gas can coexist. This single pressure and temperature is called triple point. In other words, the coexistence of three phases solid, liquid and gas is possible only in one unique state which is called triple point.

Critical Point

Liquid and vapour phases can exist together at the critical point and the corresponding values of pressure (p), temperature (T) and specific volume (α) are the critical pressure, critical temperature and critical specific volume.

1. A Numerical example : Consider the pT-diagram of water (Fig. 6.2). Horizontal line at a pressure of 1 atm intersects freezing point curve at 0 °C, boiling point curve at 100 °C. The boiling point increases with increasing pressure up to the critical temperature of 374 °C. Solid, liquid and vapour can remain only at the triple point, where the vapour pressure is 0.006 atm and the temperature is 0.01 °C.

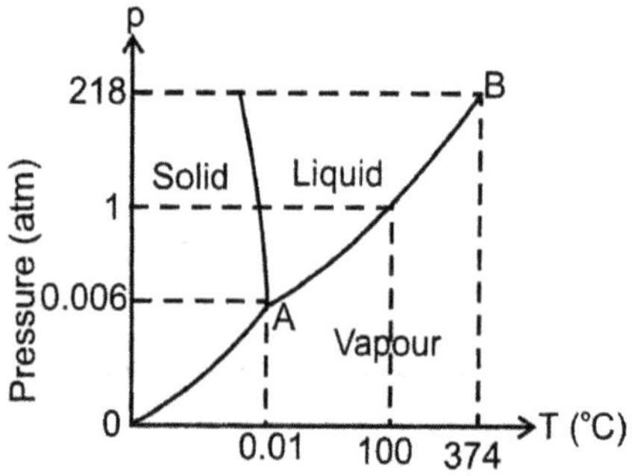

Fig. 6.2 p-T diagram change of water.

2. Another numerical example. Consider pT diagram of carbon-dioxide. The triple point temperature of CO_2 is –56.6 $\simeq$ 57 °C, corresponding pressure at p = 5.11 atm. Because of this CO_2 commonly stored in steel tanks, contain both saturated liquid and vapour. The pressure in these tanks is the vapour pressure of CO_2 at the temperature of tank. Suppose the temperature of the tank is 20 °C then the pressure inside the tank is 56 atm.

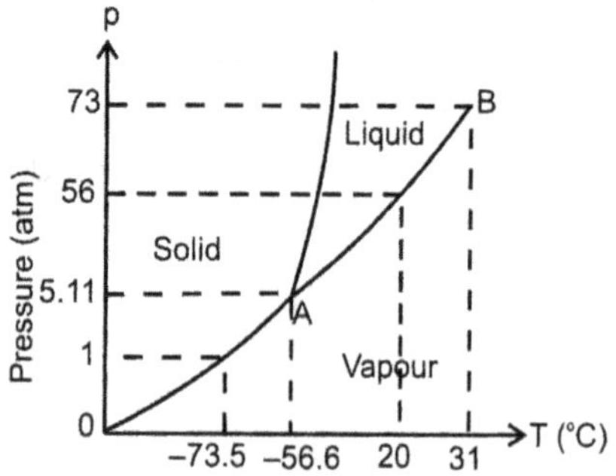

Fig. 6.3 p-T diagram for change of CO_2.

Questions

1. Define an ideal gas. State the four laws that obey ideal gases and equation of state.

2. A tank is attached to an air-compressor which has a mass of air 115 gm, temperature 30 °C, gauge pressure 4×10^5 Pa. What is the volume of air in the compressor. Find this volume at STP. Given molecular weight of dry air 28.8 gm/mol, R = 8.314 J/mol K.

3. Assuming hydrostatic equation derive a formula for variation of pressure with altitude in isothermal atmosphere.

CHAPTER 7

Thermodynamics

7.1 Zeroth Law of Thermodynamics

It states that if a body is in thermal equilibrium with a body C and another body B then they are in thermal equilibrium with one another. Thus when two or more systems are in thermal equilibrium then they are said to have same temperature. This leads to the concept of temperature.

7.1.1 Definition

Specific Heat : Specific heat of a substance is the quantity of heat required to raise the temperature of unit mass (1 gm, 1 mol, 1 kg) through one degree Celsius.

Thermal Capacity : Thermal capacity of a body is the quantity of heat required to raise the temperature of the body through one degree Celsius.

> Note : The specific heat of a body gives the thermal capacity of the body per unit.

Water Equivalent : Water equivalent of a body is numerically equal to its thermal capacity.

7.2 Specific Heat of Gases

Specific heat of a substance depends on the nature of the substance and the amount of work done on expansion of the substance by rise in temperature. The external conditions under which a gas being heated is important and hence required to be stated. Because of this we consider two specific heats of gas (i) Specific heat of gas at a constant volume and (ii) the specific heat of gas at constant pressure.

The specific heat of a gas at constant volume, is the amount of heat required to raise the temperature of one unit mass (1 gm) of gas through one degree Celsius, keeping its volume constant. It is denoted by c_v.

Similarly, the specific heat of a gas at constant pressure is the amount of heat required to raise the temperature of one unit of mass (1 gm) of gas through one degree Celsius keeping its pressure constant. It is denoted by c_p. The specific heat corresponding to a gram molecule of a gas instead of a unit mass, (such specific heats) are termed gram molecular or molar specific heats and these are denoted by C_p, C_v. They are related by c_p, c_v as follows

$$C_p > C_v$$
$$C_p = \mu\, c_p$$
$$C_v = \mu c_v$$

Where μ = molecular wt of the gas. The difference between two specific heats of a gas is given by

$$C_p - C_v = \frac{R}{J}$$

where J is mechanical equivalent of heat
R = gas constant

$$C_p - C_v = AR, \text{ where } A = \frac{1}{J}$$

$$\frac{C_p}{C_v} = \frac{c_p}{c_v} = \gamma, \text{ which is called adiabatic exponent}$$

W = work = JH, H = heat supplied,

7.3 Law of Conservation of Energy

Energy is neither created nor destroyed, it only changes from one state to another. Or the total energy of an isolated system remains constant irrespective of the process occurring in the system. This implies that the motion of matter can neither be created nor destroyed, it can change from one form to another.

7.3.1 Heat Capacity (C)

Heat capacity of a body (C) is defined by the relation

$$C = \frac{\delta Q}{dT}$$

where

δQ= Infinitesimal amount of heat transferred to the body

dT = Small change in the temperature of the body T to T + dT.

C depends upon the mass of the body, chemical composition, thermodynamic state and the process used for transfer of heat.

μ = Molecular mass or simply molecular weight

M = Mass of the gas

n = Number of moles.

M = $n\mu$.

7.3.2 The Specific Heat (c)

c is the heat capacity of a unit mass of a homogeneous substance. For homogeneous substance

$$c = \frac{C}{M}$$

where C = heat capacity

M = Mass of the substance.

$$c = \frac{C}{M} = \frac{1}{M}\frac{\delta Q}{dT}$$

or

$$\delta Q = cM\,dT$$

$$\delta Q = c \,.\, n\mu \,.\, dT \quad [C_\mu = \mu c = \text{molar heat capacity}]$$

$$\delta Q = nC_\mu\,dT$$

$$C_\mu = \frac{1}{n}\frac{\delta Q}{dT}$$

The molar heat capacity C_μ of water $\simeq$ 75.3 $Jmol^{-1}\,k^{-1}$

The molar heat capacity C_μ is the heat capacity of one kilo-mole of a substance.

7.3.3 Definition

Regelation : Ice melts under pressure (when pressure increases) and resolidifies when the pressure is decreased. This phenomena is called Regelation.

Note : The specific heat of sea water is 40% more than that of soil (land). The diurnal temperature variation of sea water is very small compared to the diurnal variation of temperature of land, which is about five times. In contrast the downward temperature transport (or penetration) in sea water is much more (due to turbulent mixing) than land where it is shallow. If Δt_l denotes diurnal variation of temperature of land and Δt_s denotes the diurnal temperature variation of sea water, then $\Delta t_l = 5\,\Delta t_s$.

7.4 First Law of Thermodynamics

Let heat be supplied to a system which is capable of doing work. Then the quantity of heat absorbed by the system is equal to the sum the external work performed by the system and increase in its internal energy.

Suppose one unit mass of gas (specific volume α) is enclosed in a cylinder which has a movable piston whose cross sections area is $d\sigma$ as shown in Fig. 7.1. Let the pressure (p) inside the cylinder be same as that of surrounding atmosphere. If the piston moves through a small distance 'dx', then the work done (δw) by expanding gas inside is given by

$$\text{pressure} = \frac{\text{Force}}{\text{area}}$$

$$[\text{Force} = \text{pressure} \times \text{area}]$$

$$[\text{Work} = \text{Force} \times \text{distance}]$$

$$\delta w = p d\sigma \times dx$$

$$\delta w = p d\alpha \qquad \text{.....(7.1)}$$

where

$$d\alpha = d\sigma \,.\, dx \text{ increase in volume.}$$

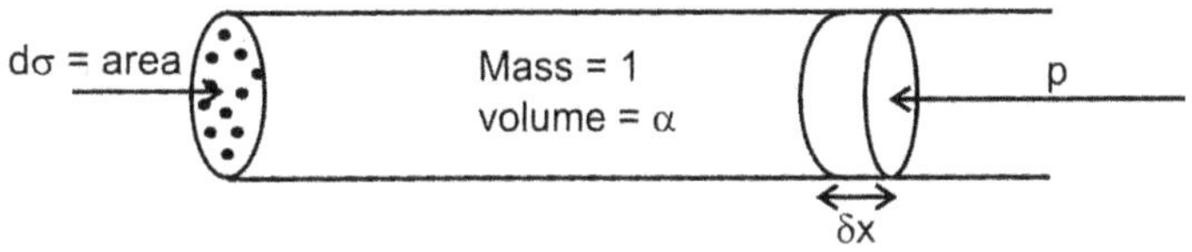

Fig. 7.1

Suppose the piston is fixed (i.e. volume is constant), but an amount of energy δQ in the form of heat is added to the gas inside cylinder. This will increase the internal energy (di) by rise in temperature dT, because we have prevented the gas from doing any work against the surroundings

$$\therefore \qquad \delta Q = C_v \, dT \qquad \text{.....(7.2)}$$

where C_v = specific heat of gas at constant volume.

Note : Internal energy of an ideal gas is proportional to the absolute temperature only, and independent of pressures and volume.

In generality suppose that the piston is free to move and we have added heat energy (δQ) to the system. This heat will be partly used to increase the internal energy (increase in temperature of the gas) and partly to do work against the surrounding (piston will move to expand gas.) i.e.,

$$\delta Q = di + \delta w \quad \text{or} \quad dH = di + dw \qquad \text{.....(7.3)}$$

Heat added to the system is equal to the sum of internal energy and the work done by the system.

where,

δQ = added energy

di = increase in internal energy

δw = work done by the gas against the surrounding.

dH = heat quantity.

For an ideal gas, which follows the equation of state ($p\alpha = R'T$)

$$di = C_v\, dT, \quad \delta w = p d\alpha$$

The eq. (7.3) becomes,

$$\delta Q = C_v\, dT + p d\alpha$$

or $$\dot{Q} = \left(\frac{dQ}{dt}\right) = C_v \frac{dT}{dt} + p\frac{d\alpha}{dt} \qquad \text{.....(7.4)}$$

[This conversion process enables the solar heat energy to drive the atmospheric motions.]

Eq. (7.4) expresses the first law of Thermodynamics or Thermodynamic energy equation.

7.4.1 Definition

Every thermodynamic system in equilibrium state possesses a state variable called internal energy (i) whose change di is given by the differential equation

$$di = dH - dw$$

We know for an ideal gas

$$pV = nRT \qquad [\rho = \frac{M}{V},\ M = n\mu,\ \alpha = \frac{1}{\rho} = \frac{V}{M},\ R' = \frac{R}{\mu}]$$

or $$p.\frac{M}{\rho} = nRT$$

or $$p\alpha = \frac{n}{M} RT$$

$$p\alpha = \frac{1}{\mu} R.T$$

$$p\alpha = R'T$$

Differentiating,

$$p d\alpha + \alpha dp = R'\, dT$$

$\therefore$ $$p d\alpha = R'\, dT - \alpha dp \qquad \text{.....(7.5)}$$

Substituting this in eq. (7.4), we have

$$\delta Q = C_v dT + R' dT - \alpha dp$$

or $$\delta Q = (C_v + R')dT - \alpha dp \qquad .....(7.6)$$

For an isobaric system dp = 0, eq. (7.6) gives

$$\delta Q = (C_v + R')\, dT \qquad ....(7.7)$$

For an isobaric system by definition

$$\delta Q = C_p dT \qquad .....(7.8)$$

From eq. (7.7) and eq. (7.8) we have

$$C_p dT = (C_v + R')\, dT$$

$$C_p - C_v = R' \qquad .....(7.9)$$

From eq. (7.4)

$$\delta Q = C_v dT + p d\alpha$$

or $$\frac{\delta Q}{dt} = C_v \frac{dT}{dt} + p\frac{d\alpha}{dt} \qquad .....(7.10)$$

$\frac{d}{dt}$ stands for differentiation with reference to time.

If H denotes amount of heat added per unit of mass and per unit of time then the above equation becomes

$$H = \frac{\delta Q}{dt} = C_v \frac{dT}{dt} + p\frac{d\alpha}{dt} \qquad .....(7.11)$$

$$dh = C_v\, dT + p d\alpha; \quad \text{where } H = \frac{dh}{dt}$$

$$p d\alpha = R'\, dT - \alpha dp \quad \text{(eq. 7.5)}$$

$$p\,\frac{d\alpha}{dt} = R'\frac{dT}{dt} - \alpha\frac{dp}{dt}$$

or $$p\,\frac{d\alpha}{dt} = \left(C_p - C_v\right)\frac{dT}{dt} - \alpha\frac{dp}{dt} \quad \left(\because\ \dot{Q} = p\frac{d\alpha}{dt} + C_v\frac{dT}{dt}\right)$$

$\therefore$ $$\dot{Q} = H = C_p\frac{dT}{dt} - \alpha\frac{dp}{dt} \qquad .....(7.12)$$

or $$dh = C_p dT - \alpha dp$$

Eq. (7.11) and eq. (7.12) represent the first law of thermodynamics which are used in meteorology.

Note : $C_v = \left(\frac{dh}{dT}\right)$ α constant;

$C_p = \left(\frac{dh}{dT}\right)$ p constant;

For dry air $C_v = 717 J\ kg^{-1} K^{-1}$ and $R = 287 J\ kg^{-1}\ {}^oK^{-1}$

∴ $C_p = 717 + 287$ $(\because\ C_p = C_v + R)$

$Cp = 1004\ J\ kg^{-1}\ {}^oK^{-1}$

7.5 Adiabatic Process

We know

$\delta Q = C_v dT + pd\alpha$ eq. (7.4)

$dh = C_v dT + pd\alpha$ eq. (7.11)

and $dh = C_p dT - \alpha dp$ eq. (7.12)

In an adiabatic process neither heat enters nor leaves the process i.e.

Q or H = Constant. i.e., dh = 0

From eq. (7.11) $0 = Cv\ dT + pd\alpha$

$$p = \frac{R'T}{\alpha}$$

or $$C_v dT + \frac{R'T}{\alpha} d\alpha = 0$$

or $$\frac{C_v}{T} dT + R' \frac{d\alpha}{\alpha} = 0$$

Integrating

$$C_v \ln T + R' \ln \alpha = \ln\left(\frac{const}{\mu}\right)$$

$$C_v \ln T + R' \ln \alpha = const.$$

$$\ln T + \ln \alpha^{\frac{R'}{C_V}} = const.$$

$$\ln T(\alpha)^{\frac{R'}{C_V}} = const.$$

$$T(\alpha)^{\frac{R'}{C_V}} = const.$$

Similarly $$T_1 (\alpha_1)^{\frac{R'}{C_v}} = \text{const.}$$

$$T_1 = \text{const.} \ (\alpha_1)^{\frac{-R'}{C_v}}$$

$$\frac{T}{T_1} = \left(\frac{\alpha}{\alpha_1}\right)^{\frac{-R'}{C_v}} \qquad .....(7.13)$$

This shows that T is inversely proportional to α

Again for adiabatic process dh = 0,

and eq. (7.12) gives

$$C_p dT - \alpha dp = 0 \qquad [p\alpha = R' \ T]$$

or $$C_p dT - \frac{R' \ T}{p} \ Tdp = 0 \qquad (\alpha = \frac{R'T}{p})$$

or $$C_p \frac{dT}{T} - R' \frac{dp}{p} = 0$$

[Differential equation of variables are separable is exact differential]

Integrating

$$C_p \ln T - R' \ln p = \text{const.}$$

or $$\ln T - \frac{R'}{C_p} \ln p = \text{const.}$$

$$\ln T - \ln (p)^{\frac{R'}{C_p}} = \text{const.}$$

$$\ln \frac{T}{(p)^{\frac{R'}{C_p}}} = \text{const.}$$

or $$T = (p)^{\frac{R'}{C_p}} \times \text{const.}$$

$\therefore$ $$\frac{T}{T_1} = \left(\frac{p}{p_1}\right)^{\frac{R'}{C_p}} \qquad .....(7.14)$$

This shows T is directly proportional to pressure

From eq. (7.13) and eq. (7.14) we have

$$\left(\frac{\alpha}{\alpha_1}\right)^{\frac{-R'}{C_v}} = \left(\frac{p}{p_1}\right)^{\frac{R'}{C_p}}$$

Taking powers of $-\left(\frac{C_v}{R'}\right)$ on both sides, we have.

$$\frac{\alpha}{\alpha_1} = \left(\frac{p}{p_1}\right)^{\frac{R'}{C_p} \times \frac{-C_v}{R'}}$$

$$\frac{\alpha}{\alpha_1} = \left(\frac{p}{p_1}\right)^{\frac{-C_v}{C_p}} \qquad \text{.....(7.15)}$$

This shows that α is inversely proportional to p

From eq. (7.15) we have [taking $\frac{-C_p}{C_v}$ power both sides]

$$\frac{p}{p_1} = \left(\frac{\alpha}{\alpha_1}\right)^{\frac{-C_p}{C_v}}$$

or $\quad p \propto \alpha^{-\gamma}$ or $p = \text{const.}\ \alpha^{-\gamma}$, where $\frac{C_p}{C_v} = \gamma$

or $\quad p\,\alpha^{\gamma} = \text{const.}$ .....(7.16)

or $\quad pV^{\gamma} = \text{const.}\ (\alpha = \frac{V}{M})$ .....(7.16')

Again from eq. (7.14) we have

$$\frac{T}{T_1} = \left(\frac{p}{p_1}\right)^{k} \qquad \text{where } k = \frac{R'}{C_p} = \frac{\gamma - 1}{\gamma}\left[\text{or } \frac{T_1}{T} = \left(\frac{p}{p_1}\right)^{-k}\right]$$

Let $\quad p_1 = p_0 \simeq 1000 \text{ h Pa}, T_1 = \theta$

$$\left[\frac{C_p}{C_v} = \gamma, C_p - C_v = R', k = \frac{R'}{C_p} = 1 - \frac{C_v}{C_p}, k = 1 - \frac{1}{\gamma} = \frac{\gamma - 1}{\gamma}\right]$$

The above equation becomes

$$\frac{T}{\theta} = \left[\frac{p}{1000}\right]^{k} \qquad \text{.....(7.17)}$$

$$k = \frac{AR}{C_p}, \ [\text{where } A = \text{Thermal equivalent of work}]$$

where $k = 0.286$ and θ is called potential temperature

$$\theta = T\left(\frac{p_1}{p}\right)^{\frac{\gamma-1}{\gamma}}$$

eq. (7.17) is called Poisson equation

$$\theta = T\left(\frac{p}{1000}\right)^{-k} \qquad .....(7.17')$$

$$\theta = T\left(\frac{1000}{p}\right)^{k}$$

$$\theta = T\left(\frac{p}{p_0}\right)^{-k}$$

where $p_0 = 1000$

7.5.1 Definition

Potential Temperature (θ) : It is the temperature attained by a sample of dry air when the sample is compressed or expanded adiabatically from a given state to the pressure of 1000 h Pa (= 100 K Pa).

θ is constant during adiabatic process. i.e., θ is conservative quantity in adiabatic process.

7.6 Another Form of Thermodynamic Equation

$$\frac{T}{\theta} = \left[\frac{p}{p_0}\right]^{k} \text{ where } p_0 = 1000 \quad \text{eq. (7.17)}$$

where $k = \frac{R'}{C_p}$ and $\left(k = \frac{AR}{C_p}\right)$

Taking logarithms and differentiating w.r.t. time 't' we have

$$\ln T - \ln \theta = k\,[\ln p - \ln p_0]$$

$$\frac{1}{T}\frac{dT}{dt} - \frac{1}{\theta}\frac{d\theta}{dt} = \frac{k}{p}\frac{dp}{dt}$$

or

$$\frac{1}{\theta}\frac{d\theta}{dt} = \frac{1}{T}\frac{dT}{dt} - \frac{k}{p}\frac{dp}{dt}$$

Multiplying both sides by C_pT we have

$$C_p \frac{T}{\theta}.\frac{d\theta}{dt} = C_p \frac{dT}{dt} - \frac{kC_pT}{p}\frac{dp}{dt}$$

where $k = \frac{R'}{C_p} = \frac{\gamma - 1}{\gamma}$

$$= C_p \frac{dT}{dt} - \frac{-R'T}{p}\frac{dp}{dt}$$

$$\therefore \quad C_p \frac{T}{\theta}\frac{d\theta}{dt} = C_p \frac{dT}{dt} - \alpha \frac{dp}{dt} \qquad \text{(since } p\alpha = R'T) \qquad .....(7.18)$$

Comparing eq. (7.18) with eq. (7.12) viz,

$$H = C_p \frac{dT}{dt} - \alpha \frac{dp}{dt}$$

we have $C_p \frac{T}{\theta}.\frac{d\theta}{dt} = H$

or

$$\frac{d}{dt}(\ln\theta) = \frac{1}{C_p}\frac{H}{T} \qquad .....(7.19)$$

eq. (7.19) is another form of thermodynamic equation.

7.6.1 Definition

Polytropic Process : A thermodynamic process in which specific heat C of the gas remains constant is called polytropic process.

The quantity $\eta = \frac{C_p - C}{C_v - C}$ is called polytropic exponent.

Some properties of common thermodynamic process of ideal gases are listed below : It is presumed that the mass of the gas remains constant. Various relations between state variables are listed. Subscripts 1 and 2 refer to initial and final states,

1. Isobaric Process : (p = constant): $\frac{V}{T}$ = constant.

   Work done in the process $\delta w = pdV$ or $W = p(V_2 - V_1)$

Amount of heat transferred

$$\delta Q = C_p dT \text{ or } Q \text{ (or H)} = C_p(T_2 - T_1)$$

Change in internal energy

$$di = C_v dT \text{ or } \Delta i = C_v (T_2 - T_1)$$

Heat capacity

$$C_p = \frac{M}{\mu} \cdot \frac{\gamma R}{\gamma - 1}$$

$$C_p = \frac{M}{\mu} \cdot \frac{R'}{k}$$

Polytropic exponent $\eta = 0$, where $\gamma = \dfrac{C_p}{C_v}$

2. ***Isothermal Process*** : (T = constant.)

Relation between state variables PV = const. or it obeys Boyle's law.

work done in the process,

$$\delta w = pdV$$

or $$W = \frac{M}{\mu} RT \ln \frac{V_2}{V_1} = \frac{M}{\mu} RT \ln \frac{p_1}{p_2}$$

Amount of heat transferred

$$\delta Q = \delta w \quad \text{or} \quad Q = W$$

Change in internal energy

$$di = 0 \quad \text{or } \Delta i = 0,$$

Heat capacity $C_T = \pm \infty$

Polytropic exponent $\eta = 1$

3. ***Isochoric Process*** : (V = constant) : $\dfrac{p}{T}$ = cost. or it obeys

Gay-lussacs Law.

Work done in the process

$$\delta w = 0 \quad \text{or} \quad W = 0$$

Amount of heat transferred

$$\delta Q = C_v \, dT \text{ or } Q = C_v(T_2 - T_1)$$

Heat capacity $C_v = \dfrac{M}{\mu} \dfrac{R}{\gamma - 1}$, polytropic exponent $\eta = \pm\infty$.

4. *Adiabatic Process* : $\delta Q = 0$, pV^{γ} = const.

or $$\frac{T}{(p)^{\frac{R'}{C_p}}} = \text{const.}$$

$$\theta = \text{potential temperature} = T\left(\frac{p}{p_0}\right)^{-k}$$

where $k = \frac{R'}{C_p} = \frac{\gamma - 1}{\gamma}$ and $p_0 = 1000$ hPa.

$$p\alpha^{\gamma} = \text{constant.}$$

Work done in the process. $\delta w = pdV = -di$

or $$W = -\Delta i = C_v (T_1 - T_2) = \frac{1}{\gamma - 1} (P_1 V_1 - P_2 V_2)$$

Amount of heat transferred

$$\delta Q = 0; Q = 0;$$

Change in internal energy

$$di = C_v \, dT = -\delta w, \quad \text{or} \quad \Delta i = -W = C_v (T_2 - T_1)$$

Heat capacity Cad = 0, polytropic exponent $\eta = \gamma = \frac{C_p}{C_v}$

5. Polytropic Process : (heat capacity = C = constant), pV^{η} = const.

$$p\,T^{\frac{\eta}{1-\eta}} = \text{const.}$$

or $$\eta = \frac{C - C_p}{C - C_\upsilon}$$

$$V\,T^{\frac{1}{\eta - 1}} = \text{const.}$$

Work done in the process. $\delta w = pdV$,

$$W = \frac{1}{\eta - 1} (p_1 V_1 - p_2 V_2)$$

Amount of heat transferred $\delta Q = CdT$ or $Q = C(T_2 - T_1)$

Change in internal energy $di = C_v dT$ or $\Delta i = C_v (T_2 - T_1)$

Heat capacity $C = \frac{M}{\mu}\frac{R(\eta-\gamma)}{(\eta-1)(\gamma-1)}$

Polytropic exponent.

$$\eta = \frac{C - C_p}{C - C_v}$$

7.7 Relationships for a Polytropic Process of an Ideal Gas

(i) Heat capacity

$$C = \frac{M}{\mu}\frac{R(\eta-\gamma)}{(\gamma-1)(\eta-1)}$$

$$\frac{\delta Q}{dV} = \frac{\gamma - \eta}{\gamma - 1}p$$

$$\frac{\delta Q}{dp} = \frac{\eta - \gamma}{\gamma - 1}V$$

$$\frac{\delta w}{\delta Q} = \frac{\gamma - 1}{\gamma - \eta}$$

$$\frac{di}{\delta Q} = \frac{\eta - 1}{\eta - \gamma};$$

$$\frac{dH}{\delta Q} = \frac{\gamma(\eta - 1)}{\eta - \gamma}$$

The lines representing isobaric, isothermal, isochoric, adiabatic and polytropic process in any thermodynamic diagrams are respectively called isobars, isotherms, isochores, adiabats and polytropis. The following Figures. of an ideal gas in V-P, T-P and V-T diagrams shows the isobars, isotherms, isochores and adiabats with initial state 1. of the gas is taken the same for all the processes.

Note :

$$\eta = \frac{C_p - C}{C_v - C}$$

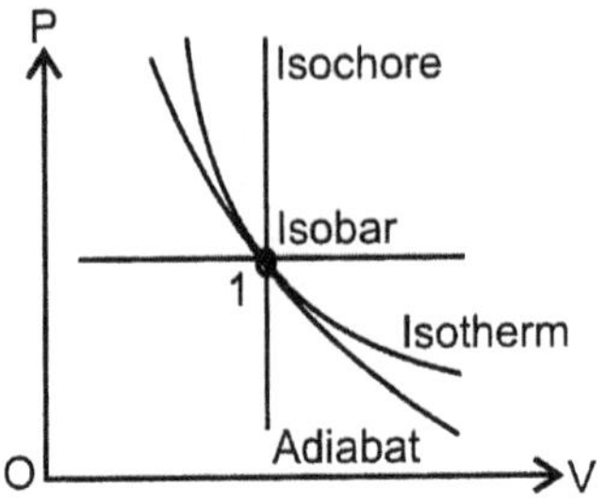

Fig. 7.2

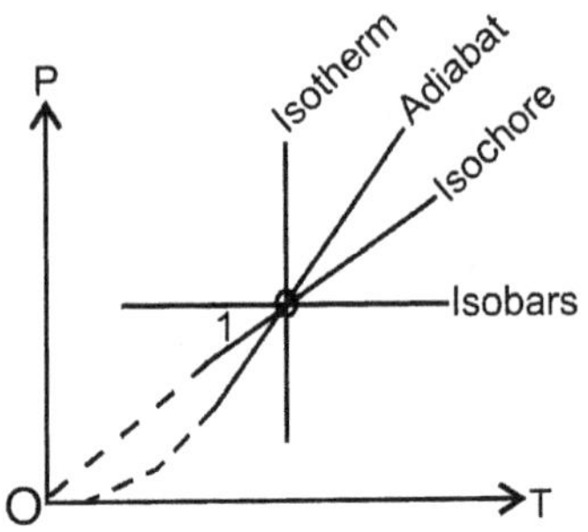

Fig. 7.3

Isobars, isotherms, Adiabats, Isochores of an ideal gas in V-P, T-P, and V-T diagrams and η-c relation. p = pressure, T = temperature, V = volume, C = heat capacity, η = polytropic exponent, (not to any scale)

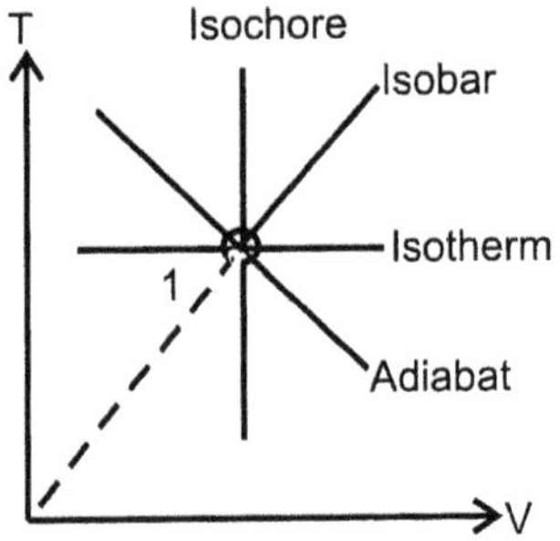

Fig. 7.4

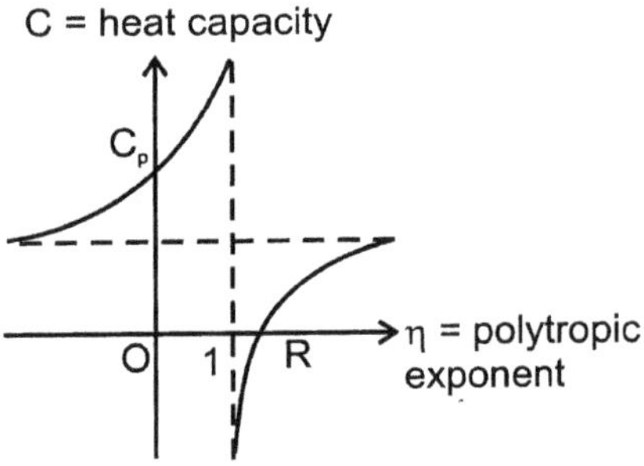

Fig. 7.5

Where C = polytropic thermal heat capacity. η is dimensionless parameter.

If C = C_p polytropic process reduces to isobaric process

If C = C_v polytropic process reduces to isochoric process

If C = 0 polytropic process reduces to adiabatic process

If C = ∞ polytropic process reduces to isothermal process

7.8 Entropy and the Second Law of Thermodynamic

Reversible Process : A thermodynamic process that allows the system to return to its initial state without any changes in its environmental medium is called a reversible process. In reversible process each state of the system is in a equilibrium state.

Irreversible Process : A thermodynamic process that does not allow the system to return to its initial state without any changes in its environmental medium is called an irreversible process. It means that a system can return to its initial state but the system plus its environment cannot be restored.

All real processes proceed at a finite rate. They are followed by friction, diffusion, heat exchange with its environment because of differences in their temperatures. Hence they are all irreversible and proceed in one direction.

Note : A thermodynamic process which proceeds with one of its state variable being constant is called Iso-Process.

Isobaric Process : It is a thermodynamic process that proceeds at constant pressure.

An *isochoric (isovolumic) process* is a thermodynamic process that proceeds at constant volume.

An *adiabatic process* is a thermodynamic process in which no heat exchange takes place with external bodies.

A system that is disturbed from a thermodynamic equilibrium returns back to equilibrium state is called *relaxation process.*

A *polytropic process* is a thermodynamic process in which specific heat (c) of the gas remains constant.

A quantity is called *state variable* if the integral of that quantity around any closed path is zero.

If $\oint d\phi = 0$, that ϕ is a state variable.

(A necessary and sufficient condition that a differential be exact is that its integral around a closed path is zero.)

Entropy may be regarded as disorder or randomness of a process. Addition or removal of heat to a process causes increase or decrease in the disorder or randomness. A natural process tends to proceed toward a state of greater disorder. In statistical mechanics the disorder and entropy is connected by the relation

$$\phi = k \, \ln \omega$$

where k = Boltzman's constant

ϕ = entropy of the system

ω = disorder parameter

Note : A state variable has a value that is characteristic of the state of the system irrespective of the fact how that state was arrived at.

If $$\oint d\phi = 0 = \int_{1-\alpha}^{\beta} d\phi + \int_{2-\beta}^{\alpha} d\phi = 0$$

integral along path 1 + integral along path 2 = 0

$$\int_{1-\alpha}^{\beta} d\phi + \int_{2-\beta}^{\alpha} d\phi = 0 \quad \text{i.e.,} \quad \int_{1-\alpha}^{\beta} d\phi = \int_{2-\alpha}^{\beta} d\phi$$

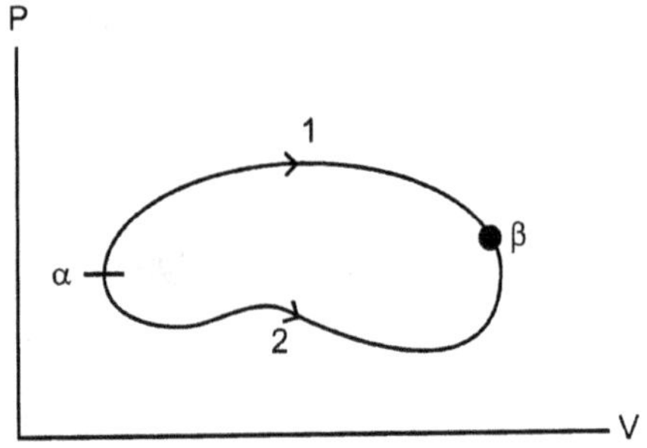

Fig. 7.6 α-β two arbitray points on the cycle 1 and 2 reversible paths joining them.

Entropy: The physical quantity which describes the ability of a system to do work is termed entropy of the system.

If $$\oint \frac{dQ}{T} = 0, \text{ or } \oint d\phi = 0$$

where $$d\phi = \frac{dQ}{T}$$

then ϕ is called entropy.

Change in entropy of any reversible process from an equilibrium state 'a' to b' is

$$\Delta\phi = \int_a^b \frac{dQ}{T} = \int_a^b d\phi$$

$$\Delta\phi = \phi_b - \phi_a = \frac{Q}{T}$$

where $\Delta\phi$ reduced heat delivered to the system

7.8.1 Definition

Isentropic Process : A thermodynamic process in which entropy of the system remains constant (does not change) is called isentropic process.

In a reversible adiabatic process $\delta Q = 0$ and hence ϕ = constant.

The common units of entropy are 1 J °K^{-1}, 1 cal °K^{-1}

Entropy of an Ideal Gas : According to first law of Thermodynamics

$$dh = \delta Q = di + pdv$$

and $\quad di = C_v\, dT;\ C_p - C_v = R$

Since $\quad pv = RT \qquad$ (general equation of gas)

This gives $pdv + vdp - R\,dT = 0$

Substituting $C_v = C_p - R$, $pdv = RdT - vdp$ in the first equation

we have

$$dh = C_v dT + pdv = (C_p - R)dT + (RdT - vdp)$$

$$\therefore \quad dh = \delta\, Q = C_p dT - vdp$$

dividing by T both side of this equation we have

$$\frac{dh}{T} = C_p \frac{dT}{T} - \frac{v}{T} dp$$

$$\frac{dh}{T} = C_p \frac{dT}{T} - \frac{R}{p} dp \qquad \left(\because \frac{v}{T} = \frac{R}{p}\right)$$

$$\frac{dh}{T} = C_p d\ln T - Rd\,(\ln p) = d\phi \text{ say [an exact differential]}$$

$$\oint \frac{dh}{T} = \oint d\phi = 0$$

where $\quad d\phi = C_p d\,(\ln T) - Rd\,(\ln p) \qquad$ .....(7.20)

Ex : Compute the change in entropy, when one Kg of ice is melted to water at 273 °K and boiled to 373 °K.

Sol : In the first step, the temperature remains constant (melting at 273 °K)

∴ The quantity of heat required to melt 1kg ice at 0 °C is

$$334 \times 10^3 \text{ J.}$$

i.e., $\quad \phi_2 - \phi_1 = \dfrac{Q}{T} = \dfrac{334 \times 10^3 \text{ J}}{273\,°\text{K}} = 1223 \text{ J } °\text{K}^{-1}$

i.e., the increase in entropy in the first process is 1223 J °K^{-1}.

Note : In any isothermal reversible process

$$\text{entropy} = \frac{\text{Amount of heat added}}{\text{Absolute temp}}$$

In the second step, 0 °C water is heated to 100 °C, here temperature is not constant.

$$dQ = \text{mass} \times \text{specific heat} \times \text{change in temperature}$$

$$(\Delta\phi) = \phi_2 - \phi_1 = \int_{T_1}^{T_2} \frac{dQ}{T} = \text{Mass} \times \text{sp.ht.} \int_{T_1}^{T_2} \frac{dT}{T}$$

$$= \text{mass} \times \text{sp ht.ln} \frac{T_2}{T_1}$$

$$= (1000 \text{ gm}) (4.19 \text{ Jgm}^{-1}\,{}^{\circ}\text{K}^{-1}).ln \frac{373\,{}^{\circ}\text{K}}{273\,{}^{\circ}\text{K}}$$

$$= 1308 \text{ J } {}^{\circ}\text{K}^{-1}$$

Note :

(i) If an ideal gas is allowed to expand adiabatically and reversibly, the change in entropy is zero, because in an adiabatic process Q = constant, $\delta Q = 0$

$$d\phi = \int_1^2 \frac{dQ}{T} = 0$$

(ii) The change in entropy in a reversible cycle is zero.

Every reversible adiabatic process is isentropic.

Ex : One kg of water at 100 °C is placed in thermal contact with 1 kg of water at 20 °C. Find the total change in entropy.

Sol : Assume the specific heat capacity of water is constant in this temperature range.

Heat lost by hot water = Heat gained by cold water.

Let x °C be the resultant temperature in thermal equilibrium.

then $\quad 1000 \times s(100 - x) = 1000 \times s(x - 20)$

$\quad 100 - x = x - 20 \quad 2x = 120 \quad x = 60$ °C

The entropy change of the hot water is $\Delta\phi_1$

$$T_1 = 373\,{}^{\circ}\text{K} = 100\,{}^{\circ}\text{C}$$

$$T_2 = 60\,{}^{\circ}\text{C} = 333\,{}^{\circ}\text{K}$$

$$\Delta\phi_1 = 1000 \times 1 \times \int_{T_1}^{T_2} \frac{dT}{T} = 1000 \int_{373}^{333} \frac{dT}{T}$$

$$= 1000 \left[\ln \frac{T_2}{T_1} \right]_{T_1=373}^{T_2=333}$$

$$= 1000 \ln \frac{333\,{}^{\circ}\text{K}}{373\,{}^{\circ}\text{K}} = -\text{ve}$$

The entropy change of the cold water is $\Delta\phi_2$

$$\Delta\phi_2 = 1000 \int_{T_1}^{T_2} \frac{dT}{T}$$

$$T_1 = (20\ ^{o}C) = 293\ ^{o}K$$

$$T_2 = (60\ ^{o}C) = 333\ ^{o}K$$

$$\Delta\phi_2 = 1000 \ln \frac{T_2}{T_1} = 1000 \ln \frac{333}{293} = +ve$$

$\therefore$ Total change in entropy = $\Delta\phi_1 + \Delta\phi_2$

7.9 Relation between Entropy (ϕ) and Potential Temperature (θ)

$$\frac{T}{\theta} = \left(\frac{p}{1000}\right)^k, \text{ Poissons equation}$$

where

$$k = \frac{AR}{C_p} = \frac{R'}{C_p}$$

Taking logarithm and then differentiating we have

$$\ln T - \ln\theta = k\,(\ln p - \ln 1000)$$

$$C_p \frac{dT}{T} - C_p \frac{d\theta}{\theta} = R' \frac{dp}{p}$$

or $\quad C_p\, d(\ln T) - R' d(\ln p) = C_p d\,(\ln\theta)$

i.e., $\quad d\,\phi = C_p d\,(\ln\theta) \qquad$ using (7.20)

$\therefore \quad \phi = C_p \ln\theta + \text{const.} \qquad$ .....(7.21)

This shows that entropy is proportional to logarithm of potential temperature.

Ex : 1 kg of ice at 0 °C melts (reversibly) to water at that same temperature, given that the latent heat of melting of ice 80 cal/gm. Find the change in entropy.

Sol :

Given that the process is reversible. The quantity of heat required to melt ice to water at 0 °C is

$$Q = 1000 \text{ gm} \times 80 \text{ cal/gm} = 8 \times 10^4 \text{ Cal.}$$

$$1\text{Cal} = 4.186 \text{ J}$$

$$\phi_{water} - \phi_{ice} = \frac{Q}{T} = \frac{8\times10^4}{273} \text{ Cal } ^{o}K^{-1}$$

$$= 292 \text{ Cal } ^{o}K^{-1} \quad \text{or} \quad 1220 \text{ J } ^{o}K^{-1}$$

Ex : Calculate the entropy change when an ideal gas expands from volume V_a to V_b in a reversible isothermal expansion.

Sol :

We know

$$di = dQ - pdV$$

because dw = pdV. (First law of Thermodynamics).

In an isothermal process di = 0

$$\therefore \quad dQ = pdV$$

Dividing by T

$$\frac{dQ}{T} = \frac{pdV}{T} \qquad \text{[gas Law. } pV = nRT]$$

$$d\phi = \frac{dQ}{T} = nR\frac{dV}{V} \qquad [\frac{p}{T} = \frac{nR}{V}]$$

$$\therefore \quad \int_a^b d\phi = \int_a^b nR\frac{dV}{V}$$

Change in entropy

$$= \phi_b - \phi_a = nR \ln\left(\frac{V_b}{V_a}\right)$$

Ex : A thermodynamic process for an isolated parcel in the atmosphere proceeds as shown in the Fig. 7.7. In the path ab of the process 143.4 Cal of heat added into the process and in the path bc 47.80 Cal heat are added.

Find the internal energy change

(i) in the process path ab

(ii) in the process path abc

(iii) and total heat advection in the process path adc

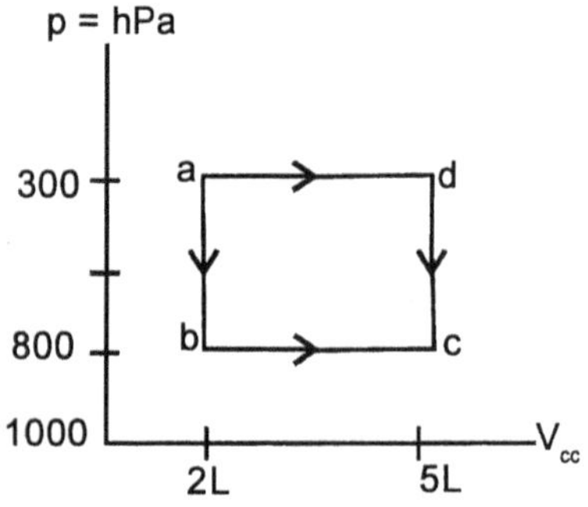

Fig. 7.7

Sol :

(i) The process ab path is isosteric (Volume constant), so work done ($d\omega$) is zero.

$$dQ = di + d\omega$$

given $dQ = 143.4 \text{ Cal} = 600 \text{ J}$ $\quad (\because 1 \text{ Cal} = 4.186 \text{ J})$

$\therefore$ $di = dQ$

$= 600 \text{ J}$

(ii) The process bc path is isobaric, so the work done is

$p = 0.8 \text{ bars}$

$= 0.8 \times 10^5 \text{ Pa}$

given $V_1 = 2000 \text{ cm}^3$

$V_2 = 5000 \text{ cm}^3$

$dw = pdV$ or $w = p(V_2 - V_1)$

$= 8 \times 10^4 \text{ Pa } (5000 - 2000) \text{ cm}^3$

$= 8 \times 10^4 \text{ Pa } (5000 - 2000) \times 10^{-6} \text{ m}^3$

$= 240 \text{ J}$

Total work done for the process abc path is $0 + 240 = 240$ J,

Total heat added $= 600 \text{ J} + 200 \text{ J} = 800 \text{ J}$ [since 47.80 cal = 200J]

$dQ = di + d\omega$

$800 \text{ J} = di + 240 \text{ J}$

$\therefore$ $di = 800 - 240 = 560$ J for the process abc.

(iii) Change in internal energy for the process path abc = Change in internal energy for process path adc

$\therefore$ di = internal energy for path process adc = 560 J. The total work done for the path adc is

$dw = pdV$ or $w = p(V_2 - V_1)$

$w = 300 \times 100 \text{ Pa } (5000 - 2000) 10^{-6} \text{ m}^3$

$+ dV$ for path ad is zero

$= 90\text{J}$

We know $dQ = di + d\omega$

total advection

$dQ = 560 \text{ J} + 90 \text{ J}$

$= 650\text{J}$

It may be noted here that though the internal energy is same for the path process abc and adc but work done and heat added are different for these path processes.

Note : If a substance receives heat ΔQ at temperature T, then it acquires entropy $\frac{\Delta Q}{T}$, but if it loesses heat ΔQ at the temperature T, then it loses entropy $\frac{\Delta Q}{T}$.

1. *Note :* The statistical definition of entropy $\phi = k \ln \omega$ relates the second law of thermodynamics in terms of statistics. A natural process directs towards higher entropy is determined by the laws of probability namely towards a more probable state. Thermodynamically, an equilibrium state is the state of maximum entropy which is the most probable state statistically. Thus second law of thermodynamics tends to show the most probable outcome of the events and not the only possible out come.

2. *Note :* A physical system comes to a state of least energy if the system is (a car rolls down a hill, hot body cools down etc) able to reduce its high energy molecules or its total energy.

Ex : Show that the work dw of expansion of an isolated parcel ascending adiabatically and quasi-statistically is given by the relation

$$dw = \frac{C_v}{C_p} g \frac{T}{\overline{T}} dz$$

Sol :

We are required to calculate the work (dw) of expansion of the parcel ascending adiabatically and statistically through a height dz.

Let us consider a unit mass. Then

$$dw = p d\alpha \qquad .....(i)$$

$$[\because \quad p\alpha = RT \text{ gas law}$$

$$p d\alpha + \alpha dp = RdT$$

$$p d\alpha = RdT - \alpha dp]$$

$$\therefore \qquad d\omega = RdT - \alpha dp \qquad .....(ii)$$

we know $\delta Q = C_p dT - \alpha dp$ [eq. (7.12) of thermodynamics]

For adiabatic process $\delta Q = 0$

i.e., $0 = C_p dT - \alpha dp$ .....(iii)

or $dT = \frac{\alpha}{C_p} dp$

From (ii) and (iii) we have

$$dw = R \cdot \frac{\alpha}{C_p} dp - \alpha dp \qquad R = C_p - C_v$$

$$= \left(\frac{R}{C_p} - 1\right) \alpha dp \qquad \frac{R}{C_p} - 1 = \frac{-C_v}{C_p}$$

$$= \frac{-C_v}{C_p} \alpha dp \qquad \text{.....(iv)}$$

Since the process is quasi-static, we have

$$dp = -g\bar{\rho} dz = -g\frac{p}{R\bar{T}} dz. \text{ (Hydrostatic eq.)} \qquad \text{.....(v)}$$

$$[p\bar{\alpha} = R\bar{T} \text{ or } \frac{p}{R\bar{T}} = \bar{\rho}]$$

$$dp = -g\bar{\rho} dz = -g\frac{p}{R\bar{T}} dz.$$

From (iv) and (v) we have

$$d\omega = -\frac{C_v}{C_p} \alpha . \frac{-gp}{R\bar{T}} dz$$

$$= g\frac{C_v}{C_p} \frac{T}{\bar{T}} dz \qquad \left(\because \frac{p\alpha}{R} = T\right)$$

$$dw = g\frac{C_v}{C_p} \frac{T}{\bar{T}} dz$$

which is the required result.

Questions

1. Define specific heat, specific heat of a gas at constant volume (C_υ), and constant pressure (C_p). Write the relation between C_p, C_υ and R.

2. Define heat capacity of a body. Derive an expression for the first law of thermodynamics as used in meteorology

$$\delta q = C_v\, dT + p\, d\alpha$$

and $$\dot{Q} = C_p \frac{dT}{dt} - \alpha \frac{dp}{dt}.$$

3. Show that in an adiabatic process

(i) $$\frac{T}{T_1} = \left(\frac{\alpha_1}{\alpha}\right)^{\frac{R'}{C_v}} \quad \text{or} \quad T(\alpha)^{\frac{R'}{C_v}} = \text{constant}$$

and (ii) $$\frac{T}{T_1} = \left(\frac{p}{p_1}\right)^{\frac{R'}{C_p}} \quad \text{or} \quad T = (p)^{\frac{R'}{C_p}} \times \text{const.}$$

4. Show that for an adiabatic process, $P V^\gamma$ = const. where $\gamma = \frac{C_p}{C_\upsilon}$

5. Derive Poisson's equation

$$\frac{T}{\theta} = \left(\frac{p}{1000}\right)^k \quad \text{where } k = \frac{\gamma - 1}{\gamma}.$$

Define potential temperature θ.

6. Using Poisson's equation derive the second form of thermodynamic equation as

$$\frac{d}{dt}(\ln \theta) = \frac{1}{C_p}\frac{H}{T}$$

7. Write briefly the following thermodynamic process for ideal gas with important properties. (i) Isobaric process, (ii) Isothermal process, (iii) Isochoric process and (iv) adiabatic process.

8. Define entropy (ϕ) and isoentropy. Find an expression of entropy of an ideal gas as $\oint d\phi = 0$

   Where $d\phi = C_p \, d(\ln T) - R \, d\,(\ln p)$

9. Find a relation between entropy (ϕ) and potential temperature (θ). or show that $\phi = C_p \ln \theta + \text{const.}$

10. One Kg of water at 100 °C is placed in thermal contact with one Kg of water at 40 °C. Find the total change in entropy.

11. Show that the work (dw) of expansion of an isolated parcel ascending adiabatically and quasi-statically is given by the relation

$$dw = \frac{C_v}{C_p} g \frac{T}{\overline{T}} dz$$

   Where $\overline{T}$ is quasistatic temperature (or average).

CHAPTER 8

The Operator ∇ (del)

∇ is also called gradient (grad), ascendent vector, nebla and Hamiltonian operator.

$$\nabla = i\frac{\partial}{\partial x} + j\frac{\partial}{\partial y} + k\frac{\partial}{\partial z}$$

Let b be any arbitrary scalar quantity. Then

$$\nabla b = i\frac{\partial b}{\partial x} + j\frac{\partial b}{\partial y} + k\frac{\partial b}{\partial z}$$

is called gradient b or grad b.

Let b = b (x, y, z, t) be a scalar variable and let the position of the particle p be at time $t = t_0$ and $t = t_0 + dt$ be at (x_0, y_0, z_0) and $(x_0 + dx, y_0 + dy, z_0 + dz)$ respectively.

Let 'db' be the change in the time interval 'dt' in the property of b. Then $db = b\,(x_0 + dx, y_0 + dy, z_0 + dz, t_0 + dt) - b(x_0, y_0, z_0, t_0)$

By Taylors expansion we have

$$db = \left[b(x_0, y_0, z_0, t_0) + \frac{\partial b}{\partial x}dx + \frac{\partial b}{\partial y}dy + \frac{\partial b}{\partial z}dz + \frac{\partial b}{\partial t}dt + \text{higher order terms}\right]$$

$$-[b\,(x_0, y_0, z_0, t_0)] \qquad \text{.....(8.1)}$$

Neglecting higher order terms and dividing by dt we have

$$\frac{db}{dt} = \frac{\partial b}{\partial x}\frac{dx}{dt} + \frac{\partial b}{\partial y}\frac{dy}{dt} + \frac{\partial b}{\partial z}\frac{dz}{dt} + \frac{\partial b}{\partial t}$$

$$\frac{db}{dt} = u\frac{\partial b}{\partial x} + v\frac{\partial b}{\partial y} + w\frac{\partial b}{\partial z} + \frac{\partial b}{\partial t}$$

where $u = \frac{dx}{dt}, \; v = \frac{dy}{dt}, \; w = \frac{dz}{dt}$

$$\frac{db}{dt} = \frac{\partial b}{\partial t} + \vec{V}.\nabla b \qquad \text{where } \vec{V} = iu + jv + kw$$

8.1 Interpretation of ∇ b

Consider the change δb in b between two different points P_1 (x, y, z) and P_2 (x + δx, y + δy, z + δz) at the same time, then

$$\delta b = b\,(x + \delta x,\; y + \delta y,\; z + \delta z) - b\,(x, y, z)$$

$$= \left[b(x,y,z) + \frac{\partial b}{\partial x}\delta x + \frac{\partial b}{\partial y}\delta y + \frac{\partial b}{\partial z}\delta z \text{ higher order terms}\right] - b(x,y,z)$$

$$= \frac{\partial b}{\partial x}\delta x + \frac{\partial b}{\partial y}\delta y + \frac{\partial b}{\partial z}\delta z$$

omitting second and higher order terms in Taylor's expansions.

$$\delta b = \nabla b \,.\, \vec{\delta r} \text{ where } \vec{\delta r} = i\delta x + j\delta y + k\delta z = \left(\overrightarrow{P_1 P_2}\right)$$

Now two cases arise.

Case (i) : The two points p_1, p_2 may lie on the equiscalar surface of b [an equiscalar surface is a surface on which b = const]. Then $\delta b = 0$. i.e., $\nabla b.\vec{\delta r} = 0$. But $\vec{\delta r}$ is a vector in the equiscalar surface and since the scalar product is zero implies that ∇b is perpendicular to the equiscalar surface as shown in the Fig. 8.1.

Case (ii) : Let the two points P_1, P_2 lie on the same normal to the equiscalar surface as shown in Fig. 8.2.

We have

$$\delta b = \nabla b \,.\, \vec{\delta n}$$

where

$\vec{\delta n}$ (= $\vec{\delta r}$) is the vector from point P_1 to P_2 along ∇b.

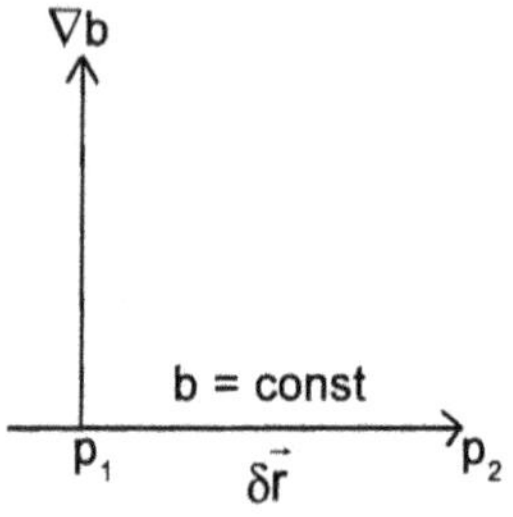

Fig. 8.1

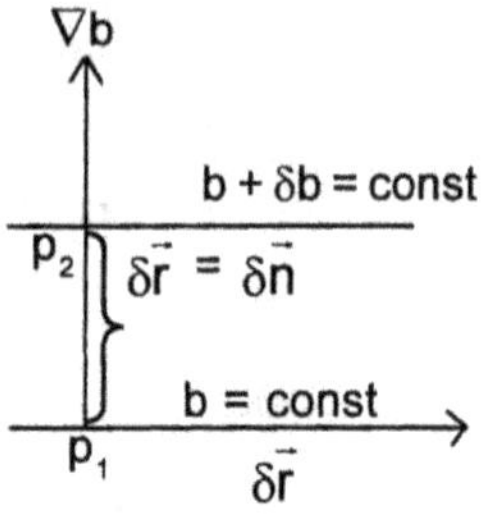

Fig. 8.2

The two vectors ∇b and $\delta\vec{n}$ have the same direction and hence $\delta b = \nabla b . \delta\vec{n}$ may be written as $|\nabla b| = \frac{\delta b}{\delta n}$. This last relation is used to evaluate the magnitude of ∇b.

Questions

1. Define del operator (∇).

   If $b = b(x,y,z,t)$ be any scalar variable, show that

   $$\frac{db}{dt} = \frac{\partial b}{\partial t} + \vec{V}.\nabla b$$

   where $\vec{V} = iu + j\upsilon + kw$.

2. If $b = b(x,y,z)$, δb be the difference between two points $P_1(x,y,z)$ and $P_2(x + \delta x, y + \delta y, z + \delta z)$ show that $\delta b = \nabla b . \vec{\delta r}$

   where $\vec{\delta r} = i\delta x + j\delta y + k\delta z = \overrightarrow{P_1P_2}$

   Discuss two cases when the two points (i) on a equiscalar surface and (ii) on the same normal to the equiscalar surface.

CHAPTER 9

The Continuity Equation

The equation of continuity expresses the conservation of mass, that is, mass is neither created nor destroyed. It expresses that the rate of change of generation of mass within a given volume is balanced by an equal outflow of mass from the volume or it means that there are no sources or sinks of mass anywhere in the atmosphere.

Consider a box of volume element $\delta V = \delta x.\ \delta y.\ \delta z$ as shown in the Fig. 9.1. where δx, δy, δz are the sides of the box, ρ density of the fluid. The mass of the air (fluid) in the box is $\rho\delta V$ (mass = volume × density). The change of mass per unit of time is

$$\frac{\partial \rho}{\partial t}\delta V \qquad \text{.....(i)}$$

where $\frac{\partial \rho}{\partial t}$ is the local density change.

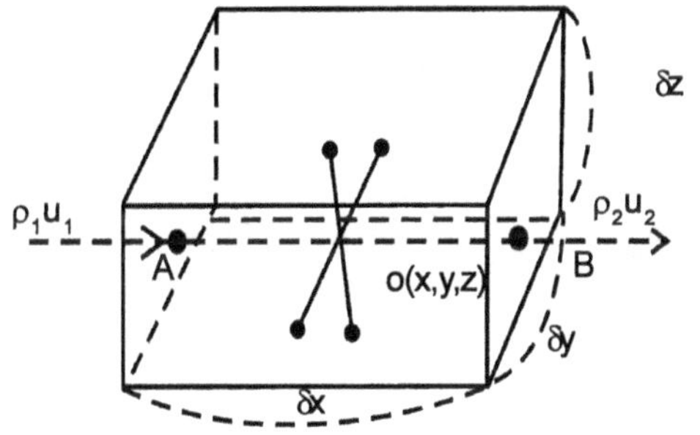

Fig. 9.1

Let the coordinates of the center of box 'o' be (x, y, z) and let u, u_1, u_2 and ρ, ρ_1, ρ_2 be the velocities and densities at the centre O of the box and at the center points A and B of the left face and right face respectively.

The coordinates of the center of the left side box A is

$$\left(x - \frac{\delta x}{2}, y, z\right)$$

and that of the center of the right side box B is

$$\left(x + \frac{\delta x}{2}, y, z\right)$$

Then the net inflow into the box in x-direction is

$$\rho_1 u_1 \delta y \delta z - \rho_2 u_2 \delta y \delta z$$

$$= \left[\rho_0 u_0 - \frac{1}{2}\left(\frac{\partial \rho u}{\partial x}\right)_0 \delta x\right]\delta y \delta z - \left[\rho_0 u_0 + \frac{1}{2}\left(\frac{\partial \rho u}{\partial x}\right)_0 \delta x\right]\delta y \delta y$$

$$= -\left[\frac{\partial \rho u}{\partial x}\right]_0 \delta x \delta y \delta z$$

$$= -\left[\frac{\partial (\rho u)}{\partial x}\right]_0 \delta V \qquad .....(ii)$$

Similarly the net inflow into the box in y-direction, z-direction are given by

$$-\left[\frac{\partial (\rho v)}{\partial y}\right]_0 \delta V \quad \text{and} \quad -\left[\frac{\partial (\rho \omega)}{\partial z}\right]_0 \delta V \qquad .....(iii)$$

where 'v, w' are the velocities in y, z directions

∴ From (i), (ii) and (iii) the equation of continuity is given by

$$\frac{\partial \rho}{\partial t}\delta V = -\left[\frac{\partial (\rho u)}{\partial x} + \frac{\partial (\rho v)}{\partial y} + \frac{\partial (\rho w)}{\partial z}\right]\delta V$$

or

$$\frac{\partial \rho}{\partial t} = -\left[\frac{\partial (\rho u)}{\partial x} + \frac{\partial (\rho v)}{\partial y} + \frac{\partial (\rho w)}{\partial z}\right]$$

i.e.,

$$\frac{\partial \rho}{\partial t} = -\nabla \cdot \left(\rho \vec{V}\right) \qquad .....(9.1)$$

where $$\nabla = i\frac{\partial}{\partial x} + j\frac{\partial}{\partial y} + k\frac{\partial}{\partial z}$$

$$\vec{V} = iu + jv + kw$$

$$\rho\vec{V} = i\rho u + j\rho v + k\rho w$$

Note :

$$\nabla \cdot \left(\rho\vec{V}\right) = \left(i\frac{\partial}{\partial x} + j\frac{\partial}{\partial y} + j\frac{\partial}{\partial z}\right) \cdot (i\rho u + j\rho v + k\rho w)$$

$$= \frac{\partial \rho u}{\partial x} + \frac{\partial \rho v}{\partial y} + \frac{\partial \rho w}{\partial z}$$

$\nabla \cdot \left(\rho\vec{V}\right)$ = mass divergence in three dimensional space

eq. (9.1) can be written as

$$\frac{\partial \rho}{\partial t} + \nabla \cdot \left(\rho\vec{V}\right) = 0$$

or $$\frac{\partial \rho}{\partial t} + \vec{V} \cdot \nabla\rho + \rho\nabla \cdot \vec{V} = 0$$

or $$\frac{d\rho}{dt} + \rho\nabla \cdot \vec{V} = 0 \quad \text{or} \quad \frac{1}{\rho}\frac{d\rho}{dt} + \nabla \cdot \vec{V} = 0 \qquad .....(9.2)$$

since $$\frac{d\rho}{dt} = \frac{\partial \rho}{\partial t} + \vec{V} \cdot \nabla\rho \text{ (implified from } \frac{d}{dt} = \frac{\partial}{\partial t} + \vec{V} \cdot \nabla)$$

Note : $\nabla \cdot \left(\rho\vec{V}\right)$ denotes the mass divergence in three dimensional space

$\nabla \cdot \vec{V}$ is called dialation or expansion.

$$\frac{d}{dt} = \frac{\partial}{\partial t} + \vec{V} \cdot \nabla \text{ denotes that}$$

the individual time derivative is equal to local time derivative plus (+) the convective acceleration.

$$\frac{d}{dt} = \text{rate of change or individual time derivative.}$$

$$\frac{\partial}{\partial t} = \text{local change or local time derivative}$$

$$\vec{V} \cdot \nabla = \text{velocity advection or convective acceleration}$$

eq. (9.1) and (9.2) are called continuity equations.

Note : If the fluid is incompressible (i.e., ρ = constant), then

$$\nabla\rho = 0, \frac{\partial\rho}{dt} = 0,$$

The equation continuity reduces to

$$\nabla.\vec{V} = 0, \text{ i.e., } \frac{\partial u}{\partial x} + \frac{\partial v}{\partial y} + \frac{\partial w}{\partial z} = 0$$

Relation between individual and local time derivatives of a scalar element ϕ is given by

$$\frac{d\phi}{dt} = \frac{\partial\phi}{dt} + u\frac{\partial\phi}{\partial x} + v\frac{\partial\phi}{\partial y} + w\frac{\partial\phi}{\partial z} = \frac{\partial\phi}{\partial t} + \vec{V}.\nabla\phi$$

$$= \frac{\partial\phi}{dt} + \vec{V}_H \nabla_H\phi + w\frac{\partial\phi}{\partial z}$$

where $\vec{V}_H$ = horizontal wind

$\nabla_H\phi$ = horizontal gradient of the element ϕ

This can also be written as

$$\frac{d\phi}{dt} = \frac{\partial\phi}{\partial t} + V\frac{\partial\phi}{\partial n}\cos\theta + w\frac{\partial\phi}{\partial z}$$

since $\vec{A}.\vec{B} = AB\cos\theta$

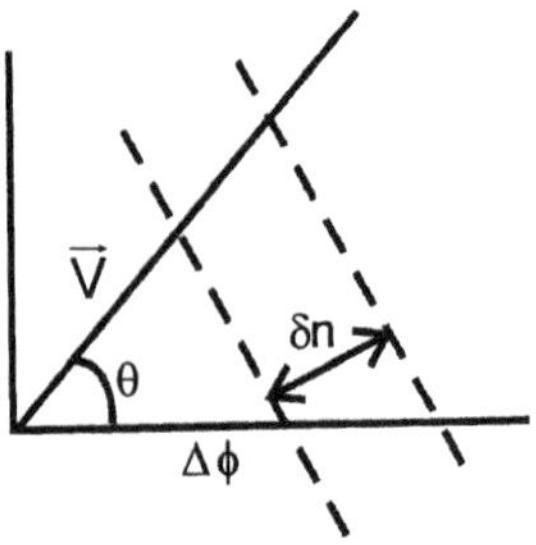

Fig 9.2

where

V = magnitude of the wind speed $|\vec{V}_H|$

$\frac{\partial\phi}{\partial n} = |\nabla_H\phi|$ = magnitude of the horizontal gradient of the element ϕ

θ = The angle between the direction of the wind $|\vec{V}_H|$ and horizontal

gradient of the element $|\nabla_H \phi| = \frac{\partial \phi}{\partial n}$

Consider an isolated moving small air parcel of constant mass δM, cross-section area A, vertical thickness δz, density ρ.

Then $\delta M = \rho A \, \delta z$

$$\ln(\delta M) = \ln \rho + \ln A + \ln \delta z$$

Differentiating w.r.t. time t

$$\therefore \quad \frac{1}{\delta M}\frac{d(\delta M)}{dt} = \frac{1}{\rho}\frac{d\rho}{dt} + \frac{1}{A}\frac{dA}{dt} + \frac{1}{\delta z}\frac{d(\delta z)}{dt}$$

But $$\frac{d(\delta M)}{dt} = 0$$

$$\therefore \quad -\frac{1}{\rho}\frac{d\rho}{dt} = \frac{1}{A}\frac{dA}{dt} + \frac{1}{\delta z}\frac{dz}{dt}$$

i.e., $$\nabla . \vec{V} = \frac{1}{A}\frac{dA}{dt} + \frac{1}{\delta z}\frac{dz}{dt}$$

or $$\nabla_H . \vec{V} + \frac{\partial w}{\partial z} = \frac{1}{A}\frac{dA}{dt} + \frac{1}{\delta z}\frac{dz}{dt}$$

this implies $$\nabla_H . \vec{V} = \frac{1}{A}\frac{dA}{dt}$$

and $$\frac{\partial w}{\partial z} = \frac{1}{\delta z}\frac{dz}{dt}$$

i.e., Horizontal divergence = Fractional rate of areal expansion

Vertical divergence = Fractional rate of thickness expansion.

Equation of continuity may be interpreted as, if a mass of air is compressed vertically it will be expanded horizontally or increase in density or both.

9.1 Geopotential (Ψ)

Acceleration due to gravity 'g' varies from place to place. The geopotential Ψ may be interpreted as the potential energy (mgh) imparted to a unit mass when it is lifted from sea level to a height z. It is defined as

$$d\Psi = g dz \quad \text{or} \quad \Psi = \int_0^z g dz$$

When $z = 0, \Psi = 0.$

or $\vec{g}_a = -\nabla\Psi_a, \; \Psi_a = \dfrac{-GM}{r} + \text{constant}$

The value of g = 9.8 m/sec^2, The geopotential of a unit mass, 1 meter above sea level is about 9.8 m^2 s^{-2} (mgh, m = 1, g = 9.8 ms^{-2}, h = 1m) implies that Ψ is nearly 9.8 times the geometric height of 1 meter. As a consequence a unit geopotential is defined and called geopotential meter, such that height in geopotential meter is

$$\Psi = \frac{1}{9.8}\int_0^z g dz$$

Ex : Convert 2998 m into geopotential meters where g = 9.806 ms^{-2}

Sol :

$$\Psi = \frac{1}{9.8}\int_0^z gdz$$

$$= \frac{1}{9.8}\int_0^{2998} 9.806\, dz$$

$$= \frac{9.806}{9.8}[Z]_0^{2998} = \frac{98.06}{98} \times 2998$$

$$\Psi = 3000 \text{ gpm}$$

9.2 An Alternative Derivation of Equation of Continuity

Consider an isolated box of volume δv of sides δx, δy, δz of fixed mass δM, which moves with the fluid as shown in the Fig.9.3.

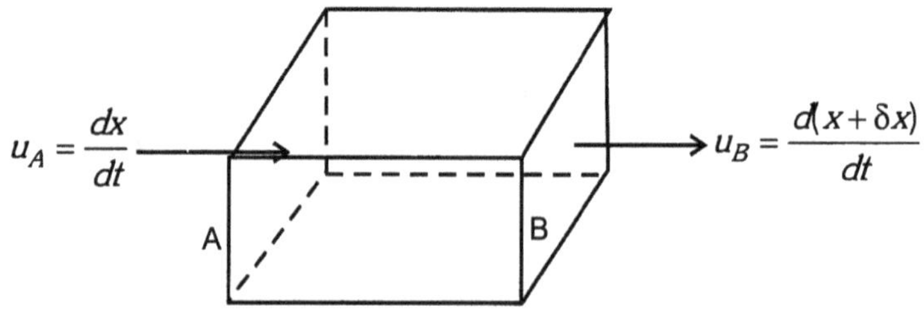

Fig 9.3

Let ρ is the density of the fluid. Then we have

$$\delta V = \delta x.\ \delta y.\ \delta z$$

and $$\delta M = \rho.\ \delta V = \rho\ .\ \delta x\ .\ \delta y.\ \delta z \qquad .....(9.3)$$

taking logarithm

$$\ln \delta M = \ln \rho + \ln \delta x + \ln \delta y + \ln \delta z \qquad .....(9.4)$$

Differentiating w.r.t time 't' we have

$$\frac{1}{\delta M}\frac{d}{dt}(\delta M) = \frac{1}{\rho}\frac{d\rho}{dt} + \frac{1}{\delta x}\frac{d(\delta x)}{dt} + \frac{1}{\delta y}\frac{d(\delta y)}{dt} + \frac{1}{\delta z}\frac{d(\delta z)}{dt} \qquad .....(9.5)$$

Let $U_A = \frac{dx}{dt}$, $U_B = \frac{d(x+\delta x)}{dt}$

be the speeds of fluid in x-direction, that is perpendicular to the face A and B respectively. Then difference of speed in x-direction in these faces is

$$\delta u = U_B - U_A = \frac{d(x+\delta x)}{dt} - \frac{dx}{dt}$$

$$= \frac{d(\delta x)}{dt}$$

similarly y, z directions we have

$$\delta v = \frac{d(\delta y)}{dt}, \delta w = \frac{d(\delta z)}{dt}$$

Substituting these values in eq. (9.5) we have

$$\frac{1}{\delta M}\frac{d(\delta M)}{dt} = \frac{1}{\rho}\frac{d\rho}{dt} + \frac{\delta u}{\delta x} + \frac{\delta v}{\delta y} + \frac{\delta w}{\delta z}$$

In the limit as δx, δy, δz tends to zero, and conservation of mass δM = constant, we have $\frac{d(\delta M)}{dt} = 0$ and the above equation reduces to

$$\frac{1}{\rho}\frac{d\rho}{dt} + \frac{\partial u}{\partial x} + \frac{\partial v}{\partial y} + \frac{\partial w}{\partial z} = 0$$

or $$\frac{1}{\rho}\frac{d\rho}{dt} + \nabla.\vec{V} = 0 \qquad \text{(eq. (9.2))}$$

where $\vec{V} = iu + jv + kw$.

which is the required equation of continuity.

Note : Equation of continuity expresses that vertical compression results in horizontal expansion and /or increase in density. Alternately convergence near the ground leads to vertical motion at some distance above the ground or divergence in the upper atmosphere implies vertical motion below in the lower atmosphere.

9.2.1 Some Definitions

Fluid : A fluid is a substance that can flow. It is either a liquid or a gas and either compressible or incompressible. A compressible fluid is a gas which is also a function of pressure. An incompressible fluid is either a gas or a liquid whose density does not depend on pressure.

A barotropic fluid is one whose density is a function of pressure alone.

A Barotropic Atmosphere : An atmosphere in which the isobaric surfaces (surfaces of constant pressure) are also surfaces of constant density (isopycnic surfaces). In barotropic atmosphere density is a function of pressure that is $\rho = \rho(p)$.

In case of ideal gas [$p\,\alpha = RT$ or $p = \rho RT$], the isobaric surfaces are also isothermal surfaces if the atmosphere is barotropic.

An Autobarotropic Atmosphere : An atmosphere which remains barotropic for all the time is called autobarotropic atmosphere.

A Baroclinic Atmosphere : An atmosphere which is not barotropic is called Baroclinic. In baroclinic atmosphere isobaric surfaces interact with the isopycnic (constant density) surfaces. In baroclinic atmosphere density is a function of pressure and temperature, $\rho = \rho\,(p, T)$.

Newtonian Fluids : Fluids whose viscosity is independent of velocity gradient is called Newtonian fluid. e.g., Air, water, Kerosine, thin lubricating oils.

Atmospheric Pressure : Pressure at any point of the atmosphere is the weight of the vertical column of air above it, extending to the top of the atmosphere, on a unit horizontal area around point (The point being at the center).

Fluid Particle : A fluid particle is an infinitesimal volume (strictly a geometrical point) whose linear dimensions are disregarded for the purpose of finding its velocity and acceleration. The infinitesimal volume whose size, however, is many times larger than the intermolecular distance (of the order 10^{-8} m in gases at standard conditions).

An ideal (perfect or inviscid) fluid is a continuous fluid substance that cannot exert any shearing stress however small. Atmosphere is regarded as a continuous fluid medium or continuum. The physical quantities (pressure, density, temperature, and velocity) are assumed to have unique values at any point in the atmosphere. These field variables and their derivatives are assumed to be continuous functions of space and time.

9.3 Hydrostatic Equation

If a fluid is in equilibrium, then every portion of the fluid is in equilibrium. Consider a small element of area A = δx δy, volume δV = δx . δy . δz as shown in the Fig. 9.4. Let ρ be the density of the fluid. The mass of the element is ρAdz and its weight dw = ρgAdz. At each of the surface, the forces exerted on the element by the fluid are perpendicular to it. The resultant horizontal forces due to fluid pressure is zero. Since the fluid element is at rest (static) the resultant of vertical forces is also zero. In the vertical direction pressure force and weight of the fluid acts.

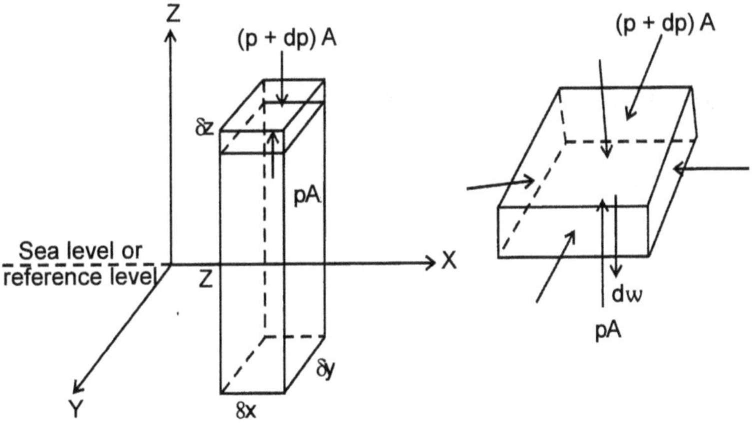

Fig. 9.4 Hydrostatic equilibrium.

Let p be the pressure on the lower face and p + dp be the pressure on the upper face. Then the upward force is pA, while the downward force is (p + dp)A, plus weight of the element dw.

∴ the vertical equilibrium is given by

$$pA = (p + dp)A + dw$$

or $$pA = (p + dp)A + \rho g A dz$$

or $$0 = dp + \rho\, g dz$$

i.e., $$dp = -g\rho dz \quad \text{or} \quad \delta p = -g\rho\delta z \qquad .....(9.6)$$

This is the fundamental statical equation in meteorology.

Integrating eq. (9.6) in the limits p_0 to p corresponding $z_0 = 0$ to z we have

$$\int_{P_0}^{P} dp = -g\rho \int_0^z dz$$

$$p - p_0 = -g\rho\,(z - 0)$$

i.e., $$p = p_0 - g\rho z \qquad .....(9.7)$$

This is the basic equation of hydrostatics for incompressible fluids.

Between any two levels corresponding to z_1, z_2, p_1, and p_2

we have

$$p_2 - p_1 = -g\rho\,(z_2 - z_1) \qquad .....(9.8)$$

In hydrostatic equilibrium, pressure at any point is equal to the weight of the column of air on a unit cross section area over that point.

Eq. (9.8) shows that pressure depends on depth and not on the shape of the surrounding. This fact was first declared by French scientist Blaise Pascal as follows : Pressure applied to an enclosed fluid is transmitted undiminished to every portion of the fluid and the walls of the containing vessel.

9.4 Variation of Density with Height

$$dp = -g\rho dz \qquad (1) \quad \text{Hydrostatic equation}$$

$$p = \rho RT \qquad (2) \quad \text{(gas equation } pv = RT\text{)}$$

Taking logarithms

$$\ln p = \ln \rho + \ln R + \ln T$$

Differentiating w.r.t z.

$$\frac{1}{p}\frac{dp}{dz} = \frac{1}{\rho}\frac{d\rho}{dz} + \frac{1}{T}\frac{dT}{dz}$$

$$-\frac{g\rho}{p} = \frac{1}{\rho}\frac{d\rho}{dz} + \frac{1}{T}\frac{dT}{dz} \qquad \left(\text{since } \frac{dp}{dz} = -g\rho\right)$$

or

$$\frac{1}{\rho}\frac{d\rho}{dz} = -\frac{g\rho}{p} - \frac{1}{T}\frac{dT}{dz}$$

$$\frac{1}{\rho}\frac{d\rho}{dz} = -\frac{g\rho}{\rho RT} - \frac{1}{T}\frac{dT}{dz} \qquad (\because\ p = \rho RT)$$

$$\frac{1}{\rho}\frac{d\rho}{dz} = -\frac{1}{T}\left(\frac{g}{R} + \frac{dT}{dz}\right) \qquad .....(9.9)$$

If ρ = constant, (i.e., Homogeneous atmosphere)

the above eq. (9.9) reduces to

$$\frac{g}{R} + \frac{dT}{dz} = 0 \text{ or } \frac{dT}{dz} = -\frac{g}{R} = \frac{-9.8 \times 1000}{287} = -34.1\ ^\circ\text{C/km}.$$

where $g = 9.8$ m/s^2, $R = 287$ m^2/s^2

$$\Gamma_H = \frac{dT}{dz} = \text{lapse rate of homogenous atmosphere} = 34.1\ ^\circ\text{C/km}$$

where Γ (= rate of fall of temperature with height) is called temperature lapse rate.

The temperature lapse rate of homogeneous atmosphere $\Gamma_H \simeq 34.1$ °C/km is very large and is about six times larger than the normal lapse rate of the atmosphere

Γ_H = 34.1 °C/km is called auto convective lapse rate.

Γ_d = dry adiabatic lapse rate $\simeq$ 9.8 °C/km.

Γ_s = Saturated adiabatic lapse rate $\simeq$ 5 °C/km.

9.5 Homogeneous Atmosphere

Assume a hypothetical case where the density of the atmosphere is constant throughout. Let H be the height of the homogeneous atmosphere.

Then $dp = -g\rho dz$ (1) [Hydrostatic equation].

$z = 0$ at the sea level, where $p = p_0$ sea level pressure. and $z = H$ height of homogeneous atmosphere, where $p = 0$ (at the top of the atmosphere). Integrating (1) we

$$\int_{p_0}^{0} dp = -g\rho \int_0^H dz$$

$$-p_0 = -g\rho H \quad \text{or} \quad p_0 = g\rho H$$

For dry atmosphere

T_0 = 273 °K at sea level, $g = 9.8\ ms^{-2}$, $R = R_d = 287\ m^2 s^{-2}$,

$$H = \frac{p_0}{\rho g} = \frac{RT_0}{g}$$

$$H = \frac{287 \times 273}{9.8} = 8000\ m$$

This is the height of the homogeneous atmosphere at temperature 0 °C

Γ_H lapse rate of homogeneous atmosphere = 34.1 0 °C/km. derived below.

$$\left(T = T_0 - \frac{g}{R}Z\right) \qquad \text{(see eq. 9.19)}$$

9.6 Isothermal Atmosphere

Assume that there is no change in temperature throughout the atmosphere

That is $$\frac{dz}{dT} = 0$$

$$dp = -g\rho dz, \quad \text{Hydrostatic equation}$$

$$p = \rho RT, \quad \text{gas equation}$$

$$\therefore \quad \frac{dp}{p} = \frac{-g}{RT} dz \qquad .....(i)$$

Let $p = p_0$ when $z = 0$ and $p = p$ when $z = z$

In isothermal atmosphere T = constant and $\frac{dT}{dz} = 0$

Integrating (i) we have

$$\int_{p_0}^{p} \frac{dp}{p} = \frac{-g}{RT} \int_0^z dz$$

$$\ln \frac{p}{p_0} = \frac{-g}{RT} z$$

$$\text{i.e.,} \quad \frac{p}{p_0} = e^{\frac{-gz}{RT}} = e^{\frac{-z}{H}}$$

$$\text{i.e.,} \quad p = p_0\, e^{\frac{-z}{H}} \qquad .....(9.10)$$

where $H = \frac{RT}{g}$ height of homogeneous atmosphere

eq. (9.10) shows that pressure decreases with height exponentially and there is no upper boundary of the atmosphere, because $p = 0$ if $e^{-\infty} = 0$, implies $z = \infty$ (i.e.,$p \to 0$ as $z \to \infty$)

If $z = H$, then eq. (9.10) becomes $p = p_0\, e^{-1}$ or $p = \frac{1}{e} \times p_0$

i.e. pressure (p) at $z = H$ (=8 Km) is $\frac{1}{e} p_0 = \frac{1}{e}$ times pressure at sea level

$$\text{i.e.,} \quad \frac{p}{p_0} = \frac{1}{e} = e^{-1} \qquad .....(9.11)$$

From eq. (9.10) and eq. (9.11) we have $e^{-1} = e^{\frac{-gz}{RT}}$,

$$\therefore \quad \frac{gz}{RT} = 1, \text{ or } z = \frac{RT}{g} = H$$

This is called scale height and denoted by Z_s.

$$\therefore \quad z_s = H$$

9.7 Uniform Lapse Rate Atmosphere

Let us assume that the temperature falls with height uniformly , that is the lapse rate is constant, $\frac{dT}{dz}$ = constant = Γ (say). T is a function of z. This implies if T is the temperature at height z and at sea level (z = 0) T_0 then we have.

$$\int_{T0}^{T} dT = -\int_{0}^{z} \Gamma dz$$

$$T - T_0 = -\Gamma z \qquad .....(9.12)$$

$$\text{or } T = T_0 - \Gamma z \Rightarrow dT = -\Gamma dz \qquad .....(9.13)$$

$dp = -g\,\rho dz$ Hydrostatic equation

$p = \rho RT$ gas equation

$$\text{then } \frac{dp}{p} = \frac{-g}{R}\frac{dz}{T} \qquad .....(9.14)$$

Let $p = p_0$ when $z = 0$, and $T = T_0$

and $p = p$ when $z = z$ and $T = T$

Integrating eq. (9.14)

$$\int_{p_0}^{p} \frac{dp}{p} = \frac{-g}{R}\int_{T_0}^{T} \frac{dz}{T}$$

or

$$\int_{p_0}^{p} \frac{dp}{p} = \frac{-g}{R}\int_{T_0}^{T} -\frac{1}{T}\frac{dT}{\Gamma} = \frac{g}{\Gamma R}\int_{T_0}^{T} \frac{dT}{T} \quad \text{(using 9.13)}$$

$$\ln \frac{p}{p_0} = \frac{g}{\Gamma R} \ln \frac{T}{T_0} \text{ or } \ln \left(\frac{T}{T_0}\right)^{\frac{g}{\Gamma R}}$$

$$\text{i.e., } p = p_0 \left(\frac{T}{T_0}\right)^{\frac{g}{\Gamma R}} \qquad .....(9.14')$$

Eq. (9.14') implies p is a function of T = function of z.

9.8 Relation Between Density ρ and Temperature T

We Know

$$p = \rho RT, \quad p_0 = \rho_0 RT_0$$

$$\rho = \frac{p}{RT} = \frac{p_0\left(\frac{T}{T_0}\right)^{\frac{g}{\Gamma R}}}{RT_0\left(\frac{T}{T_0}\right)} \quad \text{(using 9.14)}$$

$$\rho = \frac{p_0}{RT_0}\cdot\left(\frac{T}{T_0}\right)^{\frac{g}{\Gamma R}-1}$$

$$\rho = \rho_0\left(\frac{T}{T_0}\right)^{\frac{g}{\Gamma R}-1} \quad .....(9.15)$$

It is seen from $T = T_0 - \Gamma z$ eq (9.13), that the constant lapse rate atmosphere has finite height say Z_Γ, defined by $Z_\Gamma = \frac{T_0}{\Gamma}$. $\left[\because \int_{T_0}^{0} dT = -\Gamma\int_0^{Z_\Gamma} dz\right]$

For $\Gamma = 0.65 \times 10^{-2}\ °Km^{-1}$, $T_0 = 273\ °K$

and that $$Z_\Gamma = \frac{273}{0.65\times10^{-2}\ °K\,m^{-1}} = \frac{273\times10^4\,m}{65}$$

$$= 4.2 \times 10^4\ m = 42\ Km.,$$

which is the height of constant lapse rate atmosphere

At $Z = Z_\Gamma$ we have $T = 0$, $\rho = 0$ and $p = 0$.

9.9 Temperature in the Homogeneous Atmosphere

$p = \rho_H RT$ gas equation, where ρ_H = density of the homogeneous atmosphere

$$T = \frac{p}{R\rho_H} \quad .....(9.16)$$

At sea level

$$T_0 = \frac{p_0}{R\rho_H}$$ [T_0 be the temperature, p_0 = pressure at the sea level] .....(9.17)

$$dp = -g\rho_H dz \qquad \text{Hydrostatic equation}$$

Integrating $\int_{p_0}^{p} dp = -g\rho_H \int_0^z dz$

$$p - p_0 = -g\,\rho_H z \text{ or } p = p_0 - g\rho_H z \qquad .....(9.18)$$

Substituting eq. (9.18) in eq. (9.16), we have

$$T = \frac{(p_0 - g\rho_H z)}{R\rho_H} = \frac{p_0}{R\rho_H} - \frac{g}{R}z$$

$$T = T_0 - \frac{g}{R}z \quad \text{(using (9.17))} \qquad .....(9.19)$$

Eq. (9.19) shows that in a homogeneous atmosphere temperature decreases linearly with height.

Differentiating eq. (9.19) w.r.t z we have

$$\frac{dT}{dz} = \frac{-g}{R}$$

or $$\Gamma_H = -\frac{dT}{dz} = \frac{g}{R} = \frac{9.8\text{ms}^{-2}}{287\text{m}^2\text{s}^{-2}} = \frac{98}{28.7 \times 100\text{m}}$$

$$= 3.41°\text{K}/100 \text{ m}$$

$$\Gamma_H = -\frac{dT}{dz} = 34.1 \text{ °K/km.} \qquad \text{or } 34.1 \text{ °C/km.}$$

9.10 Adiabactic Atmosphere

Assume there is no change in potential temperature throughout the atmosphere i.e. θ = constant

or $$\frac{d\theta}{dz} = \frac{\partial\theta}{\partial z} = 0$$

$$\theta = T\left(\frac{p}{p_0}\right)^{-k} = T\left(\frac{p}{p_0}\right)^{\frac{-R}{c_p}} \quad \text{Poisson's equation} \qquad .....(i)$$

Taking logarithms we have

$$\ln\theta - \ln T = \frac{-R}{c_p}[\ln p - \ln p_0]$$

p = pressure at elevation

p_0 = pressure at sea level, constant.

Differentiting w.r.t. z we have

$$\frac{1}{\theta}\frac{d\theta}{dz} - \frac{1}{T}\frac{dT}{dz} = \frac{-R}{c_p}\left[\frac{1}{p}\frac{dp}{dz}\right]$$

$$-\frac{1}{T}\frac{dT}{dz} = -\frac{R}{C_p}\left[\frac{1}{p}\frac{dp}{dz}\right] \quad [\because \text{ for adiabatic atmosphere } \frac{d\theta}{dz} = 0] \quad .....(ii)$$

we know

$dp = -g\,\rho dz$ Hydrostatic equation

$p = \rho RT$ Gas equation

$$\frac{dp}{p} = \frac{-g}{R}\frac{dz}{T}$$

or $$\frac{1}{p}\frac{dp}{dz} = \frac{-g}{RT} \quad .....(iii)$$

Substituting this relation in (ii) we have

$$-\frac{1}{T}\frac{dT}{dz} = \frac{-R}{C_p} \times \frac{-g}{RT} = \frac{g}{C_pT}$$

$$\therefore \quad -\frac{dT}{dz} = \frac{g}{C_p} \quad \text{i.e., } \Gamma_a = \frac{g}{C_p}$$

$$\Gamma_a = \frac{9.8 \text{ ms}^{-2}}{1004 \text{ Jkg}^{-1}\text{°K}^{-1}} \simeq 0.098 \text{ °K/m} = 9.8 \text{ °C/km or } 9.8 \text{ °K/km} \quad .....(9.20)$$

From the constant lapse rate atmosphere, the height of the atmosphere

$$\left(Z_{\Gamma a} = \frac{T_0}{\Gamma_a}\right)$$

can be obtanied

$$z = \frac{273}{0.01} = 27300 \text{ m} = 27.3 \text{ km.} \quad (\text{where } \Gamma_a \simeq 10 \text{ °C/Km})$$

Therefore the height of the dry adiabatic atmosphere is roughly 27.3 km.

9.11 Standard Atmosphere

The hypothetical static atmospheres described above do not fit as good approximations of real atmosphere. However these concepts are used as models. Based on these, a standard atmosphere has been advocated as model with the following two simple criteria. In this model the lapse rate of temperature from MSL to an altitude of 10 km is 6.5 C/km. This layer is termed troposphere. Above troposphere the temperature is assumed to be constant (i.e. isothermal) and this layer is termed stratosphere. The boundary between these two layers is termed tropopause.

9.12 The NACA or US standard Atmosphere

For various practical uses in aviation, National Advisory Committee for Aeronautics in America defined a standard atmosphere with the following specifications up to an altitude of 32 km.

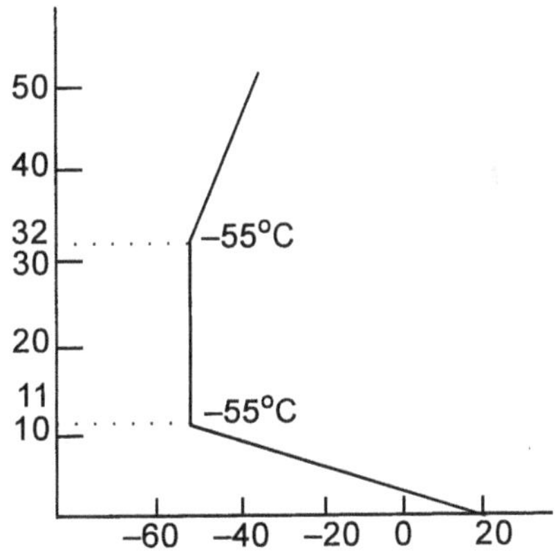

Fig 9.5 NACA standard atmosphere

1. Air is assumed to be dry and obeys gas law
2. At sea level surface air temperature 15 °C, and pressure 1013.25 h Pa
3. Temperature lapse rate 6.5 °C/km up to an altitude of 11 km. This region is called troposphere.

   Above 11 km up to 32 km the atmosphere is assumed to be isothermal (lapse rate zero) with constant temperature of –55 °C. This layer is called stratosphere.
4. The region, about 11 km elevation, separating troposphere and stratosphere is called tropopause.
5. The value of g is taken as constant = 9.86066 ms^{-2}.
6. Above 32 km temperature increase of 1 °C/km is taken.

   The above criteria is also used by ICAO (international civil aviation organization).

9.13 Hydrostatic Relation

Hydrostatic relation in terms of measured quantities of pressure, temperature.

$$dp = -g\rho dz \qquad \text{Hydrostatic equation}$$

or

$$dz = \frac{-dp}{g\rho}$$

$$p\alpha = RT \text{ or } \frac{1}{\rho} = \frac{RT}{p} \qquad \text{gas equation}$$

$$\therefore\ dz = -\frac{RT}{gp} dp$$

Integrating $\int_{z_1}^{z_2} dz = -\frac{-R}{g}\int_{p_1}^{p2} T\frac{dp}{p}$ when $p = p_1$, $z = z_1$ and $p = p_2$, $z = z_2$

$$= -\frac{R}{g}\int_{p_1}^{p_2} Td(1np)$$

or

$$z_2 - z_1 = -\frac{R}{g}\int_{p_1}^{p_2} Td(1np) \qquad .....(9.21)$$

Let T_m = mean temperature with reference to logarithmic pressure

i.e.,

$$T_m = \frac{1}{\ln\left(\frac{p_1}{p_2}\right)} \times \int_{p_1}^{p_2} Td(1n\,p) \qquad .....(9.22)$$

Substituting this value in (9.21) we have

$$z_2 - z_1 = -\frac{R}{g} \times -T_m\, 1n\left(\frac{p_2}{p_1}\right)$$

i.e.,

$$z_2 - z_1 = \frac{R}{g} T_m\, 1n\left(\frac{p_2}{p_1}\right) \qquad .....(9.23)$$

Equation (9.23) is useful in practical analysis of charts.

$z_2 - z_1$ = relative geopotential = Thickness between two isobaric surfaces p_2 and p_1.

This shows [eq. (9.23)] that thickness is proportional to the mean temperature T_m.

From hydrostatic equation and gas equation we have

$$dp = \frac{-gp}{RT}\,dz \qquad [\,dp = -g\,\rho dz \quad \rho = \frac{p}{RT}\,]$$

or $$\frac{dp}{p} = -\frac{g}{R}\frac{1}{T}dz$$

Integrating [as p_1 changes to p_2, correspondingly z_1 changes to z_2]

$$\int_{p_1}^{p_2}\frac{dp}{p} = -\int_{z_1}^{z_2}\frac{g}{R}\frac{dz}{T}$$

$$\ln\left(\frac{p_1}{p_2}\right) = \frac{g}{R}\int_{z_1}^{z_2}\frac{dz}{T} \qquad \text{.....(9.24)}$$

Since T is a function of height z i.e. T =T(z) from eq. (9.24) we can calculate pressure when z_1, z_2 are given.

EX : Calculate geo-potential difference at p = 1000 h Pa, dp = –5h Pa, T = 273 °K where g = 9.8 m s^{-2}. Assume R = 287 $m^2\ s^{-2}\ K^{-1}$

Sol :

We know

$$dp = -g\rho dz \text{ and } \frac{1}{\rho} = \frac{RT}{p}$$

$$\therefore\ dz = \frac{-RT}{gp}\,dp \qquad \text{.....(9.25)}$$

Substituting the values we have

$$dz = \frac{-287\times 273}{9.8\times 1000} \times -5 \simeq 40 \text{ m.}$$

For better approximation, suppose the pressure and temperatures were $p_0 + \delta p$, $T_0 + \delta T$ where T_0 = 273 °K, p_0 = 1000h Pa, then we can write height difference dz0 + d(dz) where dz_0 is the height difference for dp, $p = p_0$, and $T = T_0$.

$$dz_0 = \frac{-R\,T_0}{g\,p_0}\,dp \qquad \text{[From eq. (9.25)]}$$

$$dz_0 + \delta(dz) = d(z_0 + dz) = -\frac{R\,(T_0+\delta T)}{g\,(p_0+\delta p)}dp$$

$$\delta(dz) = -\frac{R}{g}dp\left[\frac{T_0+\delta T}{p_0+\delta p} - \frac{T_0}{p_0}\right] = -\frac{R}{g}dp\left[\frac{p_0(T_0+\delta T)-T_0(p_0+\delta p)}{p_0(p_0+\delta p)}\right]$$

$$\delta(dz) = -\frac{R}{g}dp.\frac{P_0\ \delta T - T_0\ \delta p}{p_0^2}$$, neglecting $p_0\delta p << p^2_0$

$$\delta(dz) = -\frac{R}{g}\frac{dp}{p_0^2}\ (p_0\ \delta T \ \square\ T0\delta\pi)$$

Using this relation we have

$$\delta(dz) = -\frac{287}{9.8} \times \frac{5}{(1000)^2}\ [1000\ .\ \delta T - 273\ \delta p]$$

$$\delta(dz) = 0.40\ \delta p - 0.15\ \delta T$$

In practice δp is large say 40 h Pa, if $\delta T = 0$, then we have $\delta(dz) = 1.6$ m. If $\delta T = 30^\circ$ and $\delta p = 0$ then $\delta(\delta z) = -0.15 \times 30 = -4.5$ m. Hence $\delta(\delta z)$ calculation error may be very large. In nature however, δp and δT have the same sign because the center of high pressure are warm while that of low pressure are cold.

9.14 International Tropical Reference Atmosphere (ITRA)

Based on US standard Atmosphere (1976) Ananthasayanam and Narsimha (1985) have proposed ITRA extending to 80 km asl. The salient features of this atmosphere are as follows.

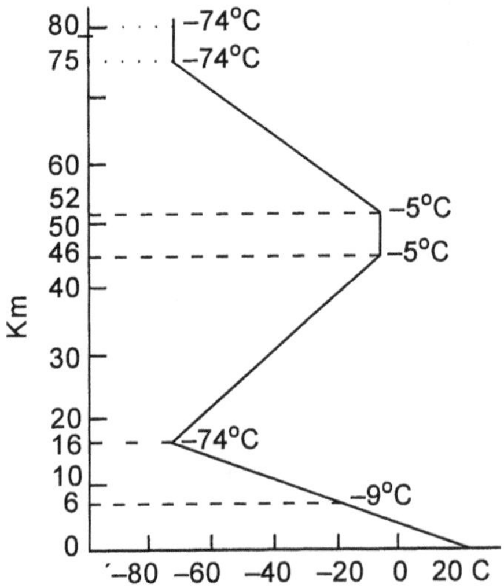

Fig 9.6 ITRA

1. ITRA is applicable between tropic of Cancer (lat 23½ °N) to Tropic of Capricorn (Lat 23½ °S)
2. Air is assumed to be dry and obeys gas law, as in USSA. The molecular weight, ratios of specific heats of air, gas constant and other constants have same values as in USSA (1976). Density of air at sea level 1.225 kg/m^3
3. At sea level surface air temperature 27 °C (300 °K) and pressure 1010 hPa
4. The acceleration due to gravity g = 9.7885 m s^{-2}
5. Temperature lapse rate $\frac{\partial T}{\partial Z}$

   (a) 6 °C/km from sea level to 6 km altitude,

   (b) 6.5 °C/km between 6 to 16 km asl,

   (c) Tropopause height 16 km with temperature –74 °C,

   (d) –2.3 °C/km (inversion) between 16 to 46 km altitudes

   (e) Stratopause at altitude 46 km and temperature –5 °C,

   (f) Isothermal layer (lapse rate zero) between altitudes 46 to 52 km with temperature –5 °C,

   (g) Lapse rate of 3 °C/km between 52 to 75 km altitudes

   (h) Mesopause near 75 km asl, temperature –74 °C

   (i) Above Mesopause isothermal layer ($\frac{\partial T}{\partial Z} = 0$), 75 to 80 km with temperature –74 °C.

Questions

1. Derive equation of continuity in the form $\frac{d\rho}{dt} + \rho \nabla . \vec{V} = 0$. Interpret meaning of this equation. Deduce the continuity equation for incompressible fluid.

2. Considering an isolated moving air parcel of constant mass, derive an expression for divergence. Using this expression interpret horizontal divergence and vertical divergence of motion.
3. What is geopotential? Define geopotential meter (gpm). Convert height 3920 m into gpm where (i) g = 9.81 m/s^2, (ii) g = 9.79 m/s^2. Find the gpm difference between these places.
4. Considering an isolated box (or parcel) with fixed mass which moves in the fluid derive equation of continuity.
5. Derive the basic hydrostatic equation for incompressible fluids. Deduce the Blaise Pascals law interpretation.
6. Assuming hydrostatic equation derive an expression for the variation of density with height. Deduce the lapse rate and height of homogeneous atmosphere taking MSL temperature 0 °C, R = 287 m^2/s^2 and g = 9.8 m/s^2.
7. Show that, in an isothermal atmosphere, pressure decreases with height exponentially.
8. For a uniform lapse rate atmosphere, derive an expression which shows pressure is a function of temperature and hence a function of height.
9. Derive an expression relating density and temperature and hence find the height of constant lapse rate atmosphere.
10. Show that in a homogeneous atmosphere temperature decreases linearly with height. Find the lapse rate of homogeneous atmosphere.
11. Show that in an adiabatic atmosphere the lapse rate of temperature is given by the expression

$$\Gamma_\alpha = -\frac{dT}{dz} = \frac{g}{C_p}$$

$$= 9.8\ °C/Km$$

Hence find the height of adiabatic atmosphere.
12. Write the salient features of US standard atmosphere.
13. Using hydrostatic equation, derive hydrostatic relation between height, temperature and pressure as

$$z_2 - z_1 = \frac{-R}{g}\int_{p_1}^{p_2} Td(\ln p)$$

How this expression used in practical analysis namely thickness is proportional to mean temperature.
14. Define ITRA with its principal features.

CHAPTER 10

Mathematical Equations of Motion

Newton's second law of motion can be expressed in the following different ways.

1. The first derivative of the momentum of a particle $\vec{p}_i$ (w.r.t time) is equal to the force $\vec{F}_i$ acting on the particle

$$\frac{d\vec{p}_i}{dt} = \vec{F}_i \quad \text{or} \quad \frac{d\left(m_i \vec{V}_i\right)}{dt} = \vec{F}_i \qquad .....(10.1)$$

$\vec{p}_i$ = momentum = m $\vec{V}i$, mi is the mass of the particle, $\vec{V}i$ = velocity of the particle.

2. A small change in momentum of a particle is equal to the impulse of the force acting on it.

Impluse = Force × time, J = $\vec{F}_i t$

$$\overrightarrow{dp_i} = \vec{F}_i\, dt, \quad \text{or } d(m_i\ \vec{V}i) = \vec{F}_i\, dt \qquad ....(10.2)$$

(where m_i = mass, is constant)

where $\vec{F}_i\,dt$, or $\vec{F}_i\ \delta t$ is called an impulse of force $\vec{F}_i$.

J = $\vec{F}(t_2 - t_1)$, implulse during the period t_1 to t_2.

3. The acceleration ($\vec{a}_i$) of a particle is directly proportional to the force ($\vec{F}_i$) acting on it and inversely proportional to the mass (mi) of the particle and coincides with the direction of the force.

$$\vec{a}_i \left(= \frac{d\vec{V}i}{dt}\right) = \frac{\vec{F}_i}{m_i}$$

It follows from eq. (10.1)

$$\frac{d(m_i \vec{V}i)}{dt} = \vec{F}_i \text{ i.e., } m_i \frac{d\vec{V}i}{dt} = \vec{F}_i$$

or $$\frac{d\vec{V}i}{dt} = \frac{\vec{F}_i}{mi} \qquad .....(10.3)$$

If $\vec{F}_i = iF_x + jF_y + kF_z$ where i, j, k unit vectors along x, y, z directions, F_x, F_y, F_z are components of vector $\overline{F}_i$ in these directions.

$$\vec{V}i = iu + jv + kw, \qquad \vec{r} = ix + iy + kz$$

$$u = \frac{dx}{dt}, \qquad v = \frac{dy}{dt}, \qquad w = \frac{dz}{dt}$$

$$\text{Then velocity} = \vec{V}i = \frac{d\vec{r}}{dt},$$

$$\text{Acceleration} = \vec{a}_i = \frac{d\vec{V}i}{dt} = \frac{\vec{F}_i}{m_i} \qquad .....(10.4)$$

or $$\frac{dm_i(iu + jv + kw)}{dt} = iF_x + jF_y + kF_z$$

or $$m_i \left[i\frac{d^2x}{dt^2} + j\frac{d^2y}{dt^2} + k\frac{d^2z}{dt^2} \right] = iF_x + jF_y + kF_z$$

$$\Rightarrow m_i \frac{d^2x}{dt^2} = F_x, \; m_i \frac{d^2y}{dt^2} = F_y, \; m_i \frac{d^2z}{dt^2} = F_z \qquad .....(10.4')$$

eq. (10.4) and (10.4') expressing acceleration $\vec{a}_i$ and force $\vec{F}_i$ are called differential eq. of motion of a particle.

Note : In general three types of coordinate axes of reference are used :

1. Rectangular cartesian (x, y, z) Fig. 10.1 (a)
2. Cylindrical (ρ, ϕ, z) Fig. 10.1 (b) and
3. Spherical (r, ϕ, θ), Fig. 10.1 (c)

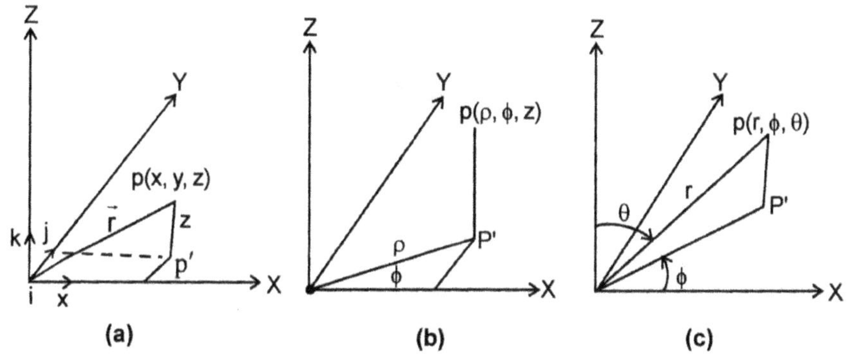

Fig. 10.1

1. Conversion of cartesian to cylindrical coordinates and vice versa are given below.

$$\rho = \sqrt{x^2 + y^2} \qquad x = \rho \cos \phi,\ y = \rho \sin \phi$$

$$\phi = \tan^{-1} \frac{y}{x} \qquad z = z$$

2. Conversion of cartesian to spherical coordinates and vice versa are given below

$$r = \sqrt{x^2 + y^2 + z^2}$$

$$\phi = \tan^{-1} \frac{y}{x}$$

$$\theta = \tan^{-1} \frac{\sqrt{x^2 + y^2}}{z}$$

$$x = r \sin \theta \cos \phi$$

$$y = r \sin \theta \sin \phi$$

$$z = r \cos \theta$$

10.1 Velocity Acceleration Speed

Let $p(\vec{r})$, $p_1(\vec{r_1})$ be the positions of a particle at time t and $t_1 = t + \Delta t$, which moved in small time period of Δt as shown in the Fig. (10.2).

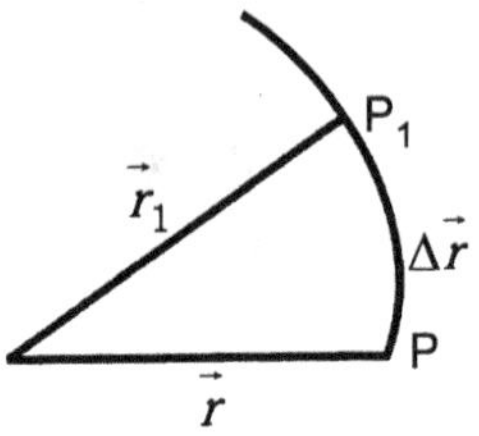

Fig. 10.2

Then velocity $\quad \vec{V} = \dfrac{\Delta\vec{r}}{\Delta t}$

or $\quad \vec{V} = \dfrac{d\vec{r}}{dt}$ (in the limit as $\Delta t \to 0$)

and acceleration $\quad \vec{a} = \dfrac{d\vec{V}}{dt} = \dfrac{d^2\vec{r}}{dt^2}$.

If $\quad \vec{r} = ix + jy + kz$, then

velocity $\quad \vec{V} = \dfrac{d\vec{r}}{dt} = i\dfrac{dx}{dt} + j\dfrac{dy}{dt} + k\dfrac{dz}{dt}$

speed $\quad V = |\vec{V}| = \sqrt{\left(\dfrac{dx}{dt}\right)^2 + \left(\dfrac{dy}{dt}\right)^2 + \left(\dfrac{dz}{dt}\right)^2}$

Note : $V = |\vec{V}| = \sqrt{\vec{V}.\vec{V}}, \dot{\vec{r}} = \dfrac{d\vec{r}}{dt}, \quad \ddot{\vec{r}} = \dfrac{d^2\vec{r}}{dt^2}$

If s is the path length of the particle moved in time t_0 to t_1 then

$$s = \int_{t_0}^{t_1} |\vec{V}| dt = \int_{t_0}^{t_1} \sqrt{\left(\frac{dx}{dt}\right)^2 + \left(\frac{dy}{dt}\right)^2 + \left(\frac{dz}{dt}\right)^2}\, .dt.$$

10.1.1 Definition

Streamline : A stream line is a curve drawn in a fluid such that , at any instant of time, the direction of the tangent at any point of the curve coincides with the direction of the velocity of the fluid particle at that point.

With the usual notation, the differential equations of streamlines are given by

$$\frac{dx}{u} = \frac{dy}{v} = \frac{dz}{w}$$

where $\vec{V} = iu + jv + kw$

$\vec{r} = ix + jy + kz$

$$\left[u = \frac{dx}{dt}, v = \frac{dy}{dt}, w = \frac{dz}{dt}\right]$$

Path line : Path line or trajectory is a curve which a particular fluid particle describes during its motion.

The differential equations of the path line are given by

$$\frac{dx}{dt} = u, \quad \frac{dy}{dt} = v, \quad \frac{dz}{dt} = w$$

It may be noted here that, stream lines reveal how each fluid particle is moving at a given instant, while the path line shows how a given particle is moving at each instant.

Euler's Approach to the hydrodynamical (or fluid dynamical) problems consists in observing the changes in velocity, density and pressure as the fluid passes through a fixed point in space occupied by the fluid or local time rate of change.

Lagrange's Approach to hydrodynamical (or fluid dynamical) problems consists in observing the changes in velocity , density and pressure as a typical particle itself moves about or individual time rate of change.

Note : 1. Streamlines do not cross one another, however they join at isolated points such as col region or infinite velocity. Flow will be along a streamline but not across it. Streamline spacing varies inversely with velocity. Converging streamlines indicate accelerated flow in that direction.

Streakline : A streakline is the instantaneous distribution of all fluid particles that pass through a fixed point in the flow of field. A line formed by the smoke particle ejected into the atmosphere from a fixed nozzle forms a streakline.

10.2 Theorem

A vector of constant length is normal to its time derivative.

Let $\vec{b} = \vec{b}(t)$ is a vector function of time t.

Then $\sqrt{\vec{b}.\vec{b}}$ = constant (= length of the vector)

i.e. $\vec{b}^2 = \vec{b}.\vec{b}$ = constant

Differentiating with respect to time 't' we have

$$2\vec{b}.\frac{d\vec{b}}{dt}=0 \text{ i.e., } \vec{b}.\frac{d\vec{b}}{dt} = 0, \text{ this proves the statement}$$

Note: $\vec{A}.\vec{B} = ab\cos\theta$, $\vec{A}\times\vec{B} = ab\sin\theta\,\hat{n}$

Where $a = |\vec{A}|$, $b = |\vec{B}|$, θ is the angle between the vectors $\vec{A}$ and $\vec{B}$, $\vec{n}$ unit vector perpendicular to the plane containing $\vec{A}$ and $\vec{B}$.

Let $\vec{T}$ be a unit vector in the direction of the tangent on the trajectory, then

$$\vec{V}=V\vec{T} \qquad \text{where } V = |\vec{V}| \qquad .....(10.5)$$

Note: Velocity vector lies along the tangent at any point of the trajectory

$$\vec{a}=\frac{d\vec{V}}{dt}=\frac{d}{dt}(V\vec{T})=\frac{dV}{dt}\vec{T}+V\frac{d\vec{T}}{dt} \qquad .....(10.6)$$

10.3 Theorem

Acceleration on curved path is equal to sum of tangential acceleration and normal acceleration.

Let p, p_1 be the positions of a particle on the trajectory at time $t = t_0$ and $t_1 = t_0 + \Delta t$, and let $\vec{T}$, $\vec{T_1}$ be the unit vectors along the tangents at P and P_1 as shown in the Fig. (10.3). Let the normals at P and P_1 intersect at C. Then CP = R is the radius of curvature.

From the Fig. $\vec{T} + \Delta\vec{T} = \vec{T_1}$

$$\overrightarrow{PS} = \overrightarrow{P_1Q_1} = \vec{T_1},$$

$$\overrightarrow{QS} = \overrightarrow{PS} - \overrightarrow{PQ}$$

or $\vec{T_1} - \vec{T} = \Delta\vec{T}$

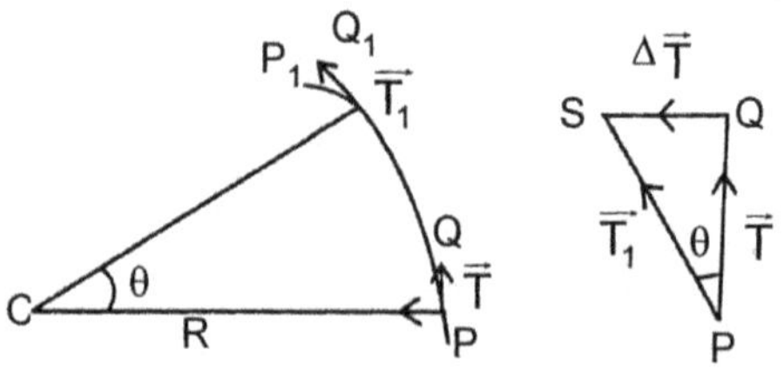

Fig. 10.3

ΔCPP_1 is similar to Δ PQS

$$\therefore \qquad \frac{|\Delta\vec{T}|}{|\vec{T}|} = \frac{PP_1}{CP} = \frac{\Delta s}{R} \quad \text{where } \Delta s = PP_1 \text{ and } CP = R$$

or

$$\frac{|\Delta\vec{T}|}{\Delta t.|\vec{T}|} = \frac{\Delta s}{R\,\Delta t}$$

as $\Delta t \to 0$, the above relation becomes

$$\left|\frac{d\vec{T}}{dt}\right| = \frac{1}{R}\frac{ds}{dt} = \frac{V}{R}, \quad (\because \frac{ds}{dt} = V) \qquad .....(10.7)$$

We note that the vector $\Delta\vec{T} = \overrightarrow{QS}$ in the limit is directed along the normal $\vec{N}$ (pointing towards the center of curvature C), this follows from the above theorem, as $\vec{T}$ is of constant length.

The above relation (10.7) may be written as

$$\frac{d\vec{T}}{dt} = \frac{V}{R}\vec{N} \qquad .....(10.8)$$

Further, from $\vec{a} = \frac{dV}{dt}\vec{T} + V\frac{d\vec{T}}{dt}$ (eq. 10.6)

$$\vec{a} = \frac{dV}{dt}\vec{T} + V\frac{V}{R}\vec{N}, \quad \text{using the above eq. (10.8)}$$

$$\vec{a} = \frac{dV}{dt}\vec{T} + \frac{V^2}{R}\vec{N} \qquad .....(10.9)$$

i.e., Acceleration = Tangential acceleration + normal acceleration

eq. (10.9) may be stated as: 'The acceleration is composed of tangential acceleration $\left(\frac{dV}{dt}\vec{T}\right)$ [which is equal in magnitude to the time derivative of the speed $\left(\frac{dV}{dt}\right)$] and a normal acceleration $\frac{V^2}{R}\vec{N}$. The magnitude of normal acceleration $\frac{V^2}{R}$ is also called as centripetal acceleration.

Ex : To compute velocity and position vector when acceleration is given.

Let

$$\vec{r} = \vec{A} + \vec{B}t \qquad ......(10.10)$$

be the position vector, where $\vec{A}$, $\vec{B}$ are vectors which are independent of time 't'. Differentiating (10.10) we have

$$\text{velocity} \qquad = \vec{V} = \frac{d\vec{r}}{dt} = \vec{B}$$

$$\text{acceleration} \qquad = \vec{a} = \frac{d\vec{V}}{dt} = \frac{d\vec{B}}{dt} = 0,$$

Conversely by let $\vec{a} = 0$ .....(10.11)

Then by integrating with respect to time 't' we have

$$\vec{V} = \int \vec{a}dt = \int 0.dt = \text{Constant say } \vec{V}_0 \text{ vector} \qquad (\because \vec{a} = 0)$$

Again integrating velocity we have,

$$\vec{r} = \int \vec{V}dt = \int \vec{V}_0 dt \qquad \text{(from above)}$$

$$\vec{r} = \vec{V}_0 t + \vec{r}_0 \qquad \text{(where } \vec{r}_0 \text{ is a constant vector)}$$

Thus we have (i) $\vec{V} = \vec{V}_0$ a constant vector

and (ii) $\vec{r} = \vec{V}_0 t + \vec{r}_0 =$ a constant vector

i.e., a motion along a straight line with constant speed has zero acceleration.

Ex : Suppose $\vec{a} = \vec{a}_0$ (a constant vector).

Integrating we have

$$\vec{V} = \int \vec{a}_0 dt = \vec{a}_0 t + \vec{V}_0 \text{ (a constant vector)}$$

Again integrating we get position vector $\vec{r}$

$$\vec{r} = \int \vec{V}dt = \int (\vec{a}_0 t + \vec{V}_0)dt$$

$$= \vec{a}_0 \frac{t^2}{2} + \vec{V}_0 t + \vec{r}_0 \text{ (where } \vec{r}_0 \text{ is a constant vector)}$$

i.e. (i) $\vec{V} = \vec{a}_0 t + \vec{V}_0$

and (ii) $\vec{r} = \vec{a}_0 \frac{t^2}{2} + \vec{V}_0 t + \vec{r}_0$

Questions

1. Express mathematically Newton's second law of motion in three different ways.
2. Define a streamline and a pathline. Give differential equations satisfying them. State Euler's and Lagranges approach to the hydrodynamical problems.
3. Show that a vector of constant length is normal to its time derivative.
4. Show that the acceleration of a particle on a curved path is equal to the sum of tangential and normal accelerations.

CHAPTER 11

Kinematics of Rotating Motion

Classical mechanics is the study of the motion of objects (in meteorology air) and deals with the relations of force, motion and matter. Mechanics comprises of (i) Statics (ii) Kinematics and (iii) dynamics.

Statics deals with the laws of composition of forces under equilibrium condition (acceleration is zero). Kinematics deals with the motion of bodies (air) without dealing with the forces that are responsible for it. Dynamics deals with the motion of bodies (or air) with the forces that are responsible for it.

Suppose a particle (or point) P rotates about an axis with an angular velocity $\vec{\Omega}$ (angular speed Ω), where $\vec{\Omega}$ is a vector pointing along the axis of rotation as shown in the Fig.11.1. Let OP = $\vec{R}$, and P makes an angle θ with the axis. Then the speed of the particle P is given by

$$V = \Omega R \sin\theta \qquad .....(11.1)$$

because, arc = radius × radian

$$\text{Radian} = \Omega, \quad \text{Radius} = R\sin\theta$$

The velocity can be written as

$$\text{velocity} = \vec{V} = \vec{\Omega}\times\vec{R} \qquad .....(11.2)$$

This representation is inconformity with the definition of cross product of vectors.

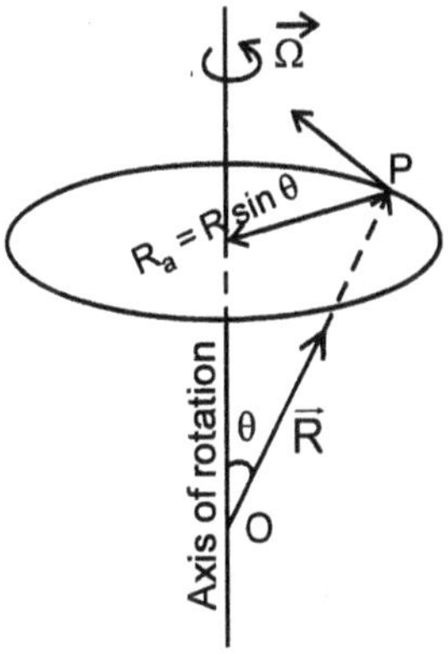

Fig. 11.1

If the speed (V) is constant, then from the acceleration theorem

$$\left(\vec{a} = \frac{dV}{dt}\vec{T} + \frac{V^2}{R}\vec{N}\right)$$

it follows that $$\vec{a}_c = \frac{V^2}{R_a}\vec{N} \qquad \left(\because \frac{dV}{dt} = 0\right)$$

$$\vec{a}_c = \frac{\Omega^2 R^2 \sin^2\theta}{R \sin\theta}\vec{N} \qquad .....(11.3)$$

or $$\vec{a}_c = \vec{\Omega} \times (\vec{\Omega} \times \vec{R}) \qquad (11.3')$$

where $\vec{a}_c$ = circular accleration, R_a = Radius about the axis = R sin θ.

Question

1. Suppose a particle P rotates about an axis with angular velocity $\vec{\Omega}$, with a radius vector OP = $\vec{R}$ (O is a point on the axis) which makes an angle θ with the axis. Find the velocity of the particle ($\vec{V}$) and circular acceleration ($\vec{a}$) in terms of $\vec{\Omega}$, $\vec{R}$ and θ.

CHAPTER 12

Absolute and Relative Velocity

Suppose a particle p moves with a velocity $\vec{V}_r$ relative to a point M (some geometrical point). Suppose the point M rotates about an axis with constant angular velocity $\vec{\Omega}$. Then

$$\vec{V}_M = \vec{\Omega} \times \vec{R} \qquad .....(12.1)$$

as described in the rotating motion.

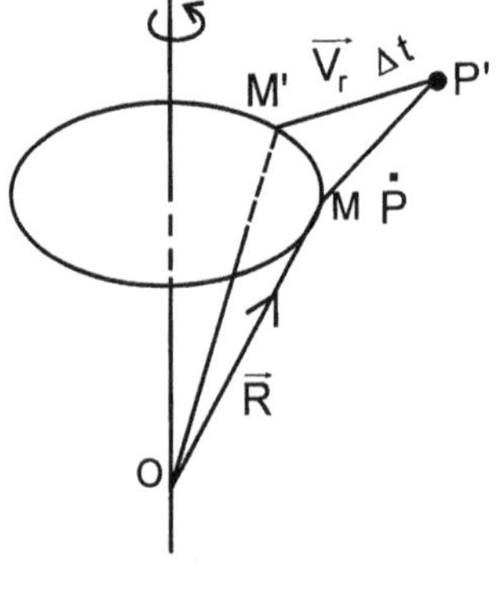

Fig. 12.1 (a)

Suppose during a small interval of time Δt, M moves to M' so that $MM' = \vec{V}_M \Delta t$.

During this same interval of time Δt, suppose P has moved to P' so that $P P' = \vec{V}_r \Delta t$.

The total or absolute displacement

$$\vec{MP'} = \left(\vec{V}_M + \vec{V}_r\right)\Delta t \qquad ......(12.2)$$

From eq. (2.2) we have dividing by Δt and taking limit as $\Delta t \to 0$

$$\frac{\vec{MP'}}{\Delta t} = \vec{V}_M + \vec{V}_r$$

$$\vec{V}_a = \vec{V}_r + \vec{V}_M = \vec{V}_r + \vec{\Omega} \times \vec{R} \text{ [using (12.1)]} \qquad .....(12.3)$$

Equation (12.3) states that ($\vec{V}_a$) absolute velocity (or velocity in one system) is equal to ($\vec{V}_r$) the relative velocity (or velocity in the second system) plus the velocity of the second system with reference (or relative) to the first (or absolute).

Since $\vec{V}_a = \left(\frac{d\vec{R}}{dt}\right)_a$, $\vec{V}_r = \left(\frac{d\vec{R}}{dt}\right)_r$ by definition of velocity

[where suffixes a and r indicate absolute and relative reference axes.]

eq. (12.3) can be written as

$$\left(\frac{d\vec{R}}{dt}\right)_a = \left(\frac{d\vec{R}}{dt}\right)_r + \vec{\Omega} \times \vec{R} \qquad .....(12.4)$$

Consider an arbitrary vector $\vec{Q}$ such that $\vec{Q} = \vec{R_2} - \vec{R_1}$ where $\vec{R_2}$ is the position vector of end point of $\vec{Q}$ and R_1 is the position vector of beginning point of $\vec{Q}$ as shown in the Fig. 12.1(b).

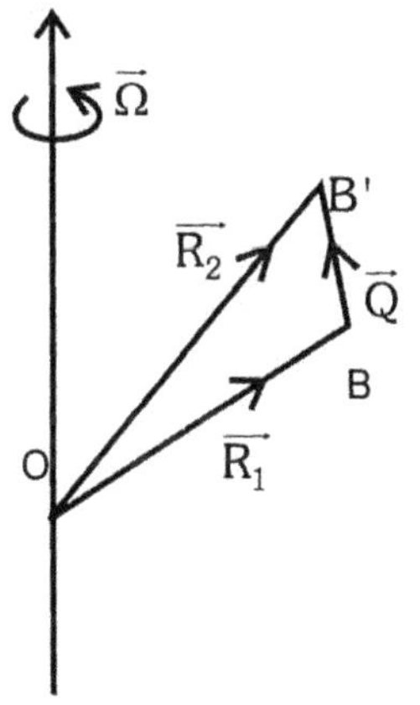

Fig. 12.1 (b)

Using above relation eq. (12.4) we have for $\vec{R_1}$, $\vec{R_2}$ vectors

$$\left(\frac{d\vec{R_2}}{dt}\right)_a = \left(\frac{d\vec{R_2}}{dt}\right)_r + \vec{\Omega}\times\vec{R_2} \qquad \text{.....(12.5)}$$

$$\left(\frac{d\vec{R_1}}{dt}\right)_a = \left(\frac{d\vec{R_1}}{dt}\right)_r + \vec{\Omega}\times\vec{R_1} \qquad \text{.....(12.6)}$$

Substracting (12.6) from (12.5), we have

$$\left(\frac{d\vec{R_2}}{dt}\right)_a - \left(\frac{d\vec{R_1}}{dt}\right)_a = \left(\frac{d\vec{R_2}}{dt}\right)_r - \left(\frac{d\vec{R_1}}{dt}\right)_r + \vec{\Omega}\times(\vec{R_2}-\vec{R_1})$$

or $$\left[\frac{d}{dt}(\vec{R_2}-\vec{R_1})\right]_a = \left[\frac{d}{dt}(\vec{R_2}-\vec{R_1}\right]_r + \vec{\Omega}\times(\vec{R_2}-\vec{R_1})$$

i.e., $$\left(\frac{d\vec{Q}}{dt}\right)_a = \left(\frac{d\vec{Q}}{dt}\right)_r + \vec{\Omega}\times\vec{Q} \qquad \because \vec{Q} = \vec{R_2}-\vec{R_1} \qquad \text{.....(12.7)}$$

It follows from eq. (12.7) that for any artitrary vector $\vec{Q}$ the relation (12.4) is true.

12.1 Absolute and Relative Acceleration

We Know

$\vec{V_a} = \vec{V_r} + \vec{\Omega}\times\vec{R}$, eq. (12.3) of absolute and relative velocity.

In the equation (12.7) above put $\vec{Q} = \vec{V_a}$, then we have

$$\left(\frac{d\vec{V}_a}{dt}\right)_a = \left(\frac{d\vec{V}_a}{dt}\right)_r + \vec{\Omega}\times\vec{V}_a \qquad \text{.....(12.8)}$$

Substituting $\vec{V_a} = \vec{V_r} + \vec{\Omega}\times\vec{R}$, on the right side of this equation we have

$$\left(\frac{d\vec{V}_a}{dt}\right)_a = \left[\frac{d}{dt}(\vec{V}_r + \vec{\Omega}\times\vec{R}\right]_r + \vec{\Omega}\times[\vec{V}_r + \vec{\Omega}\times\vec{R}]$$

$$= \left(\frac{d}{dt}\vec{V}_r\right)_r + \vec{\Omega}\times\left(\frac{d\vec{R}}{dt}\right)_r + \vec{\Omega}\times\vec{V}_r + \vec{\Omega}\times\vec{\Omega}\times\vec{R}$$

$$\left(\frac{d\vec{V}_a}{dt}\right)_a = \left(\frac{d\vec{V}_r}{dt}\right)_r + 2\vec{\Omega}\times\vec{V}_r + \vec{\Omega}\times\vec{\Omega}\times\vec{R},$$

$$\because \quad \left(\frac{d\vec{R}}{dt}\right)_r = \vec{V}_r \qquad .....(12.9)$$

Where $\left(\frac{d\vec{V}_a}{dt}\right)_a$ = Absolute acceleration

This is the acceleration of a particle in Newton's second Law.

$\left(\frac{d\vec{V}_r}{dt}\right)_r$ = Relative acceleration

$2\vec{\Omega}\times\vec{V}_r$ = Coriolis acceleration. This is perpendicular to the relative velocity. This exists only when the relative motion is non-zero.

$\vec{\Omega}\times\vec{\Omega}\times\vec{R}$ = Centripetal acceleration $\vec{a_c}$. This depends on position only. Its values is Ω^2 R sin θ derived in kinematic of rotating motion eq. (11.3) where θ co-latitude or $\Omega^2 R \cos\phi$, where ϕ is latitude. $\theta + \phi = 90°$. The maximum value of the magnitude of $\vec{a_c}$ occurs at equator, where $\theta = 90$ or $\phi = 0$.

If $\qquad \Omega = 7.29 \times 10^{-5} s^{-1}$, $R = \frac{2}{\pi} \times 10^7$ m

Then $\qquad \Omega^2 R = 0.03 m/s^2$

In meteorology we call $-2\vec{\Omega}\times\vec{V}_r$ as Coriolis force and $-\vec{\Omega}\times\vec{\Omega}\times\vec{R}$ Centrifugal force.

The earth revolves about it axis in every 23 h 56 min 42s = 86164s

$$\therefore \quad \Omega = \frac{2\pi^c}{T} = \frac{2\times 22}{7\times 86164} = 7.29 \times 10^{-5} \text{ c/s}$$

(π^C = π – radiance) or c/s = radiance per second

12.2 The Friction or Viscosity Force

There are two types of friction–internal and external. Internal friction (or viscosity) is the result of friction between layers of the gas or liquid which flows parallel to each other at different speeds. The cause of viscosity is the transfer of momentum by the molecules from one layer to another. Tangential forces which obstructs the displacement of portions of liquids or gases with respect to each other. The external friction develops due to interaction between bodies and acts at their surfaces of contact and obstructs in their relative displacement. The later one is also named as contact friction. In meteorology both these types act.

Suppose the motion is horizontal i.e., u, v exists and w = 0. Further let us assume that the spatial variation is in the z-direction only. Let

$$\vec{V} = iu_z + jv_z \qquad .....(12.9)$$

Fig. (12.2(a)) shows the simplest case where v = 0 (variation only in zx-plane). In this case fluid at one height impart acceleration to the fluid below. Conversely the fluid below may slow down the fluid above if the above layer moves faster than the lower layer. Basically we assume that

$$(\vec{V_2} - \vec{V_1}) \propto \text{accelerating force.}$$

Where $\vec{V_2}$, $\vec{V_1}$ are velocities just above and below a given level as shown in Fig 12.2(a) by broken line. Supposing the velocity variation is continuous we have

$$(\vec{V}_2 - \vec{V}_1) = \frac{d\vec{V}}{dz} dz \qquad .....(12.10)$$

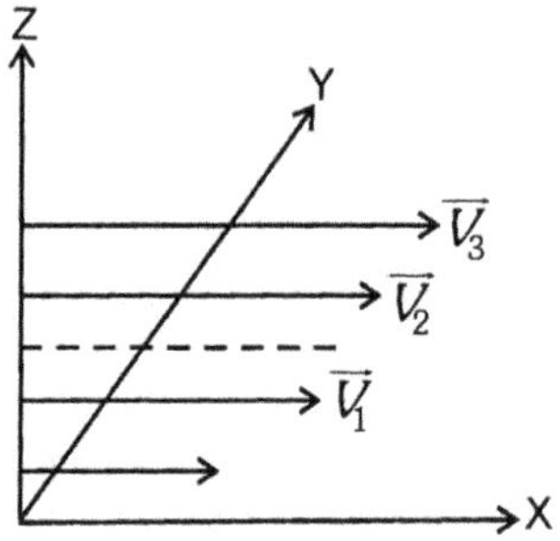

Fig. 12.2 (a)

If $\vec{\tau}$ is the frictional force per unit area, then we may write

$$\vec{\tau} = \mu \frac{d\vec{V}}{dz} \qquad .....(12.11)$$

where $\mu = \mu(z)$ = coefficient of viscosity. $\vec{\tau}$ acts along the surface while pressure acts perpendicular to the surface.

Consider a box of unit area of base and top and thickness dz as shown in Fig. 12.2 (b). The force on the topside is $\vec{\tau}(z + dz)$ while on the bottom side is $-\vec{\tau}(z)$.

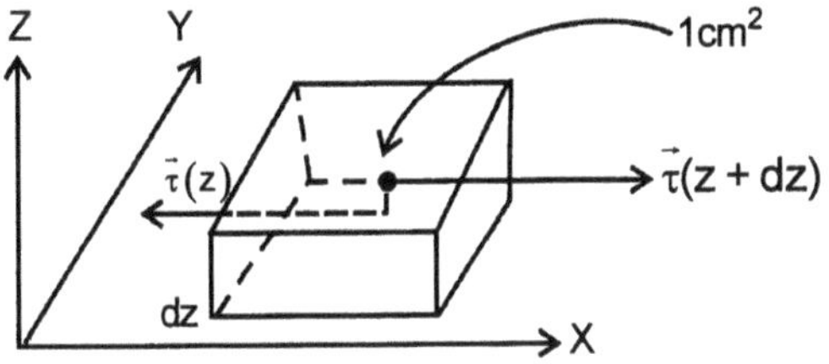

Fig. 12.2 (b)

The resulting force is given by

$$\vec{\tau}(z+dz)-\vec{\tau}(z)=\frac{d\vec{\tau}}{dz}.dz \qquad .....(12.12)$$

As shown, the volume of the box is (A × h) = 1 × dz.
[volume= base area × height, mass = density × volume].

∴ the mass of the air in the box = ρdz.

Dividing eq. (12.12) by ρdz both sides we have

$$\frac{\vec{\tau}(z+dz)-\vec{\tau}(z)}{\rho dz}=\frac{d\vec{\tau}}{dz}.dz\times\frac{1}{\rho dz}$$

i.e. The force per unit mass is

$$\frac{1}{\rho}\frac{d\vec{\tau}}{dz}=\frac{1}{\rho}\frac{d}{dz}\left(\mu\frac{d\vec{V}}{dz}\right), \text{ [using (12.11) i.e., } \tau=\mu\frac{d\vec{V}}{dz}$$

.....(12.13)

Strictly the above equation is valid with partial derivatives as we have considered v = 0 and w = 0

Note : eq. (12.13) gives the component only under the assumption horizontal motion of wind given by eq. (12.9) and v = 0. When we consider three dimensional motion including vertical component w, viscous force becomes very complicated. In practice it is observed main variation of horizontal wind is in the vertical direction and eq. (12.13) is sufficient.

The dimensions of $\vec{\tau}$ is force per unit area.

For air at 273 °K the value of μ is taken

$= 1.7 \times 10^{-10}\ N\ m^{-2} s^{-1}$

Note :

1. The property of fluids (liquids or gases) which acts tangentially on the layers of the liquid in motion and tries to prevent the motion is called viscosity. Or simply a backward dragging force is called viscosity.
2. *Newton's formula for viscous force.* The viscous force between two parallel layers each of area A and having velocity gradient $\frac{dV}{dx}$ is given by $F = -\eta A \frac{dV}{dx}$, where η = coefficient of dynamic viscosity of the fluid.

   In eq. (12.11) above μ may be replaced by η.
3. *Coefficient of dynamic viscosity* (η) defined as the tangential viscous force which maintains a unit velocity gradient $\left(\frac{dV}{dx} = 1\right)$ between two parallel layers each of unit area (A = 1).
4. Some viscosity values are given below.

Temp °C	Viscosity of air µp	Viscosity of water cp	Viscosity of castor oil p
0	171	1.792	53
20	181	1.005	9.86
100	218	0.284	0.17

where p = poise, µp = micropoise or 10^{-6}p, c.p. = centi poise = 10^{-2}p

Units : CGS unit is Poise (dyn.s. cm^{-2}), SI unit 1 Ns m^{-2};

1 decapoise = 10 poise.

1 poise = 1dyns cm^{-2} = 10^{-1} Ns m^{-2}

Stoke's Law : The backward dragging force acting on small sphere of radius r moving with uniform velocity V through a fluid medium of coefficient of viscosity η is given by $F = 6\pi\eta rv$. This is Stoke's Law. When the falling body attains terminal velocity [= constant= V_T], the weight of the body becomes equal to the sum of back ward dragging force and the upthrust due to the medium.

Viscous retarding force with terminal velocity + buoyant force or upthrust due to fluid medium = weight of the sphere.

If σ = density of the fluid medium, ρ = density of the material of the sphere, then terminal velocity is given by

$$6\,\pi\eta r\,V_T + \frac{4}{3}\,\pi r^3\,\sigma\,g = \frac{4}{3}\,\pi r^3 .\,\rho.g$$

$$V_T = \frac{4}{3}\pi r^3 g(\rho - \sigma) \times \frac{1}{6\pi\eta r}$$

$$v_T = \frac{2}{9}\frac{r^2(\rho - \sigma)g}{\eta}$$

12.3 The Force of Gravitation

According to universal law of gravitation, every particle (or body) of mass m_1 in the universe attracts every other particle of mass m_2 with a force ($\vec{F}$) directly proportional to the product of their masses ($m_1 \times m_2$) and inversely proportional to the square of the distance (R) between them.

i.e., $$\vec{F} \propto \frac{m_1 m_2}{R^2} \quad \text{or} \quad \vec{F} = -G\frac{m_1 m_2}{R^3}\vec{R} \qquad .....(12.14)$$

where G = universal gravitational constant = $6.66 \times 10^{-11}\ Nm^2kg^{-2}$

If $m_1 = M$ is the mass of the earth $\simeq 6 \times 10^{24}$ kg and $m_2 = 1$ unit mass (kg) and $\vec{R}$ = distance between unit mass and the centre of the earth 6380 km, then the force of earth gravitation $\vec{g_a}$ on unit mass is given by

$$\vec{g}_a = -G \times \frac{6 \times 10^{24} \times 1}{(6.38 \times 10^6]^2}$$

$$|\vec{g}_a| = \frac{6.66 \times 10^{-11} \times 6 \times 10^{24} \times 1}{[6.38 \times 10^6]^2} = \simeq 9.82\ ms^{-2}$$

It was defined earlier that geo-potential ψ is a gradient of scalar potential function of $\vec{g}_a$ and is given by the relation

$$d\Psi = gdz$$

or $$\Psi = \int_0^z gdz$$

where

z = altitude,

Ψ = potential energy gained by unit mass when lifted from sea level to the height z.

when $z = 0, \Psi = 0$

or $$\vec{g_a} = -\nabla \Psi_a \qquad .....(12.15)$$

$$\Psi_a = -\frac{GM}{R} + \text{constant.}$$

We have proved that the centrifugal force (CF), which depends on the position is given by $\vec{\Omega} \times \vec{\Omega} \times \vec{R}$ where $|\vec{R}|$ = radius of the earth, and $\vec{\Omega}$ = angular velocity of the earth. Since both $\vec{g_a}$ and CF, depend on position on the earth, these may be combined into a single force called gravity as shown in Fig. 12.3

$$\vec{g} = \vec{g_a} - \vec{\Omega} \times \vec{\Omega} \times \vec{R} \qquad .....(12.16)$$

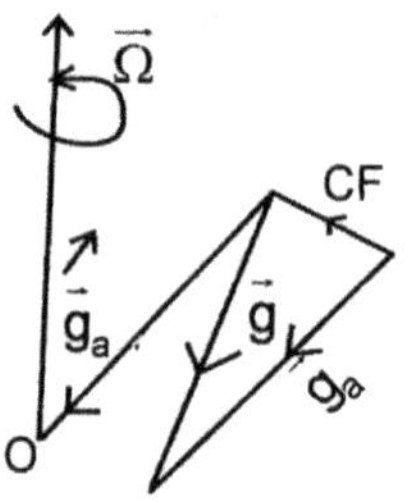

Fig. 12.3

12.4 The Pressure Gradient Force

The pressure gradient force or simply pressure force is the result of changes of pressure is defined as the force per unit area $\left(p = \frac{F}{A}\right)$. Pressure at a point in the fluid is the same in all directions (Pascals law) at any given instant and is independent of direction.

12.4.1 Definition

Atmospheric pressure : Pressure at any point of the atmosphere is the weight of the vertical column of air above it (extended to the top of the atmosphere) on unit area, the point being at the centre.

Consider a box of volume element $\delta v = \delta x . \delta y . \delta z$, the coordinates of whose centre 0 is x, y, z as shown in Fig. 12.4. We know that due to random motion of molecules momentum transfer per unit area, per unit time is pressure exerted on the sides. Therefore w.r.t to the pressure at the centre, the pressure at the centres of the walls can be expressed by Taylor's expansion.

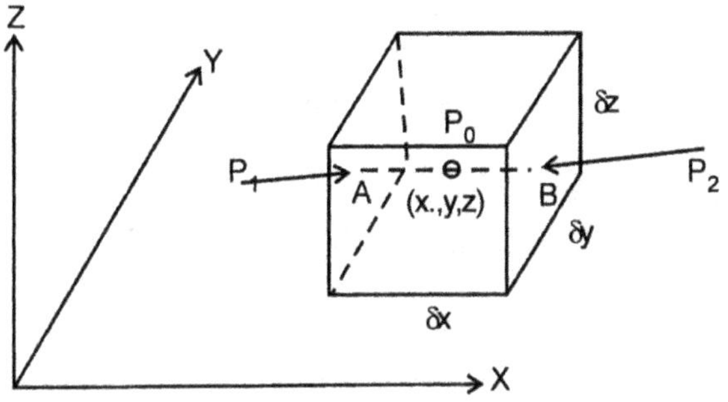

Fig. 12.4

Let p_0, p_1, p_2 be the pressure forces acting at the centre of the box, at centre of the left side face A, and the right side face B of the box respectively.

The net force acting in the x-direction is given by

$$p_1\delta y\delta z - p_2\delta y\delta z = \left\{\left[p_0 - \frac{\partial p_0}{\partial x}\frac{\delta x}{2} + \ldots\right] - \left[p_0 + \frac{\partial p_0}{\partial x}\frac{\delta x}{2} + \ldots\right]\right\} \times \delta y\, \delta z$$

$$= -\frac{\partial p_0}{\partial x}\, \delta x\delta y\delta z$$

$$= -\frac{\partial p_0}{\partial x}\, \delta v$$

Therefore the net pressure force per unit mass in x - direction is

$$= \frac{1}{\rho\delta v} \times -\frac{\partial p_0}{\partial x}\delta v = -\frac{1}{\rho}\frac{\partial p_0}{\partial x}$$

($\because$ mass of the air in the box $= \rho\delta v$)

Similarly the net pressure forces per unit mass in y-direction, z-direction are respectively given by

$$-\frac{1}{\rho}\frac{\partial p_0}{\partial y},\ -\frac{1}{\rho}\frac{\partial p_0}{\partial z}$$

Therefore the net pressure force per unit of mass

$$= -\frac{1}{\rho}\left[i\frac{\partial p}{\partial x} + j\frac{\partial p}{\partial y} + k\frac{\partial p}{\partial z}\right] \text{ (where } p_0 \text{ is replaced by p)}$$

$$= -\frac{1}{\rho}\nabla p$$

As the gradient of the pressure (∇p) is normal to the isobaric surfaces and is directed from lower to higher pressure, the pressure force is also normal to the isobaric surfaces but directed from higher to lower pressure.

12.5 The Equations of Motion

From Newton's second law we have

$$\frac{d\vec{V}_a}{dt} = \sum_{i=1}^{n} \vec{F}_i \qquad .....(12.17)$$

i.e., "The acceleration $\frac{d\vec{V}_a}{dt}$ of the particle is equal to the sum of the forces (ΣF_i) acting on it". The LHS of eq. (12.17) with reference to the absolute or inertial coordinates.

We have derived the relation

$$\left(\frac{d\vec{V}_a}{dt}\right)_a = \left(\frac{d\vec{V}}{dt}\right)_r + 2\vec{\Omega} \times \vec{V}_r + \vec{\Omega} \times \vec{\Omega} \times \vec{R}.$$

The RHS of eq. (12.17) indicate, the sum of the forces acting on the particle of unit mass, namely pressure force $\left(-\frac{1}{\rho}\nabla p\right)$, gravitational force $\vec{g_a}$ and friction force $\vec{\tau} = \mu\frac{d\vec{V}}{dt}$.

Substituting these (relations) values in eq. (12.17), omitting suffix - r, we have

$$\frac{d\vec{V}}{dt} + 2\vec{\Omega} \times \vec{V} + \vec{\Omega} \times \vec{\Omega} \times \vec{R} = -\frac{1}{\rho}\nabla p + \vec{g_a} + \frac{1}{\rho}\frac{d\vec{\tau}}{dz}$$

or $$\frac{d\vec{V}}{dt} = -\frac{1}{\rho}\nabla p - 2\vec{\Omega} \times \vec{V} + \vec{g_a} - \vec{\Omega} \times \vec{\Omega} \times \vec{R} + \frac{1}{\rho}\frac{d\vec{\tau}}{dz}$$

$$\frac{d\vec{V}}{dt} = -\frac{1}{\rho}\nabla p - 2\vec{\Omega} \times \vec{V} + \vec{g} + \frac{1}{\rho}\frac{d\vec{\tau}}{dz} \qquad .....(12.18)$$

Eq. (12.18) is the required equations of motion, where $2\vec{\Omega} \times \vec{V}$ = coriolis force, $\vec{\Omega} \times \vec{\Omega} \times \vec{R}$ = centrifugal force. $\vec{g} = \vec{g_a} - \vec{\Omega} \times \vec{\Omega} \times \vec{R}$ = is gravity force or simply gravity.

Eq. (12.18) contains five dependent variables, viz u, v, w, p, ρ while $\vec{g}$ and $\vec{\Omega}$ are known. (eq. 12.18) has five variables and to solve, we use hydrostatic equation, gas equation, continuity equation and first law of thermodynamics, which were discussed earlier.

12.6 Equation of Motion in Local Cartesian Coordinates

Suppose geographical east as the x-axis, north the y-axis and z-axis local zenith (upwards or opposite to the local direction of gravity) as shown in Fig.12.5) xy-plane is normal to the zenith or the gravity vector

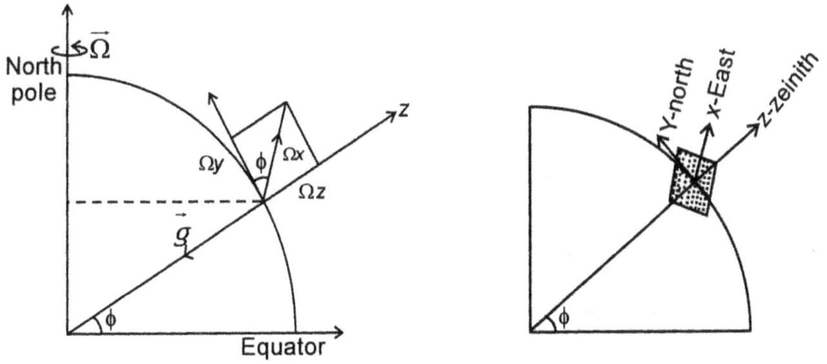

Fig. 12.5

Let $\vec{V} = iu + jv + kw$ .....(12.19)

where i, j, k are unit vectors in the direction of east, north and local zenith and u, v, w are the velocity components in these directions. φ = latitude of the place. Then

(Differentiating velocity eq. (12.19) w.r.t time t) we have

$$\frac{d\vec{V}}{dt} = i\frac{du}{dt} + j\frac{dv}{dt} + k\frac{dw}{dt} \quad .....(12.20)$$

(pressure gradient force)

$$-\frac{1}{\rho}\nabla p = \frac{-1}{\rho}\left(i\frac{\partial p}{\partial x} + j\frac{\partial p}{\partial y} + k\frac{\partial p}{\partial z}\right) \quad .....(12.21)$$

$\vec{g} = -kg$, where k unit vector in z - direction .....(12.22)

$$\frac{1}{\rho}\vec{F} = \frac{1}{\rho}(iF_x + jF_y + kF_z), \quad .....(12.23)$$

Where F_x, F_y, F_z are components of friction force in local coordinate direction.

From Fig. 12.5 we have

$\vec{\Omega}$ = Angular velocity of the earth = $i\Omega_x + j\,\Omega_y + k\Omega_z$.

where $\Omega_x = 0$, $\Omega_y = \Omega\cos\phi$, $\Omega_z = \Omega\sin\phi$

$$-2\vec{\Omega}\times\vec{V} = -2\begin{vmatrix} i & j & k \\ 0 & \Omega_y & \Omega_z \\ u & v & w \end{vmatrix} = -2\begin{vmatrix} i & j & k \\ 0 & \Omega\cos\phi & \Omega\sin\phi \\ u & v & w \end{vmatrix}$$

$$= -2i(w\,\Omega\cos\phi - v\,\Omega\sin\phi) + 2j(0 - u\Omega\sin\phi) - 2k(0 - u\,\Omega\cos\phi)$$

Substituting eq. (12.20) to (12.23) in the equation of motions.

$$\frac{d\vec{V}}{dt} = -\nabla p - 2\vec{\Omega}\times\vec{V} + \vec{g} + \frac{1}{p}\vec{F}$$

we get
$$i\frac{du}{dt} + j\frac{dv}{dt} + k\frac{dw}{dt} = -\frac{1}{\rho}\left(i\frac{\partial p}{\partial x} + j\frac{\partial p}{\partial y} + k\frac{\partial p}{\partial z}\right)$$

$$-2i(w\,\Omega\cos\phi - v\,\Omega\sin\phi) - 2ju\,\Omega\sin\phi + 2ku\,\Omega\cos\phi + kg +$$

$$\frac{1}{\rho}(iF_x + jF_y + kF_z)$$

Equating like terms we have

$$\left.\begin{aligned} \frac{du}{dt} &= -\frac{1}{\rho}\frac{\partial p}{\partial x} + fv - ew + \frac{1}{\rho}F_x \\ \frac{dv}{dt} &= -\frac{1}{\rho}\frac{\partial p}{\partial y} - fu + \frac{1}{\rho}F_y \\ \frac{dw}{dt} &= -\frac{1}{\rho}\frac{\partial p}{\partial z} + eu - g + \frac{1}{\rho}F_z \end{aligned}\right\} \quad \text{.....(12.24)}$$

where $f = 2\Omega\sin\phi$, $e = 2\Omega\cos\phi$.

Eq. (12.24) are the required equations of motion in local cartesian coordinates.

Consider the magnitude of various terms (scale analysis)

$$g \simeq 10\text{ ms}^{-2},\ |u|_{max} = 100\text{ ms}^{-1},\ \Omega = 7.29\times10^{-5}\text{s}^{-1}$$

$$|v| = 10\text{ ms}^{-1},\ |w| < 0.10\text{ ms}^{-1},\ \frac{dw}{dt} \simeq 0.10\text{ ms}^{-2}$$

$$|eu| < 2\Omega|u|_{max} = 14.58\times10^{-5}\times10^{2} = 1\times10^{-2},\ |e\omega| \ll |fv|$$

e u , ew can be neglected in comparison to other terms.

Consider KE (Kinetic energy) per unit mass.

$$KE = \frac{1}{2}\vec{V}.\vec{V} = \frac{1}{2} \ (iu + iv + kw) \cdot (iu + jv + kw)$$

$$= \frac{1}{2} \ (u^2 + v^2 + w^2)$$

Multiplying the three eq. in (12.24) by u, v, w respectively and adding we get

$$\frac{d(KE)}{dt} = u\frac{du}{dt} + \frac{dv}{dt} + w\frac{dw}{dt}$$

$$= - \ \frac{1}{\rho}\left(u\frac{\partial p}{\partial x} + v\frac{\partial p}{\partial y} + w\frac{\partial p}{\partial z}\right) - gw + \frac{1}{\rho}[uF_x + vF_y + wF_z]$$

i.e.,
$$\frac{d(KE)}{dt} = -\frac{1}{\rho}\vec{V}.\nabla p - gw + \frac{1}{\rho}\vec{V}.\vec{F} \qquad .....(12.25)$$

Work = Force × distance

$\therefore$ $-\frac{1}{\rho}\vec{V}.\nabla p =$ work per unit of time by pressure force.

$-gw$ = work per unit of time by gravity

$\frac{1}{\rho}\vec{V}.\vec{F}$ = Work per unit of time by frictional force.

$2\vec{\Omega} \times \vec{V}$ has no contribution to the rate of change of KE (because it is perpendicular to the velocity.)

$\therefore$ The eq. of motion (12.24) may be approximated to

$$\frac{du}{dt} = -\frac{1}{\rho}\frac{\partial p}{\partial x} + fv + \frac{1}{\rho}F_x$$

$$\frac{dv}{dt} = -\frac{1}{\rho}\frac{\partial p}{\partial y} - fu + \frac{1}{\rho}F_y$$

$$\frac{dw}{dt} = 0 = -\frac{1}{\rho}\frac{\partial p}{\partial z} - g \qquad .....(12.26)$$

12.7 Pressure as a Vertical Coordinate

In weather analysis pressure as a vertical coordinate is very useful. The hydrostatic equation $\delta p = -g\rho dz$ is very advantageous which provides a relation between pressure and height. So far we have considered physical and kinematic variables of atmosphere as functions of space (x, y, z) and time (t). With the help of hydrostatic equation we may consider these variables as functions of x, y, p and t. In our earlier discussion p = p(x, y, z, t), pressure is a dependent variable while z, height is independent variable. In the current discussion we use pressure p as independent variable and z dependent variable, z = z(x, y, p, t).

12.8 Individual and Local Time Derivatives Treating Pressure as Vertical Coordinate

Consider an arbitrary variable

$$b = b(x, y, p, t) \qquad .....(12.27)$$

(which is a function of x, y, p and t).

Let the position of a particle at time $t = t_0$ and $t = t_0 + dt$ be at (x_0, y_0, p_0) and $(x_0 + dx, y_0 + dy, p_0 + dp)$ respectively. db the change in b in time dt is given by $db = b(x_0 + dx, y_0 + dy, p_0 + dp, t_0 + dt) - b(x_0, y_0, p_0, t_0)$.

Using Taylor's expansion we have

$$db = \left(\frac{\partial b}{\partial x}\right)_p dx + \left(\frac{\partial b}{\partial y}\right)_p dy + \left(\frac{\partial b}{\partial p}\right)_p dp$$

$$+ \left(\frac{\partial b}{\partial t}\right)_p dt + \text{higher order trems (hot)}$$

Dividing by dt we have

$$\frac{db}{dt} = \left(\frac{\partial b}{\partial t}\right)_p + \left(\frac{\partial b}{\partial x}\right)_p \frac{dx}{dt} + \left(\frac{\partial b}{\partial y}\right)_p \frac{dy}{dt} + \left(\frac{\partial b}{\partial p}\right)\frac{dp}{dt}$$

or

$$\frac{db}{dt} = \left(\frac{\partial b}{\partial t}\right)_p + u\left(\frac{\partial b}{\partial x}\right)_p + v\left(\frac{\partial b}{\partial y}\right)_p + \omega\left(\frac{\partial b}{\partial p}\right)_p \qquad .....(12.28)$$

where

$$u = \frac{dx}{dt}, \; v = \frac{dy}{dt} \; \omega = \frac{dp}{dt},$$

The subscript p indicates that derivatives are evolved keeping p as constant.

$$\frac{db}{dt} = \text{LHS} = \text{individual time derivative, and}$$

$$\text{RHS} = \left(\frac{\partial b}{\partial t}\right)_p + u\left(\frac{\partial b}{\partial x}\right)_p + v\left(\frac{\partial b}{\partial y}\right)_p + \omega\left(\frac{\partial b}{\partial p}\right)_p$$

$$= \text{local time derivative.}$$

Replacing b by z in eq. (12.28) we have

$$\frac{dz}{dt} = \left(\frac{\partial z}{\partial t}\right)_p + u\left(\frac{\partial z}{\partial x}\right)_p + v\left(\frac{\partial z}{\partial y}\right)_p + \omega\left(\frac{\partial z}{\partial p}\right)_p$$

$$\frac{dz}{dt} = \left(\frac{\partial z}{\partial t}\right)_p + \vec{V}_h \nabla_p Z + \omega\left(\frac{\partial z}{\partial p}\right)_p \qquad .....(12.29)$$

where $\vec{V_h} = iu + jv$

Using hydrostatic, eq. (12.29) may be further written as

$$\frac{dz}{dt} = \left(\frac{\partial z}{\partial t}\right)_p + \vec{V_h}.\nabla_p Z - \frac{\alpha}{g}\omega \qquad .....(12.30)$$

$$(\because \delta p = -g\rho\delta z \text{ or } \frac{\partial z}{\partial p} = -\frac{\alpha}{g})$$

12.9 Transformation relations from z-system to p-system

$$\delta p = -g\rho\delta z, \text{ Hydrostatic equation, } \frac{\partial z}{\partial p} \simeq -\frac{\alpha}{g} \qquad .....(12.31)$$

$$d\psi = g dz, \quad \psi = \text{geopotential} \qquad .....(12.32)$$

or $\psi = gz$

Differentiating eq. (12.32) w.r.t. p. we have

$$\frac{\partial \psi}{\partial p} = g\frac{\partial z}{\partial p}$$

$$= g \times -\frac{\alpha}{g} \quad \text{using (eq. 12.31)}$$

therefore $$\frac{\partial \psi}{\partial p} = -\alpha \qquad .....(12.33)$$

12.10 General Transformation Relations

Let a constant height surface (z = const) and an isobaric surface (p = const.) intersect along a line that passes through the point 1. as shown in Fig. 12.6. (shown in two dimensions), where z = constant is a horizontal surface while p = constant is an isobaric surface. Let 2 be a point on the horizontal surface such that δx = distance between the points 1 and 2 in x-direction. Let the vertical line through the point 2 intersect the isobaric surface at a point 3 such that δz = distance between point 2 and 3. Let b be any arbitrary scalar variable.

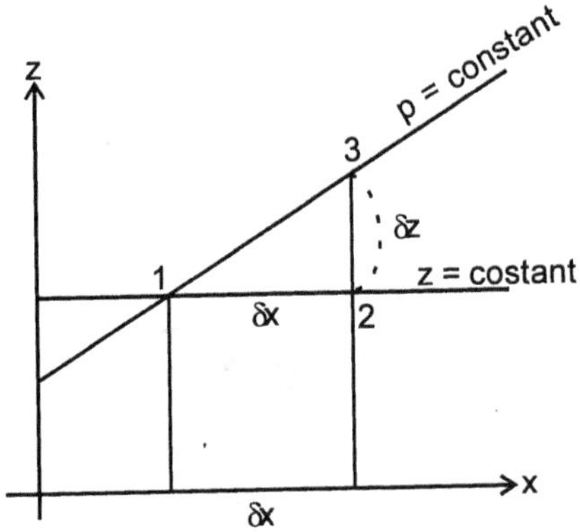

Fig. 12.6

$$\left(\frac{\partial b}{\partial x}\right)_z = \lim_{\delta x\to 0}\frac{b_2 - b_1}{\delta x} \qquad \text{.....(12.34)}$$

$$\left(\frac{\partial b}{\partial x}\right)_p = \lim_{\delta x\to 0}\frac{b_3 - b_1}{\delta x} \qquad \text{.....(12.35)}$$

From Fig. 12.6 we have

$$\left(\frac{\partial b}{\partial x}\right)_z = \lim_{\delta x\to 0}\frac{b_2 - b_1}{dx} = \lim_{\delta x\to 0}\left(\frac{b_3 - b_1}{\delta x} - \frac{b_3 - b_2}{\delta x}\right)$$

$$= \lim_{\delta x\to 0}\frac{b_3 - b_1}{\delta x} - \lim_{\delta x\to 0}\frac{b_3 - b_2}{\delta x}$$

$$= \left(\frac{\partial b}{\partial x}\right)_p - \lim_{\delta x\to 0}\frac{b_3 - b_2}{\delta x}$$

$$= \left(\frac{\partial b}{\partial x}\right)_p - \frac{\partial z}{\partial x}\frac{\partial p}{\partial z}\frac{\partial b}{\partial p}$$

$$\left(\frac{\partial b}{\partial x}\right)_z = \left(\frac{\partial b}{\partial x}\right)_p - \left(\frac{\partial z}{\partial x}\right)_p \left(\frac{\partial p}{\partial z}\right)\left(\frac{\partial b}{\partial p}\right)$$

$$\left(\frac{\partial b}{\partial x}\right)_z = \left(\frac{\partial b}{\partial x}\right)_p + g\rho\left(\frac{\partial b}{\partial p}\right)\left(\frac{\partial z}{\partial x}\right)_p \qquad .....(12.36)$$

[using hydrostatic eq. $\frac{\partial p}{\partial z} = -g\rho$]

$\left(\frac{\partial z}{\partial x}\right)_p$ = slope of isobaric surface relative to horizontal surface, hence the subscript p.

Similarly we can write for independent variables y and t

$$\left(\frac{\partial b}{\partial y}\right)_z = \left(\frac{\partial b}{\partial y}\right)_p + g\rho\left(\frac{\partial b}{\partial p}\right)\left(\frac{\partial z}{\partial y}\right)_p \qquad .....(12.37)$$

$$\left(\frac{\partial b}{\partial t}\right)_z = \left(\frac{\partial b}{\partial t}\right)_p + g\rho\left(\frac{\partial b}{\partial p}\right)\left(\frac{\partial z}{\partial t}\right)_p \qquad .....(12.38)$$

eq. (12.36) (12.37) and (12.38) are transformation relations.

Further we have

$$\frac{\partial b}{\partial z} = \frac{\partial b}{\partial p}\frac{\partial p}{\partial z}$$

or $$\frac{\partial b}{\partial z} = -g\rho\frac{\partial b}{\partial p}, \quad \text{since } \frac{\partial p}{\partial z} = -g\rho \qquad .....(12.39)$$

Denoting

$$\nabla_p b = i\left(\frac{\partial b}{\partial x}\right)_p + j\left(\frac{\partial b}{\partial y}\right)_p, \quad \text{where } \nabla_p * = i\left(\frac{\partial *}{\partial x}\right)_p + j\left(\frac{\partial *}{\partial y}\right)_p$$

$$\nabla_p z = i\left(\frac{\partial z}{\partial x}\right)_p + j\left(\frac{\partial z}{\partial y}\right)_p, \quad \text{where } \nabla_z * = i\left(\frac{\partial *}{\partial x}\right)_z + j\left(\frac{\partial *}{\partial y}\right)_z$$

where * can be replaced by scalar variable.

Then using (12.37) and (12.38) above equations, the last equation can be written as,

Viz. $\nabla_z b = i\left(\frac{\partial b}{\partial x}\right)_z + j\left(\frac{\partial b}{\partial y}\right)_z$ can be written as

$$\nabla_z b = i\left[\left(\frac{\partial b}{\partial x}\right)_p + g\rho\left(\frac{\partial b}{\partial p}\right)\left(\frac{\partial z}{\partial x}\right)_p\right] + j\left[\left(\frac{\partial b}{\partial y}\right)_p + g\rho\left(\frac{\partial b}{\partial p}\right)\left(\frac{\partial z}{\partial y}\right)_p\right]$$

or $$\nabla_z b = i\left(\frac{\partial b}{\partial x}\right)_p + j\left(\frac{\partial b}{\partial y}\right)_p + g\rho\frac{\partial b}{\partial p}\left[\left(i\frac{\partial z}{\partial x}\right)_p + j\left(\frac{\partial z}{\partial y}\right)_p\right]$$

$$\nabla_z b = \nabla_p b + g\rho\frac{\partial b}{\partial p}\nabla_p z \qquad \text{.....(12.40)}$$

This is the required general transformation (z = const to p = const.)

Note : $\nabla_p z$ is a horizontal vector perpendicular to the isolines for height in the isobaric surface.

$\nabla_p b$ is a horizontal vector = gradient in the horizontal field

$\nabla_p T$= gradient for the temperature field and perpendicular to the isotherms.

In case b is vector say $\vec{b} = ib_x + jb_y$, b_x, b_y are components of $\vec{b}$ in x, y-directions,

The divergence is

$$\nabla_z \cdot \vec{b} = \left[\left(i\frac{\partial}{\partial x}\right)_z + j\left(\frac{\partial}{\partial y}\right)_z\right] \cdot [ib_x + jb_y]$$

$$\nabla_z \cdot \vec{b} = \left(i\frac{\partial b_x}{\partial x}\right)_z + \left(\frac{\partial b_y}{\partial y}\right)_z \qquad \text{.....(12.41)}$$

in eq. (12.36) put $b = b_x$ then we have

$$\left(\frac{\partial b_x}{\partial x}\right)_z = \left(\frac{\partial b_x}{\partial x}\right)_p + g\rho\frac{\partial b_x}{\partial p}\left(\frac{\partial z}{\partial x}\right)_p \qquad \text{.....(12.42)}$$

put $b = b_y$ in eq. (12.37), then we have

$$\left(\frac{\partial b_y}{\partial y}\right)_z = \left(\frac{\partial b_y}{\partial y}\right)_p + g\rho\frac{\partial b_y}{\partial p}\left(\frac{\partial z}{\partial y}\right)_p \qquad \text{.....(12.43)}$$

From eqn. (12.41, 12.42) and eq. (12.43) we have (by adding)

$$\nabla_z \cdot \vec{b} = \nabla_p \cdot \vec{b} + g\rho \frac{\partial \vec{b}}{\partial p} \cdot \nabla_p z \quad .....(12.44)$$

eq. (12.44) is a general transformation when arbitrary variable is a vector.

12.11 Equilibrium of Static Atmosphere

Let all particles in the atmosphere are at a state of rest relative to the earth. This implies that the velocity of particles relative to the earth is zero i.e. $\vec{V}_r = 0$

$\therefore \frac{d\vec{V}_r}{dt} = 0$, coriolis force $(2\vec{\Omega} \times \vec{V})$ and the frictional force $\left(\frac{1}{\rho}\frac{d\vec{\tau}}{dz}\right)$ vanish.

Under these conditions the equations of motion

$$\left(\frac{d\vec{V}}{dt} = -\frac{1}{\rho}\nabla p - 2\vec{\Omega} \times \vec{V} + \vec{g} + \frac{1}{\rho}\frac{d\vec{\tau}}{dz}\right)$$

reduces to
$$0 = -\frac{1}{\rho}\nabla p + \vec{g} \quad .....(12.45)$$

(x, y, z) and (x + δx, y + δy, z + δz)

We know $\vec{g} = -gk = -\nabla\psi$, where ψ is geo-potential $(= \int_0^z g\,dz)$

Using this, eq. (12.45) becomes

$$\frac{1}{\rho}\nabla p = \vec{g} = -\nabla\psi \quad \text{or } \nabla\psi = -\alpha\nabla p \quad .....(12.46)$$

Let
$$\vec{dr} = i\,dx + j\,dy + k\,dz \quad .....(12.47)$$

where the end point coordinates of $\vec{r}$ (Fig. 12.7) are

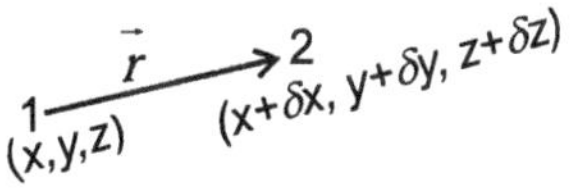

Fig 12.7

we have proved that $db = \nabla b \cdot d\vec{r}$ (in operator ∇)

where b is an arbitrary scalar

$$\therefore \quad d\psi = \nabla\psi \cdot d\vec{r} = -\alpha\nabla p \cdot (idx + jdy + kdz)$$

[using (12.46) and (12.47)]

$$= -\alpha\left(i\frac{\partial p}{\partial x} + j\frac{\partial p}{\partial y} + k\frac{\partial p}{\partial z}\right). (idx + jdy + kdz)$$

$$= -\alpha\left(\frac{\partial p}{\partial x}dx + \frac{\partial p}{\partial y}dy + \frac{\partial p}{\partial z}dz\right)$$

$$d\psi = -\alpha dp \qquad .....(12.48)$$

The equiscalor surfaces of ψ are called geo-potential surfaces (while the equiscalor pressure surfaces are called isobaric surfaces).

12.12 The Equations of Motion in Pressure Coordinate System

The equations of motion in local cartesian coordinates or the z-system are given by

$$\frac{du}{dt} = -\frac{1}{\rho}\frac{\partial p}{\partial x} + fv + \frac{1}{\rho}F_x \quad \text{where } f = 2\Omega\sin\phi, e = 2\Omega\cos\phi$$

$$\frac{dv}{dt} = -\frac{1}{\rho}\frac{\partial p}{\partial y} - fu + \frac{1}{\rho}F_y$$

$$0 = -\frac{1}{\rho}\frac{\partial p}{\partial z} - g \quad \text{or } dp = -g\rho\delta z \qquad .....(12.49)$$

From the last (Hydrostatic) equation we find that

$$\frac{1}{\rho}\left(\frac{\partial p}{\partial x}\right)_z = g\left(\frac{\partial z}{\partial x}\right)_p \qquad .....(12.50)$$

$$\frac{1}{\rho}\left(\frac{\partial p}{\partial y}\right)_z = g\left(\frac{\partial z}{\partial y}\right)_p \qquad .....(12.51)$$

We have proved in z to p transformation that (put b = p), in eq. (12.36, 12.37),

$$\left(\frac{\partial p}{\partial x}\right)_z = \left(\frac{\partial p}{\partial x}\right)_p + g\rho\frac{\partial p}{\partial p}\left(\frac{\partial z}{\partial x}\right)_p$$

$$\left(\frac{\partial p}{\partial y}\right)_z = \left(\frac{\partial p}{\partial y}\right)_p + g\rho\frac{\partial p}{\partial p}\left(\frac{\partial z}{\partial y}\right)_p$$

From individual, local time derivatives in p system we have (by putting b = u and b = v in eq. (12.28))

$$\frac{du}{dt} = \left(\frac{\partial u}{\partial t}\right)_p + u\left(\frac{\partial u}{\partial x}\right)_p + v\left(\frac{\partial u}{\partial y}\right)_p + \omega\frac{\partial u}{\partial p}$$

$$\frac{dv}{dt} = \left(\frac{\partial v}{\partial t}\right)_p + u\left(\frac{\partial v}{\partial x}\right)_p + v\left(\frac{\partial v}{\partial y}\right)_p + \omega\frac{\partial v}{\partial p}$$

Neglecting friction force and using these above two relations, the equations of motion (12.49) becomes

$$\left.\begin{aligned}
&\left(\frac{\partial u}{\partial t}\right)_p + u\left(\frac{\partial u}{\partial x}\right)_p + v\left(\frac{\partial u}{\partial y}\right)_p + \omega\frac{\partial u}{\partial p} = -\frac{\partial \psi}{\partial x} + fv \\
&\left[\because \psi = gz, \left(\frac{\partial \psi}{\partial p}\right) = -\alpha \text{ or } \partial\psi = -\frac{1}{\rho}\,\delta p\right] \\
&\left(\frac{\partial v}{\partial t}\right)_p + u\left(\frac{\partial v}{\partial x}\right)_p + v\left(\frac{\partial v}{\partial y}\right)_p + \omega\frac{\partial v}{\partial p} = -\frac{\partial \psi}{\partial y} - fu \\
\text{and} \quad &\frac{\partial \psi}{\partial p} = -\alpha
\end{aligned}\right\} \quad .....(12.52)$$

Equation (12.52) are the required equations of motion in pressure coordinates.

12.13 The Equation of Continuity in Pressure Coordinate System

Consider an isolated box (parallelopiped) with sides δx, δy, δz and with top and bottom of isobaric faces with pressure difference δp. This box is fixed in (x, y, p) space, which has A_1, A_2, A_3, A_4 as points of center in the left, right, top and bottom faces as shown in the Fig. 12.8. Let ρ be the density of the fluid. The volume of the box. is $\simeq \delta x\, \delta y\, \delta z$.

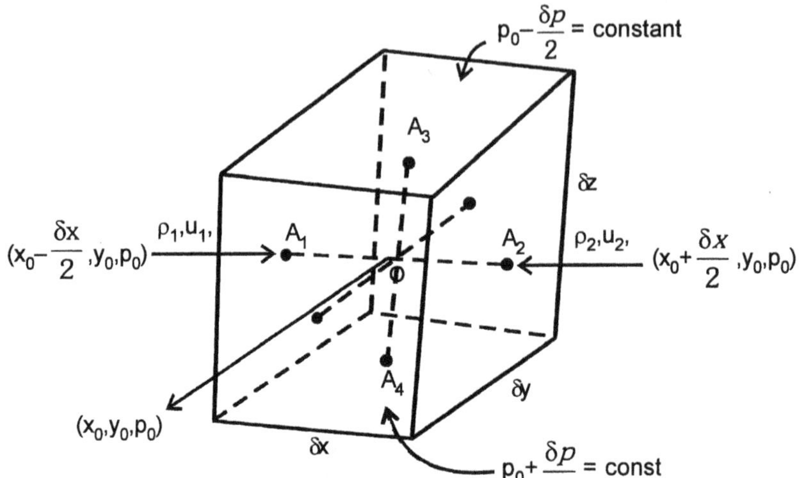

Fig. 12.8

Mass of the fluid inside the box

$$\simeq \delta x\ \delta y\ \delta z\ \rho,\ (\text{mass} = \text{volume} \times \text{density})$$

$$= \delta x.\ \delta y.\ \frac{\delta p}{g},\ (\delta p = -g\rho\delta z \text{ hydrostatice equation})$$

$$= \text{const. since } \delta x, \delta y, \delta p \text{ are constants.}$$

Let (ρ_1, u_1), (ρ_2, u_2) be the density and velocity of the fluid at A_1, A_2 in x-directions. Let the coordinates of '0', the center of the box be (x_0, y_0, p_0). where density velocity are ρ_0, u_0. The coordinates of A_1 (Center of the left side of the box) are $\left(x_0 - \frac{\delta x}{2}, y_0, p_0\right)$ and that of A_2 (center of the right side of the box) are $\left(x_0 + \frac{\delta x}{2}, y_0, p_0\right)$.

The net mass inflow into the box in x-direction is given by

$$\rho_1 u_1 \delta y \delta z - \rho_2 u_2 \delta y \delta z = u_1\ \delta y \frac{\delta p}{g} - u_2 \delta y \frac{\delta p}{g}$$

(using hydrostatic equation)

Using Taylor's expansion, this equation becomes

$$\left[u_0 - \frac{1}{2}\left(\frac{\partial u}{\partial x}\right)_0 \delta x\right]\delta y \frac{\delta p}{g} - \left[u_0 + \frac{1}{2}\left(\frac{\partial u}{\partial x}\right)_0 \delta x\right]\delta y \frac{\delta p}{g}$$

(Suffix '0' where indicates at the centre of the box)

$$= -\frac{\partial u}{\partial x}\delta x\delta y\frac{\delta p}{g} \quad .....(12.52A)$$

where '0' dropped for generality)

Similarly the contribution from the y-direction

$$= -\frac{\partial v}{\partial y}\delta x\delta y\frac{\delta p}{g} \quad .....(12.53)$$

and from the vertical direction

$$= -\frac{\partial \omega}{\partial p}\delta x\delta y\frac{\delta p}{g} \quad .....(12.54)$$

Therefore the net mass flux into the box

$$= -\left(\frac{\partial u}{\partial x}+\frac{\partial v}{\partial y}+\frac{\partial w}{\partial p}\right)\delta x\delta y\frac{\delta p}{g} \quad .....(12.55)$$

By the principle of conservation of mass (mass is neither created or destroyed), the net mass flow must be zero.

Therefore from eq. (12.55) we have

$$\frac{\partial u}{\partial x}+\frac{\partial v}{\partial y}+\frac{\partial w}{\partial p}=0 \quad .....(12.55A)$$

Which is the equation of continuity in p - system. This is the same equation for the incompressible fluid.

12.14 Thermodynamic Equations in Pressure Coordinates

First law of thermodynamics may be written as

$$H = \frac{dQ}{dt} = C_v\frac{dT}{dt}+p\frac{d\alpha}{dp}$$

$$= (C_p - R)\frac{dT}{dt}+p\frac{d\alpha}{dt}$$

$$H = C_p\frac{dT}{dt}-\alpha\frac{dp}{dt} \quad .....(12.56)$$

$$(\because p\alpha = RT \text{ gives } \alpha\frac{dp}{dt} = R\frac{dT}{dt} - p\frac{d\alpha}{dt} \text{ and } C_p - C_V = R)$$

Poisson's equation

$$\frac{T}{\theta} = \left(\frac{p}{1000}\right)^{+\frac{R}{C_p}}$$

Taking logarithms

$$\ln T - \ln \theta = + \frac{R}{C_p} (\ln p - \ln 1000)$$

$$C_p \ln T - C_p \ln \theta = + R \ln p - R \ln 1000$$

$$\frac{C_p}{T}\frac{dT}{dt} - \frac{C_p}{\theta}\frac{d\theta}{dt} = \frac{R}{p}\frac{dp}{dt} \quad \text{(Differentiating w.r.t time t)}$$

$$C_p.\frac{1}{\theta}\frac{d\theta}{dt} = -\frac{\alpha}{T}\frac{dp}{dt} + \frac{C_p}{T}\frac{dT}{dt} \qquad (\because \frac{R}{p} = \frac{\alpha}{T})$$

$$= \frac{1}{T}\left(C_p \frac{dT}{dt} - \alpha \frac{dp}{dt}\right) = \frac{H}{T}, \text{ using eq. (12.56)}$$

$$\therefore \quad \frac{1}{\theta}\frac{d\theta}{dt} = \frac{H}{C_p T} \qquad .....(12.57)$$

In p-system we have shown that

$$\frac{dz}{dt} = \left(\frac{\partial z}{\partial t}\right)_p + \vec{V_h}.\nabla_p z + \omega \left(\frac{\partial z}{\partial p}\right)_p, \quad \text{using (12.29)}$$

In this equation replacing z by ln θ we have

$$\frac{d \ln \theta}{dt} = \left(\frac{\partial \ln \theta}{\partial t}\right)_p + \vec{V_h}.\nabla_p \ln \theta + \omega \frac{\partial \ln \theta}{\partial p}$$

$[\vec{V_h} = \vec{V}$ horizontal velocity$]$

or

$$\frac{1}{\theta}\frac{d\theta}{dt} = \frac{1}{\theta}\left(\frac{\partial \theta}{\partial t}\right)_p + \frac{\vec{V}}{\theta}.\nabla_p \theta + \frac{\omega}{\theta}\frac{\partial \theta}{\partial p} \qquad .....(12.58)$$

using eq. (12.57) this equation becomes

$$\frac{H}{C_p T} = \frac{1}{\theta}\left(\frac{\partial \theta}{\partial t}\right)_p + \frac{\vec{V}}{\theta}\nabla_p \theta + \frac{\omega}{\theta}\frac{\partial \theta}{\partial p} \qquad ....(12.59)$$

Again

$$\theta = T\left(\frac{p}{p_0}\right)^{\frac{-R}{C_p}} \qquad \text{Poisson's equation}$$

Taking logarithms

$$\ln \theta = \ln T - \frac{R}{C_p}(\ln p - \ln p_0)$$

$$= \ln p + \ln \alpha - \ln R - \frac{R}{C_p}\ln p + \frac{R}{C_p}\ln p_0 \qquad \left(\because T = \frac{p\alpha}{R}\right)$$

$$\ln \theta = \ln \alpha + \ln p\left[1 - \frac{R}{C_p}\right] - \ln R + \frac{R}{C_p}\ln p_0$$

Differentiating this equation partially (treating p = const.) w.r.t t

We have

$$\left.\begin{aligned} \frac{1}{\theta}\left(\frac{\partial \theta}{\partial t}\right)_p &= \frac{1}{\alpha}\left(\frac{\partial \alpha}{\partial t}\right)_p \\ \text{and} \quad \frac{1}{\theta}\nabla_p\theta &= \frac{1}{\alpha}\nabla_p\alpha \end{aligned}\right\} \qquad .....(12.60)$$

Substituting eq. (12.60) in eq. (12.59), we have

$$\frac{H}{C_p T} = \frac{1}{\alpha}\left(\frac{\partial \alpha}{\partial t}\right)_p + \vec{V}.\frac{1}{\alpha}\nabla_p\alpha + \frac{\omega}{\theta}\frac{\partial \theta}{\partial p}$$

$$\frac{\alpha}{T}\frac{H}{C_p} = \left(\frac{\partial \alpha}{\partial t}\right)_p + \vec{V}.\nabla_p\alpha + \frac{\omega\alpha}{\theta}\frac{\partial \theta}{\partial p}$$

$$\frac{R}{p}\frac{H}{C_p} = -g\frac{\partial}{\partial t}\left(\frac{\partial z}{\partial p}\right) + \vec{V}.\nabla_p\alpha + \frac{\omega\alpha}{\theta}\frac{\partial \theta}{\partial p}$$

$$\left[\because p\alpha = RT,\ \delta p = -g\rho\delta z \text{ or } \alpha = -g\frac{\partial z}{\partial p}\right]$$

or

$$-g\frac{\partial}{\partial t}\left(\frac{\partial z}{\partial p}\right) = \frac{R}{C_p}\frac{H}{P} - \vec{V}.\nabla_p\alpha + \omega\left(\frac{-\alpha}{\theta}\frac{\partial \theta}{\partial p}\right)$$

Integrating this equation w.r.t p between p_1 to p_2 (isobaric surfaces) we have

$$g\frac{\partial}{\partial t}(z_1 - z_2) = \frac{R}{C_p}\int_{p_1}^{p_2}\frac{H}{p}dp - \int_{p_1}^{p_2}\vec{V}.\nabla_p\alpha dp + \int_{p1}^{p2}\omega\left(\frac{-\alpha}{\theta}\frac{\partial\theta}{\partial p}\right)dp$$

where z_1, z_2 corresponds to p_1, p_2, and $h = Z_1 - Z_2$

$$g\frac{\partial h}{\partial t} = \frac{R}{C_p}\int_{p_1}^{p_2}\frac{H}{p}dp - \int_{p_1}^{p_2}\vec{V}.\nabla_p\alpha dp + \int_{p1}^{p2}\omega\left(\frac{-\alpha}{\theta}\frac{\partial\theta}{\partial p}\right)dp \quad \text{......(12.61)}$$

It follows from eq. (12.61) that mean temperature changes in three ways.

1. By horizontal advection air with another temperature,
2. The coupling of the vertical p-velocity with the static stability

$$\left[-\frac{\alpha}{\theta}\frac{\partial\theta}{\partial p}\right]$$

If $\frac{\partial\theta}{\partial p} < 0$ then static stability is positive which shows subsidence ($\omega > 0$) causes rise in mean temperature in the layer, and

3. The direct effect of the heat source in the layer column.

Considering the temperature at an arbitrary point on an isobaric surface (p = const) and using Poisson's equation, we derive the following

$$\frac{T}{\theta} = \left(\frac{p}{p_0}\right)^{\frac{R}{C_p}} \quad \text{(Poissons equation)}$$

$$\ln T = \ln\theta + \frac{R}{C_p}(\ln p - \ln p_0)$$

Differentiating partially w.r.t 't' (time) treating p-constant, we have

$$\frac{1}{T}\left(\frac{\partial T}{\partial t}\right)_p = \frac{1}{\theta}\left(\frac{\partial\theta}{\partial t}\right)_p \quad \text{and} \quad \frac{1}{T}\nabla_p T = \frac{1}{\theta}\nabla_p\theta$$

Substituting these values in eq. (12.59) we have

$$\frac{H}{C_p T} = \frac{1}{T}\left(\frac{\partial T}{\partial t}\right)_p + \vec{V}.\frac{1}{T}\nabla_p T + \omega\frac{1}{\theta}\frac{\partial\theta}{\partial p}$$

or $$\frac{H}{C_p} = \left(\frac{\partial T}{\partial t}\right)_p + \vec{V}.\nabla_p T + \omega \frac{T}{\theta}\frac{\partial \theta}{\partial p} \qquad .....(12.62)$$

The three effects mentioned above is also true for the change of temperature at a point.

12.15 Bernoulli's Equation

Consider a steady non-viscous, in compressible fluid flow through a pipe whose cross section gradually increases from the end PQ to RS as shown in the Fig. 12.9. Let the area of cross sections of the pipe at the ends PQ, RS be A and B whose mean heights above reference level be h_a, h_b ($h_b > h_a$). Let P_a, V_a and P_b, V_b be the pressure and speeds of the fluid at ends of PQ and RS respectively.

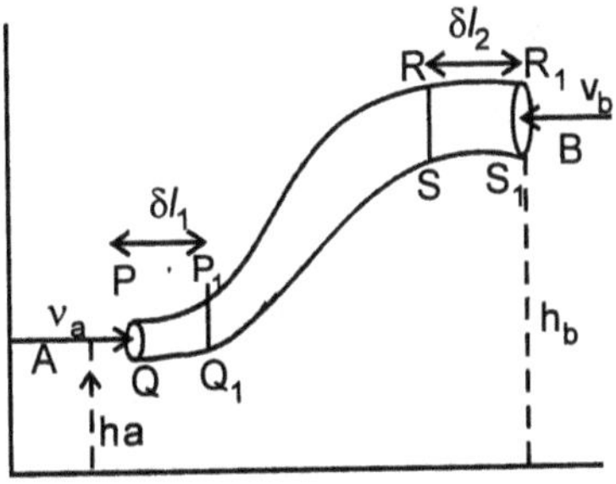

Fig. 12.9

Let in Δt time period a portion of the fluid at the end PQ be pushed through an elemental length δl_1 ($= Va\ \Delta t$) to a position P_1Q_1 through the force P_aA. Let the fluid at the end RS move through an elemental length $\delta l_2 = V_b\Delta_t$ to the position of R_1S_1.

In the first case the work done on the system $W_a = P_aA\Delta l_1$ or $W_a = P_aAV_a\Delta t$ and the work done by the system in second case $W_b = P_bB\Delta l_2 = P_bBV_b\Delta t$.

Therefore the net work done by the liquid pressure is

$$W = W_a - W_b$$

$$W = (AP_aV_a - BP_bV_b)\Delta t.$$

$$W = (P_a - P_b)\ V \qquad .....(12.63)$$

where V = volume $= AV_a\Delta t = BV_b\ \Delta t$, $AV_a = BV_b$.

Let ρ be the density of the fluid. During the time Δt the mass of the liquid ρV (mass = Volume $\times$density) changes from h_a to h_b position.

The work done on the liquid by gravitational field W_g is given by

$$W_g = -\rho V g\,(h_b - h_a) = \rho V.g\,(h_a - h_b) \qquad .....(12.64)$$

According to work-energy theorem (namely, work done by resulting force acting on the system = change in Kinetic energy of the system),

the total work done on the system = $W + W_g$ = change in KE

$$\therefore\ (P_a - P_b)\cdot V + \rho V g\,(h_a - h_b) = \frac{1}{2}\,\rho V\,(V_b^2 - V_a^2)$$

$$P_a - P_b + \rho g\,(h_a - h_b) = \frac{1}{2}\,\rho\,(V_b^2 - V_a^2)$$

$$\frac{p_a}{\rho g} + \frac{V_a^2}{2g} + h_a = \frac{p_b}{\rho g} + \frac{V_b^2}{2g} + h_b$$

or

$$\frac{P}{\rho g} + \frac{V^2}{2g} + h = (C_1) = \text{const.} \qquad .....(12.65)$$

eq. (12.65) is called Bernoulli's equation for steady, non-viscous, incompressible fluid flow.

Note :

(i) Bernoulli's equation is valid only for steady irrotational fluid flow

(ii) Flow is essentially one dimensional (along streamline)

(iii) only pressure, gravity forces are present.

Each term in this eq. (12.65) has the dimensions of length and is called a head.

$\frac{p}{\rho g}$ is called pressure or piezometric head

$\frac{V^2}{2g}$ is called velocity head and h is called gravitational head.

This eq. (12.65) says that sum of pressure head, velocity head and gravitational head is equal to constant.

eq. (12.65) can also be written as

$$p + \frac{1}{2}\rho v^2 + \rho g h (= C_2) = \text{const.} \qquad [\because \rho g = \text{const}]$$

In this equation both p and p $+\rho g h$ are called static pressure, $\frac{1}{2}\rho v^2$ is called dynamics pressure or dynamic head. In this form this equation says that static pressure + dynamic pressure is equal to constant.

12.16 Use of Bernoulli's Equation

During squalls and cyclonic storms the roofs of some houses are blown off while no damage is caused to the other parts of the house (doors, windows etc). This is not freak accident. The reason is as follows (Fig. 12.10). When strong wind blows over the roof it creates a low pressure area p_2 on the top, while under the roof lies the pressure p_1 (the atmospheric pressure) which is comparatively larger than p_2. This pressure defect (difference P1-P2) causes an upward thrust which lifts the roof. Once the roof is lifted it is blown off with the wind. This happens particularly in the path of a tornado when the doors and windows are closed and there are no ventilators in the direction of the wind flow.

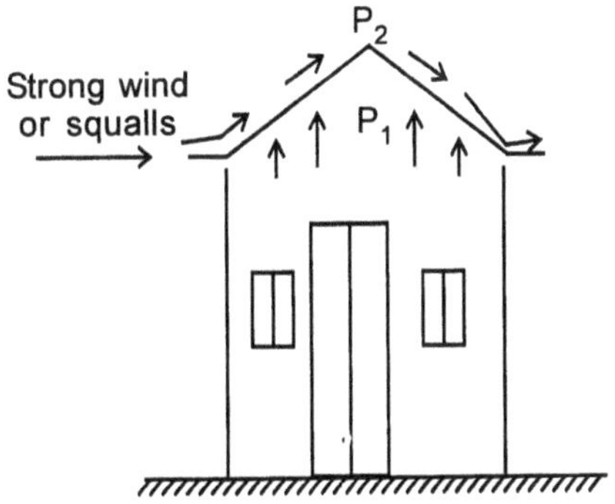

Fig. 12.10

12.17 Equation of Continuity in Isobaric Coordinates

Consider an elemental mass of fluid δM with cross-sectional area $\delta x.\ \delta y$ which is bounded by two isobaric surfaces of pressure p and $p\text{-}\delta p$ as shown in the Fig. 12.11. Let ρ be the density of the fluid.

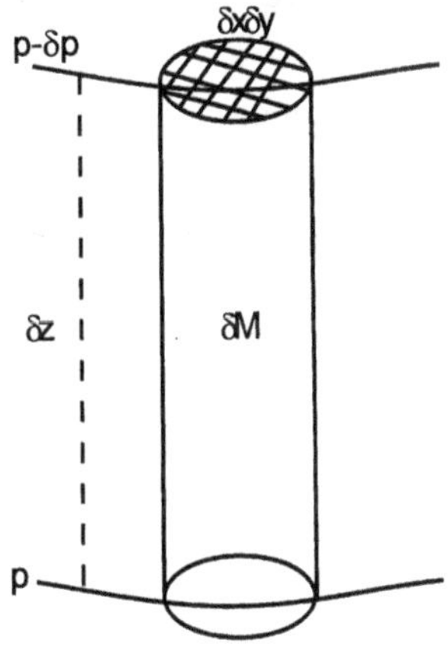

Fig. 12.11 Air column of fixed mass δM (bounded by two isobaric surfaces).

Then

$$\delta M = \rho\delta x\delta y\delta z$$

$$\delta M = \delta x\delta y \frac{\delta p}{g} \quad \text{(Using hydrostatic equation, } \delta p = g\rho dz\text{)}$$

Taking logarithms

$$\ln \delta M = \ln\left(\frac{\delta x\delta y\delta p}{g}\right)$$

Differentiating w.r.t 't' (time) we have

$$\frac{1}{\delta M}\frac{d}{dt}\delta M = \frac{g}{\delta x\delta y\delta p}\frac{d}{dt}\left(\frac{\delta x\delta y\delta p}{g}\right) = 0 \quad (\because \text{ mass is conserved})$$

$$\therefore \frac{d}{dt}(\delta x\delta y\delta p) = 0 \quad \text{(g being constant)}$$

Expanding

$$\delta x\delta y\frac{d(\delta p)}{dt} + \delta x\delta p\frac{d}{dt}(\delta y) + \delta y\delta p\frac{d}{dt}(\delta x) = 0$$

Dividing by $\delta x\delta y\delta p$ we have

$$\frac{1}{\delta p}\delta\left(\frac{dp}{dt}\right) + \frac{1}{\delta y}\delta\left(\frac{dy}{dt}\right) + \frac{1}{\delta x}\delta\left(\frac{dx}{dt}\right) = 0$$

Taking limit as δx, δy, $\delta p \to 0$ we obtain equation of continuity as,

$$\frac{\partial u}{\partial x} + \frac{\partial v}{\partial y} + \frac{\partial w}{\partial p} = 0 \quad \left(\text{where } u = \frac{dx}{dt},\ v = \frac{dy}{dt},\ \omega = \frac{dp}{dt}\right)$$

Ex : Calculate the rate of flow of glycerine of density $1.25 \times 10^3 \text{kgm}^{-3}$ through the cross-section of a pipe, if the radii of its ends are 10 cm and 4 cm and the pressure drop across its length is 10 N/m^2.

Sol :

According to continuity equation, mass in flow = mass out flow.

$$A_1 v_1 = A_2 v_2 \text{ or } \pi r_1^2 v_1 = \pi r_2^2 v_2$$

$$v_1 = \frac{r_2^2}{r_1^2} v_2 = \frac{(0.04)^2}{(0.1)^2} \times v_2 = 0.16\, v_2$$

From Bernoulli's equation we have g = const. at same place. $h_a = h_b$ being at the same level

$$\frac{p_1}{\rho} + \frac{v_1^2}{2} = \frac{p_2}{\rho} + \frac{v_2^2}{2}$$

or $$p_1 - p_2 = \frac{1}{2}\rho\,(v_2^2 - v_1^2)$$

$\therefore$ $$10 = \frac{1}{2} \times 1.25 \times 10^3\,[v_2^2 - (0.16v_2)^2]$$

$$\frac{20}{1.25 \times 10^3} = v_2^2\ (1 - 0.0256)$$

$\therefore$ $$v_2^2 = \frac{20}{1.25 \times 10^3 \times 0.9744} \simeq 1.642 \times 10^{-2}$$

or $$v_2 = 0.128\ ms^{-1}$$

Rate of flow of glycerine = volume of glycerin flowing in a second

$$V = \pi r_2^2 v_2 = \frac{22}{7} \times (0.04)^2 \times 0.128 \approx 6.43 \times 10^{-4}\ m^3 s^{-1}$$

12.18 Fundamentals of Vectors

Scalar : A quantity which can be completely specified by magnitude is called a scalar quantity. eg. length, temperature, pressure, mass etc.

Vector : A quantity which can be completely specified by magnitude and direction is called a vector. eg. velocity, acceleration, momentum etc.

A vector is represented by a letter with an arrow mark over it ($\vec{A}$). In geometry it is a directed line-segment, which has initial point and terminal point ($\overrightarrow{AB}$).

The magnitude of a vector $\vec{A}$ is represented by $|\vec{A}|$ or A. It is called absolute value or modulus (which is scalar quantity).

Base vectors or unit vectors : In rectangular coordinate system, the segments of unit length in the direction of three axes X, Y, Z are denoted by i, j, k respectively. They are also called base vectors (Fig. 12.12).

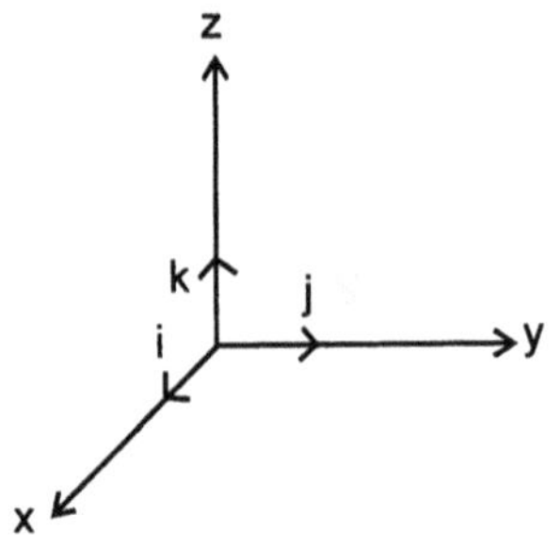

Fig. 12.12 Base Vectors.

The coordinates of a vector : The rectangular coordinates of $\vec{A}$ are the projections of the vector $\vec{A}$ on the coordinate axes. Thus if $\vec{A} = iA_x + jA_y + kA_z$ then $\overrightarrow{A_x}$, $\overrightarrow{A_y}$, $\overrightarrow{A_z}$ are the projections of $\vec{A}$ on the X, Y, Z coordinates,

$$\overrightarrow{A_x} = iA_x,\ \overrightarrow{A_y} = jA_y,\ \overrightarrow{A_z} = kA_z$$

$$|\vec{A}| = A = \sqrt{A_x^2 + A_y^2 + A_z^2}$$

Scalar product : Scalar or dot product of two vectors $(\vec{A}, \vec{B})$ is defined as $\vec{A} . \vec{B} = A\,B \cos\theta$, where θ is the angle between $\vec{A}$ and $\vec{B}$ (Fig. 12.13).

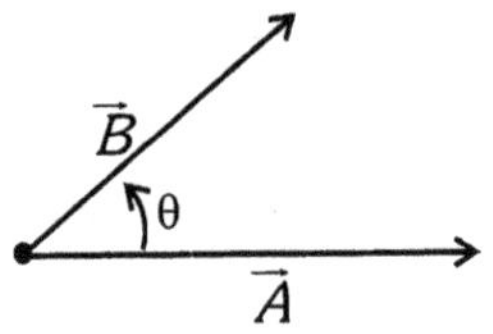

Fig. 12.13

It follows that the scalar product of two perpendicular vectors is zero.

$$\vec{A} . \vec{B} = (iA_x + jA_y + kA_z) \cdot (iB_x + jB_y + kB_z)$$
$$= A_xB_x + A_yB_y + A_zB_z$$

$\vec{A}$ and $\vec{B}$ are at right angles if $A_xB_x + A_yB_y + A_zB_z = 0$.

Thus we have $i \cdot i = j \cdot j = k \cdot k = 1$ and

$$ij = -ji = 0,$$

$$jk = -kj = 0,$$

$$ki = -ik = 0$$

Ex : Find the angle between the vectors $\vec{A} = 2i + 3j + 4k$, $\vec{B} = -i - 2j + 3k$.

Sol :

Here $A_x = 2$, $A_y = 3$, $A_z = 4$, $B_x = -1$, $B_y = -2$, $B_z = 3$

From the def. $\cos\theta = \dfrac{\vec{A}.\vec{B}}{AB}$

$$\cos\theta = \frac{A_xB_x + A_yB_y + A_zB_z}{\sqrt{A_x^2 + A_y^2 + A_z^2}\sqrt{B_x^2 + B_y^2 + B_z^2}}$$

$$= \frac{(2\times -1) + (3\times -2) + (4\times 3)}{\sqrt{4+9+16}\sqrt{1+4+9}}$$

$$= \frac{-2-6+12}{\sqrt{29}\sqrt{14}} = \frac{4}{\sqrt{406}}$$

$$\therefore \theta = \cos^{-1}\frac{4}{\sqrt{406}}$$

Note :

(i) If $\vec{F}$ is the force and $\vec{d}$ the displacement, then the work $= \vec{F}.\vec{d}$. Thus the scalar product $\vec{F}.\vec{d}$ is numerically equal to the work of the force.

(ii) The scalar product $\vec{A}.\vec{B}$ vanishes if and only if one of the factors is a null vector or if $\vec{A}$ and $\vec{B}$ are perpendicular to each other

(iii) If the vectors are parallel or collinear, then

$\vec{A}.\vec{B} = |\vec{A}||\vec{B}| = \pm AB$ (+ when they are in same direction, – when they are in opposite direction)

$$\vec{A}.\vec{A} = |\vec{A}|^2 \quad \text{or} \quad \sqrt{A^2} = |\vec{A}|$$

Vector product or cross product : The cross product of two vectors $\vec{A}$ and $\vec{B}$ is defined as $\vec{A} \times \vec{B} = AB \sin\theta\, \hat{n} = \vec{C}$ (see Fig. 12.14). It is a vector quantity in the direction perpendicular to the plane containing these vectors. $\hat{n}$ = unit vector perpendicular to the plane containing these vectors. The sign is determined by the right handed screw system.

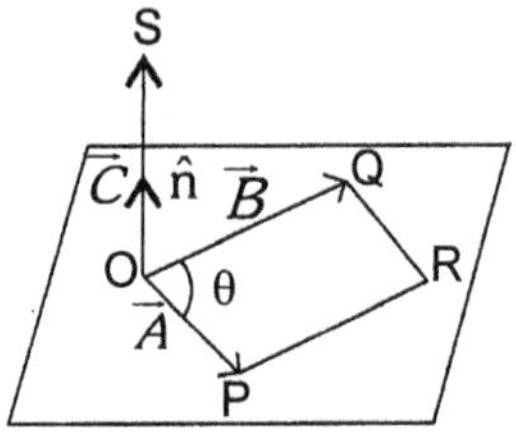

Fig. 12.14

Note:

(i) The cross product $\vec{A} \times \vec{B}$ vanishes only when $\vec{A}$ and $\vec{B}$ collinear, or one or both of them are null vectors.

$$\vec{B} \times \vec{A} = -\vec{A} \times \vec{B}$$

(ii) Absolute value of $\vec{A} \times \vec{B}$ is numerically equal to the area of a parallelogram (OPRQ) constructed on the vectors

Vector product of base vectors :

$$i \times i = j \times j = k \times k = 0$$

$$i \times j = k = -j \times i$$

$$j \times k = i = -k \times j$$

$$k \times i = j = -i \times k$$

$$\vec{A} \times \vec{B} = (iA_x + jA_y + kA_z) \times (iB_x + jB_y + kB_z)$$

$$= \begin{vmatrix} i & j & k \\ A_x & A_y & A_z \\ B_x & B_y & B_z \end{vmatrix}$$

$$= i(A_y B_z - A_z B_y) - j(A_x B_z - A_z B_x) + k(A_x B_y - A_y B_x)$$

Vector differentiation : Total differentiation of a vector $\vec{A} = \vec{A}(x, y, z)$ is given by

$$d\vec{A} = \frac{\partial \vec{A}}{\partial x}dx + \frac{\partial \vec{A}}{\partial y}dy + \frac{\partial \vec{A}}{\partial z}dz$$

Let $\vec{r}$ = position vector = ix + jy + kz

then $d\vec{r}$ = (idx + jdy + kdz)

∇ (del-operator or Hamiltonian operator) is defined as

$$\nabla = i\frac{\partial}{\partial x} + j\frac{\partial}{\partial y} + k\frac{\partial}{\partial z}$$

$$(\nabla . d\vec{r})\vec{A} = \left(\frac{\partial}{\partial x}dx + \frac{\partial}{\partial y}dy + \frac{\partial}{\partial z}dz\right)\vec{A}$$

$$= \frac{\partial \vec{A}}{\partial x}dx + \frac{\partial \vec{A}}{\partial y}dy + \frac{\partial \vec{A}}{\partial z}dz$$

$$= d\vec{A}$$

i.e., $d\vec{A} = (\nabla . d\vec{r})\vec{A}$

(i) If ϕ is a scalar function, then gradient of ϕ is denoted as grad ϕ. It is given by

$$\text{grad } \phi = \nabla\phi$$

$$= i\frac{\partial \phi}{\partial x} + j\frac{\partial \phi}{\partial y} + k\frac{\partial \phi}{\partial z}$$

(ii) If $\vec{V}$ = iu + jv + kw is a velocity vector, then the divergence of vector $\vec{V}$ is denoted by Div $\vec{V}$ and is given by

$$\text{Div } \vec{V} = \nabla \cdot \vec{V} = \left(i\frac{\partial}{\partial x} + j\frac{\partial}{\partial y} + k\frac{\partial}{\partial z}\right). (iu + jv + kw)$$

$$= \frac{\partial u}{\partial x} + \frac{\partial v}{\partial y} + \frac{\partial w}{\partial z}$$

Div $\vec{V}$ is a scalar quantity.

(iii) The rotation or curl of vector is denoted by rot $\vec{V}$ or curl $\vec{V}$ and is given by

$$\text{rot } \vec{V} = \text{Curl } \vec{V} = \nabla \times \vec{V} = \begin{vmatrix} i & j & k \\ \frac{\partial}{\partial x} & \frac{\partial}{\partial y} & \frac{\partial}{\partial z} \\ u & v & w \end{vmatrix}$$

(iv) $\nabla \cdot (\phi \vec{V}) = \nabla \phi . \vec{V} + \phi \nabla \cdot \vec{V}$, where ϕ is a scalar, $\vec{V}$ is a vector

(v) $\nabla \times (\phi \vec{V}) = \nabla \phi \times \vec{V} + \phi \nabla \times \vec{V}$

(vi) $\nabla .(\nabla \times \vec{V})$

$$= \left(i\frac{\partial}{\partial x} + j\frac{\partial}{\partial y} + k\frac{\partial}{\partial z}\right) . \left[i\left(\frac{\partial w}{\partial y} - \frac{\partial v}{\partial z}\right) + j\left(\frac{\partial u}{\partial z} - \frac{\partial w}{\partial x}\right) + k\left(\frac{\partial v}{\partial x} - \frac{\partial u}{\partial y}\right)\right]$$

$$= \frac{\partial^2 w}{\partial x \partial y} - \frac{\partial^2 v}{\partial z \partial x} + \frac{\partial^2 u}{\partial y \partial z} - \frac{\partial^2 w}{\partial x \partial y} + \frac{\partial^2 v}{\partial z \partial x} - \frac{\partial^2 u}{\partial y \partial z} = 0$$

$$\therefore \nabla .(\nabla \times \vec{V}) = 0$$

(vii) $\nabla \cdot \nabla \phi = \nabla^2 \phi = \left(i\frac{\partial}{\partial x} + j\frac{\partial}{\partial y} k\frac{\partial}{\partial z}\right) . \left(i\frac{\partial \phi}{\partial x} + j\frac{\partial \phi}{\partial y} + k\frac{\partial \phi}{\partial z}\right)$

$$= \frac{\partial^2 \phi}{\partial x^2} + \frac{\partial^2 \phi}{\partial y^2} + \frac{\partial^2 \phi}{\partial z^2}$$

$\nabla \cdot \nabla = \nabla^2 = \frac{\partial^2}{\partial x^2} + \frac{\partial^2}{\partial y^2} + \frac{\partial^2}{\partial z^2}$ is caled Laplacian

Divergence of a gradient = Laplacian

(viii) $\vec{V}.\nabla = (iu + jv + kw).\left(i\frac{\partial}{\partial x} + j\frac{\partial}{\partial y} + k\frac{\partial}{\partial z}\right)$

$$= u\frac{\partial}{\partial x} + v\frac{\partial}{\partial y} + w\frac{\partial}{\partial z}$$

(ix) $\nabla \times \nabla \phi = \left(i\frac{\partial}{\partial x} + j\frac{\partial}{\partial y} + k\frac{\partial}{\partial z}\right) \times \left(i\frac{\partial \phi}{\partial x} + j\frac{\partial \phi}{\partial y} + k\frac{\partial \phi}{\partial z}\right)$

$$= k\frac{\partial^2 \phi}{\partial x \partial y} - j\frac{\partial^2 \phi}{\partial z \partial x} - k\frac{\partial^2 \phi}{\partial x \partial y} + i\frac{\partial^2 \phi}{\partial y \partial z} + j\frac{\partial^2 \phi}{\partial z \partial x} - i\frac{\partial^2 \phi}{\partial y \partial z} = 0$$

Note :

(i) Divergence of a vector field is a scalar field.

$$\nabla.\vec{A} = \text{Div}\vec{A} = \frac{\partial A_x}{\partial x} + \frac{\partial A_y}{\partial y} + \frac{\partial A_z}{\partial z}.$$

where $\vec{A}$ = iAx + jAy +kAz .

(ii) Rotation of a vector field is vector field.

$$\text{rot } \vec{A} = \nabla \times \vec{A} = i\left(\frac{\partial A_z}{\partial y} - \frac{\partial A_y}{\partial z}\right) + j\left(\frac{\partial A_x}{\partial z} - \frac{\partial A_z}{\partial x}\right) + k\left(\frac{\partial A_y}{\partial x} - \frac{\partial A_x}{\partial y}\right)$$

$$= \begin{vmatrix} i & j & k \\ \frac{\partial}{\partial x} & \frac{\partial}{\partial y} & \frac{\partial}{\partial z} \\ A_x & A_y & A_z \end{vmatrix}$$

This is akin to circulation of a vector field or related to circulation.

$$\oint_{(L)} \vec{A}.d\vec{r} = \oint_{(L)} A_t dL \int_{(s)} \text{rot}\,\vec{A}.dS$$

where

$\vec{r}$ = ix + jy + kz; τ = unit tangent vector;

(S) = surface bounded by the contour (L) also called "pulled over L"

Ex : Show that

$$|\vec{A}\times\vec{B}|^2 = |\vec{A}|^2 |\vec{B}|^2 - (\vec{A}.\vec{B})^2$$

Sol:

We know that

$\vec{A}\times\vec{B} = |\vec{A}||\vec{B}| \sin\phi\ \hat{n}$ where ϕ is the angle between $\vec{A}$ and $\vec{B}$

$$|\vec{A}\times\vec{B}|^2 = |\vec{A}|^2 |\vec{B}|^2 \sin^2\phi = |\vec{A}|^2 |\vec{B}|^2 (1-\cos^2\phi) \quad (\because \hat{n}^2 = 1)$$

$$= |\vec{A}|^2 |\vec{B}|^2 - \left[|\vec{A}||\vec{B}|\cos\phi\right]^2$$

$$\therefore \quad |\vec{A}\times\vec{B}|^2 = |\vec{A}|^2 |\vec{B}|^2 - \left[\vec{A}.\vec{B}\right]^2 \quad \left[\because \vec{A}.\vec{B} = AB\cos\phi\right]$$

Ex : If $\vec{A} = [2,4,6], \vec{B} = [1,-3, 2]$, find the projection of $\vec{A}$ on $\vec{B}$ and the values of $\sin\phi$, $\cos\phi$.

Sol :

$\vec{A}\cdot\vec{B} = AB\cos\phi, A\cos\phi = \dfrac{\vec{A}.\vec{B}}{B}$, which is the projection of $\vec{A}$ on $\vec{B}$

$$A = |\vec{A}| = \sqrt{A^2{}_x + A^2{}_y + A^2{}_z} \text{ etc.,}$$

$$\vec{A}\ .\ \vec{B} = (2i + 4j + 6k)\ .\ (i - 3j + 2k)$$

$$= 2 - 12 + 12 = 2$$

$$A = \sqrt{2^2 + 4^2 + 6^2} = \sqrt{56}$$

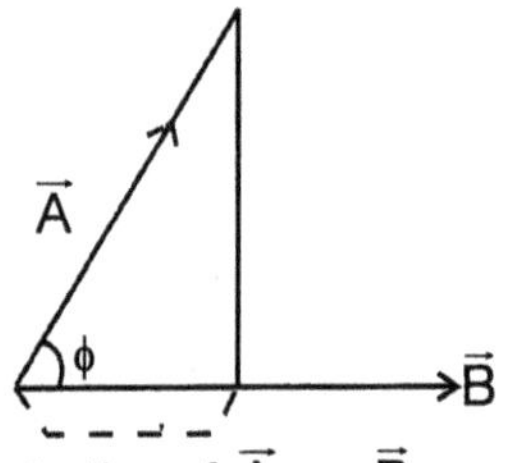

Projection of $\vec{A}$ on $\vec{B}$

Fig. 12.15

$$B = \sqrt{1+9+4} = \sqrt{14}$$

Projection of $\vec{A}$ on $\vec{B}$ is $= \dfrac{\vec{A}.\vec{B}}{B} = \dfrac{2}{\sqrt{14}}$

$$\cos\phi = \frac{\vec{A}.\vec{B}}{AB} = \frac{2}{\sqrt{14}\times\sqrt{56}} = \frac{1}{14}$$

$$\sin\phi = \sqrt{1-\cos^2\phi} = \sqrt{1-\left(\frac{1}{14}\right)^2} = \sqrt{\frac{195}{196}}$$

$$= \frac{\sqrt{195}}{14}$$

Ex : Show that

$$\vec{A}.(\vec{B}\times\vec{C}) = \begin{vmatrix} A_1 & A_2 & A_3 \\ B_1 & B_2 & B_3 \\ C_1 & C_2 & C_3 \end{vmatrix}.$$

Sol :
We know

$$\vec{B}\times\vec{C}=\begin{vmatrix} i & j & k \\ B_1 & B_2 & B_3 \\ C_1 & C_2 & C_3 \end{vmatrix}.$$

$$\therefore \vec{A}.\left(\vec{B}\times\vec{C}\right)=\left(iA_1+jA_2+KA_3\right).\begin{vmatrix} i & j & k \\ B_1 & B_2 & B_3 \\ C_1 & C_2 & C_3 \end{vmatrix}=\begin{vmatrix} A_1 & A_2 & A_3 \\ B_1 & B_2 & B_3 \\ C_1 & C_2 & C_3 \end{vmatrix}$$

Ex : Find a unit vector perpendicular to the plane of $\vec{A}$ = [2, –6, –3] and $\vec{B}$ = [4,3,-1]

Sol :

$$\therefore \vec{A}\times\vec{B}=\begin{vmatrix} i & j & k \\ A_x & A_y & A_z \\ B_x & B_y & B_z \end{vmatrix}=\begin{vmatrix} i & j & k \\ 2 & -6 & -3 \\ 4 & 3 & -1 \end{vmatrix}.$$

$$= i\,(6 + 9) - j\,(-2 + 12) + k\,(6 + 24) = 15i - 10j + 30k$$

This vector 15i – 10j + 30k is perpendicular to the plane of $\vec{A}$ and $\vec{B}$

∴ the required unit vector is

$$=\frac{15i-10j+30k}{\sqrt{15^2+10^2+30^2}}=\frac{15i-10j+30k}{35}$$

$$=\frac{3}{7}i-\frac{2}{7}j+\frac{6}{7}k.$$

Ex : Let $\phi = \phi\,(x, y, z)$ be a scalar function. Show that $\nabla\phi.d\vec{r}=d\phi$, r= ix+jy+kz.

Sol :

$$\nabla\phi.d\vec{r}=\left(i\frac{\partial\phi}{\partial x}+j\frac{\partial\phi}{\partial y}+k\frac{\partial\phi}{\partial z}\right).\left(idx+jdy+kdz\right)$$

$$=\frac{\partial\phi}{\partial x}dx+\frac{\partial\phi}{\partial y}dy+\frac{\partial\phi}{\partial z}dz=d\phi.\ \text{(total differential)}$$

$$\therefore \nabla\phi.d\vec{r}=d\phi$$

Ex : $\phi = \phi\,(x, y, z, t)$ then show that $d\phi = \nabla\phi.d\vec{r} + \frac{\partial\phi}{\partial t}dt.$

Sol :

$$d\phi = \frac{\partial\phi}{\partial x}dx + \frac{\partial\phi}{\partial y}dy + \frac{\partial\phi}{\partial z}dz + \frac{\partial\phi}{\partial t}dt \quad \text{[total differential]}$$

$$\therefore d\phi = \nabla\phi.d\vec{r} + \frac{\partial\varphi}{\partial t}dt \qquad \text{[using above example]}$$

Ex : Let $\vec{V} = iVx + jVy + kVz$ be a velocity vector emanating at the centre of the cuboid of sides $\delta x, \delta y, \delta z$ as shown in the Fig. (12.16)

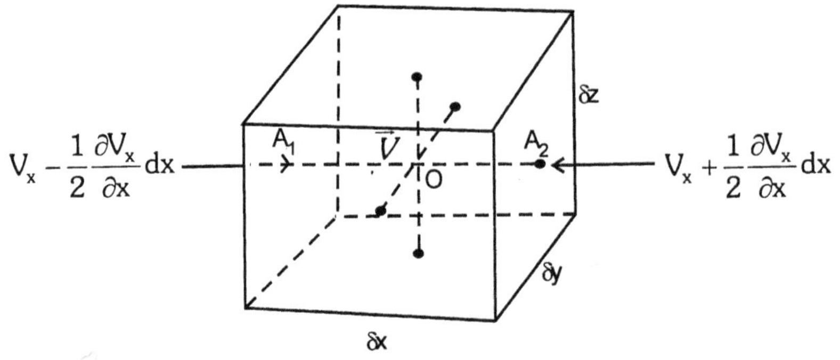

Fig 12.16

The rate of velocity at A_1 (the centre of the left face in x-direction) is given by $V_x - \frac{1}{2}\frac{\partial V_x}{\partial x}dx$ by Taylors expansion.

The rate of velocity at A_2 (the centre of right face in x-direction) is given by $V_x + \frac{1}{2}\frac{\partial V_x}{\partial x}dx$. [By Taylors expansion]

$\therefore$ the net force acting on the cuboid (or flux of flow) in x-direction is

$$F_x \simeq \left[\left(V_x - \frac{1}{2}\frac{\partial V_x}{\partial x}dx\right) - \left(V_x + \frac{1}{2}\frac{\partial V_x}{\partial x}dx\right)\right]dydz$$

$$F_x = -\frac{\partial V_x}{\partial x}dxdydz.$$

Similarly the net flux of flow across other two faces in y, z-directions (dzdx; dxdy) are given by

$$F_y = -\frac{\partial V_y}{\partial y}.dxdydz, \quad F_z = -\frac{\partial V_z}{\partial z}dxdydz.$$

∴ The net out flow across the cube = Fx + Fy + Fz

$$= -\left[\frac{\partial V_x}{\partial x} + \frac{\partial V_y}{\partial y} + \frac{\partial V_z}{\partial z}\right]dxdydz.$$

∴ The net out flow across the cuboid per unit of volume

$$\frac{F_x + F_y + F_z}{dxdydz} = -\left(\frac{\partial V_x}{\partial x} + \frac{\partial V_y}{\partial y} + \frac{\partial V_z}{\partial z}\right) = -\nabla.\vec{V} = -\text{div}\vec{V}$$

Ex : Show that Div $\vec{r}$ = 3 where $\vec{r}$ = ix + jy + kz.

Sol :

$$\text{Div}\,\vec{r} = \nabla.\vec{r} = \left(i\frac{\partial}{\partial x} + j\frac{\partial}{\partial y} + k\frac{\partial}{\partial z}\right).(ix + jy + kz)$$

$$= 1 + 1 + 1 = 3$$

Ex : Given $\vec{F} = ixyz + jx^2y^2z + kyz^3$. Find divergence of $\vec{F}$

Sol :

$$\text{Div}\,\vec{F} = \nabla.\vec{F} = \left(i\frac{\partial}{\partial x} + j\frac{\partial}{\partial y} + k\frac{\partial}{\partial z}\right)\left(ixyz + jx^2y^2z + kyz^3\right)$$

$$= yz + 2x^2\,yz + 3yz^2$$

Ex : Consider $\nabla \times \vec{V}$ where $\vec{V} = iV_x + jV_y + kV_z$

$$\nabla \times \vec{V} = \begin{vmatrix} i & j & k \\ \frac{\partial}{\partial x} & \frac{\partial}{\partial y} & \frac{\partial}{\partial z} \\ V_x & V_y & V_z \end{vmatrix}$$

$$= i\left(\frac{\partial V_z}{\partial y} - \frac{\partial V_y}{\partial z}\right) - j\left(\frac{\partial V_z}{\partial x} - \frac{\partial V_x}{\partial z}\right) + k\left(\frac{\partial V_y}{\partial x} - \frac{\partial V_x}{\partial y}\right)$$

Let $\vec{V}$ be the velocity vector at the point O. Consider an elemental rectangular plate ABCD of sides dy,dz as shown in Fig. 12.17.

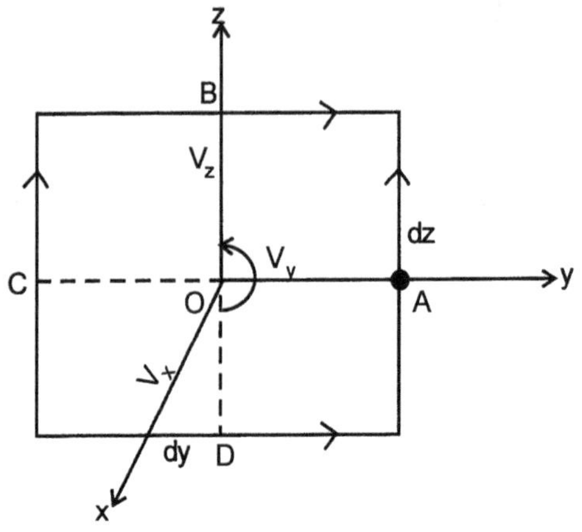

Fig 12.17

The component of velocity V_z at the point $A = V_z + \frac{1}{2}\frac{\partial V_z}{\partial y}dy$

The component of velocity of V_z at the point $C = V_z - \frac{1}{2}\frac{\partial V_z}{\partial y}dy$

∴ resultant velocity in z-direction

$$= \left(V_z + \frac{1}{2}\frac{\partial V_z}{\partial y}dy\right) - \left(V_z - \frac{1}{2}\frac{\partial V_z}{\partial y}dy\right) = \frac{\partial V_z}{\partial y}dy \qquad \text{..... (i)}$$

Similarly the resultant of velocity y-direction is

$$= \left(V_y + \frac{1}{2}\frac{\partial V_y}{\partial z}dz\right) - \left(V_y - \frac{1}{2}\frac{\partial V_y}{\partial z}dz\right)$$

$$= \frac{\partial V_y}{\partial z}dz \qquad \text{.....(ii)}$$

∴ x-component of spin is given by (using (i) and (ii))

$$= i\left(\frac{\partial V_z}{\partial y}dy - \frac{\partial V_y}{\partial z}dz\right) = \text{Curl}_x \vec{V}$$

Similarly we have y and z components of spin as

$$j\left(\frac{\partial V_x}{\partial z}dz - \frac{\partial V_z}{\partial x}dx\right) \text{ and } k\left(\frac{\partial V_y}{\partial x}dx - \frac{\partial V_x}{\partial y}dy\right)$$

which are denoted by $\text{Curl}_y \vec{V}$, $\text{Curl}_z \vec{V}$

Thus $\nabla \times \vec{V}$ gives the Curl $\vec{V}$ or rot $\vec{V}$, which is spin.

The relationship between

$$\omega = \frac{dp}{dt} \text{ and } w = \frac{dz}{dt}$$

We know in x y z coordinates

$$\omega = \frac{dp}{dt} = \frac{\partial p}{\partial t} + \frac{\partial p}{\partial x}\frac{dx}{dt} + \frac{\partial p}{\partial y}\frac{dy}{dt} + \frac{\partial p}{dz}\frac{dz}{dt}$$

$$= \frac{\partial p}{\partial t} + u\frac{\partial p}{\partial x} + v\frac{\partial p}{\partial y} + w\frac{\partial p}{\partial z}$$

$$\omega = \frac{\partial p}{\partial t} + \vec{V}.\nabla p + w\frac{\partial p}{\partial z} \qquad .....(12.66)$$

where $\vec{V}$ = iu + jv, horizontal velocity.

Again $\vec{V} = \vec{V}_g + \vec{V}$, where $\vec{V}_g^{'}$ = geostrophic wind $= \frac{\alpha}{f} k \times \nabla p$,

$\vec{V}^{'}$ is much smaller than the $\vec{V}_{g'}$

$$\vec{V}.\nabla p = \left(\vec{V}_g + \vec{V}^{'}\right).\nabla p = \vec{V}_g.\nabla p + \vec{V}^{'}.\nabla p.$$

$$\vec{V}_g.\nabla p = \frac{\alpha}{f} k \times \nabla p.\nabla p = \frac{\alpha}{f}\left[k \times \left(i\frac{\partial p}{\partial x} + j\frac{\partial p}{\partial y}\right)\right].\nabla p$$

$$= \frac{\alpha}{f}\left[j\frac{\partial p}{\partial x} - i\frac{\partial p}{\partial y}\right].\left(i\frac{\partial p}{\partial x} + j\frac{\partial p}{\partial y}\right)$$

$$= \frac{\alpha}{f}\left[\frac{\partial p}{\partial x}\frac{\partial p}{\partial y} - \frac{\partial p}{\partial x}\frac{\partial p}{\partial y}\right] = 0$$

$$\therefore \vec{V}.\nabla p = \vec{V}^{'}.\nabla p$$

Using this result

eq. (12.66) becomes

$$\omega = \frac{\partial p}{\partial t} + \vec{V}^{'}.\nabla p + w\frac{\partial p}{\partial z}$$

$$\omega = \frac{\partial p}{\partial t} + \vec{V}^{'}.\nabla p - g\rho w \quad (\therefore \delta p = -g\rho\delta z)$$

Therefore first approximation is given by

$$\omega = -g\rho w$$

since $\frac{\partial p}{\partial t}$ and $\vec{V}'.\nabla p$ are comparatively much smaller than $g\rho w$

$$g\rho w \approx 10\left(\frac{\partial p}{\partial t}\right), \quad \vec{V}'.\nabla p = 0.1\ (g\rho w)$$

Questions

1. (a) A particle P moves with a velocity $\vec{V}_r$ relative to a point M. M rotates about an axis with radius vector $\vec{R}$ with constant angular velocity $\vec{\Omega}$. Show that the absolute velocity of the particle $\vec{V}_a$ is the sum of relative velocity $\vec{V}_r$ and the velocity of the second system ($\vec{V}_M$) with respect to the first

   that is $\quad \vec{V}_a = \vec{V}_r + \vec{V}_M$

   or $\quad \vec{V}_a = \vec{V}_r + \vec{\Omega} \times \vec{R}$.

   (b) show that absolute acceleration $\frac{d\vec{V}_a}{dt}$ is given by

$$\left(\frac{d\vec{V}_a}{dt}\right)_a = \left(\frac{d\vec{V}_r}{dt}\right)_r + 2\vec{\Omega} \times \vec{V}_r + \vec{\Omega} \times \vec{\Omega} \times \vec{R},$$

   where $\left(\frac{d\vec{V}_r}{dt}\right)_r$ is relative acceleration.

2. Define viscous force, coefficient of viscosity (η). Give Newton's formula for viscous force and Stokes law and terminal velocity V_T.
3. Define Universal law of gravitation. Find the mass of the earth (M_E) if

   G = gravitational constant = 6.673×10^{-11} N m^2/kg^2

   g = gravity of earth = 9.80 m/s^2, R = radius of the earth = 6370 Km.

(use weight on earth $w = Mg = \frac{GM.M_E}{R^2}$)

4. What is pressure gradient force? Show that

Pressure force per unit of mass at a point $= -\frac{1}{\rho}\nabla p$

Where p = pressure force , ρ = density of fluid/air

5. Derive equations of motion (in vector form) using Newton's second law of motion. Deduce the equations of motion in local cartesian coordinates.

6. Treating pressure as vertical coordinate derive equation connecting individual and local time derivatives. Use hydrostatic equation.

Deduce the equation

$$\frac{dz}{dt} = \left(\frac{\partial z}{\partial t}\right)_p + \vec{V}_h.\nabla_p z - \frac{\alpha}{g}\omega$$

7. Derive the general transformation z = const to p = const as

$$\nabla_z b = \nabla_p b + g\rho\frac{\partial b}{\partial p}\nabla_p z$$

where b is any arbitrary scalar variable

Transform the above equation if $\vec{b} = i\,b_x + j\,b_y$ is a vector.

8. Define geopotential Ψ. Transform the equations of motion for equilibrium of static atmosphere.

9. Assuming the equations of motion in local cartesian coordinates and using hydrostatic equation, transform the above equations into pressure coordinates system.

10. Derive the equation of continuity in pressure coordinates.

11. Using Poisson's equation of potential temperature (θ), derive the thermodynamic relation

$$\frac{1}{\theta}\nabla_p\theta = \frac{1}{\alpha}\nabla_p\alpha$$

Hence transform the thermodynamic equation $H = \frac{d\theta}{dt} = Cv\frac{dT}{dt} + p\frac{d\alpha}{dt}$ into pressure coordinates, and show that temperature changes in three ways.

12. For steady, non-viscous, incompressible fluid flow, derive Bernoullis equation.

13. Derive equations of continuity in isobaric coordinates.

14. Define scalar product and vector product of two vectors $\vec{A} = iA_x + jA_y + kA_z$ and $\vec{B} = iB_x + jB_y + kB_z$.
Show that the scalar product of two perpendicular vectors is zero.

15. (i) Show that the divergence of a gradient ϕ is a Laplacian
(ii) Curl of a gradient ϕ is zero, where ϕ is a scalar function.

16. If $\vec{A} = 2i + 4j + 6k$ and $\vec{B} = i - 3j + 2k$ find the projection of $\vec{A}$ on $\vec{B}$ and the values of $\sin\theta$, $\cos\theta$, where θ is the angle between the vectors $\vec{A}$ and $\vec{B}$.

17. (i) If $\phi = \phi(x, y, z)$ be a scalar function, $\vec{r} = ix + jy + kz$ any vector show that $\nabla\phi . d\vec{r} = d\phi$

(ii) If $\phi_1 = \phi_1(x, y, z, t)$, then show that $d\phi_1 = \nabla\phi_1 . d\vec{r} + \frac{\partial \phi_1}{\partial t} dt$.

18. Find the first approximation relation between $\omega = \frac{dp}{dt}$ and $w = \frac{dz}{dt}$ using hydrostatic equation and $\vec{V} = \vec{V}_g + \vec{V}$ (perturbation).

CHAPTER 13

Circulation

Flow : Let $\vec{V}$ be an arbitrary velocity field, and let A and B be any two points in the fluid space (three dimensional), on the arbitrary curve L. Then the line integral

$$F = \int_A^B \vec{V}.d\vec{r}$$

is called the flow along the path from A to B. Where $d\vec{r} \simeq \delta\vec{r}$ is an infinitesimal vector along the curve (L) as shown in the Fig. 13.1 . The positive direction is along the curve A to B.

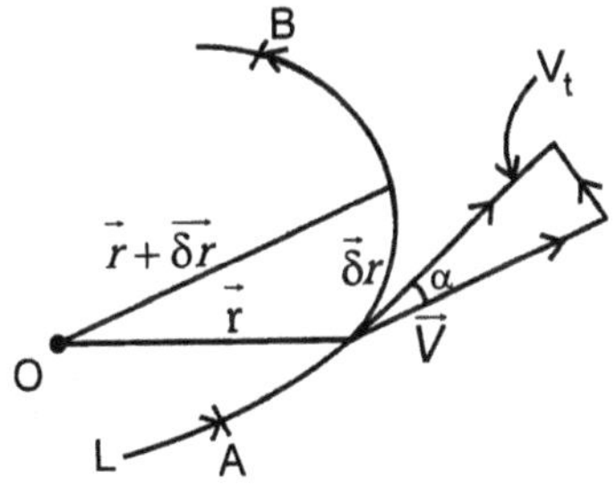

Fig. 13.1

Let V_t be the velocity component of $\vec{V}$ along the curve (along $\delta\vec{r}$) and let α be the angle between $\vec{V}$ and V_t as shown in the Fig.13.1.

Note:

$$\vec{V}.d\vec{r} \simeq V_t\delta r \text{ and } \vec{V}.d\vec{r} = udx + vdy + wdz$$

where $$d\vec{r} = idx + jdy + kdz$$

$$\vec{V} = iu + jv + kw$$

$$\therefore \quad F = \int_A^B V_t\delta r = \overline{V}_t L$$

where $\overline{V}_t$ = average velocity component over the length (AB) of the curve.

Circulation (c) : The flow round a closed curve is called circulation. Fig.13.2

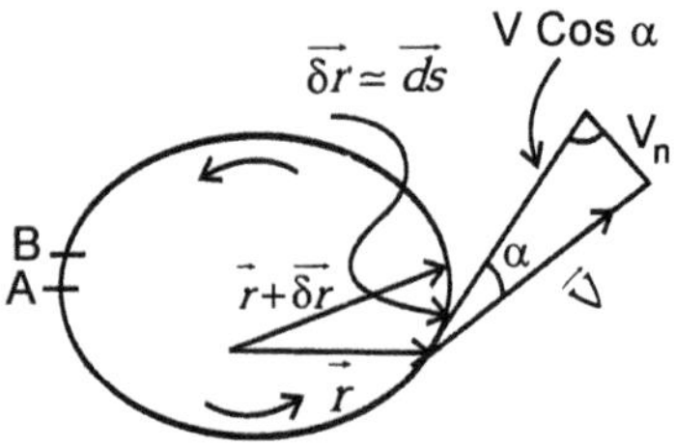

Fig. 13.2

Thus $$C = \oint \vec{V}\cdot d\vec{r} = \oint V_t dr$$

$$= \oint V \cos\alpha \, dr$$

$$= \overline{V} L$$

Where the sign $\oint$ represents the line integral taken around the closed curve.

V_t = velocity component along the tangent to the curve

$\overline{V}$ = average velocity round the closed curve, L is the length of the closed curve.

$$C = \oint \vec{V}\cdot d\vec{r} = \oint (iu + jv + kw) \,.\, (idx + jdy + kdz)$$

$$= \oint (udx + vdy + wdz)$$

Ex : If the motion is uniform and around a closed circle of radius r having tangential velocity V, then circulation around the circle $C = 2\pi rV$.

Note 1 : The motion of an infinitesimal element of fluid for a small period Δt consists of body translation, solid rotation, volume expansion or compression and shear. In Meteorology we consider mainly motion of solid rotation and circulation.

Note 2 : Circulation and vorticity are two primary measures of rotation in a fuild. Circulation is an integral quantity and is a microscopic measure of rotation for a finite area of the fluid. Vorticity is a vector field which provides a microscopic measure of rotation at any point in the fluid.

Theorem : The circulation of an ascendent (gradient) is zero along any closed curve.

Proof : We know circulation $C = \oint \vec{V} \cdot d\vec{r}$.

Replace $\vec{V}$ by $\nabla\phi$ (ascendent), then we have

$$C = \oint \nabla\phi \cdot d\vec{r}$$

where ϕ is any scalar function of (x, y, z).

But
$$\nabla\phi . d\vec{r} = \left(i\frac{\partial\phi}{\partial x} + j\frac{\partial\phi}{\partial y} + k\frac{\partial\phi}{\partial z} \right) \cdot (idx + jdy + kdz)$$

$$= \frac{\partial\phi}{\partial x} dx + \frac{\partial\phi}{\partial y} dy + \frac{\partial\phi}{\partial z} dz$$

$$= d\phi, \quad \text{total differential if } \phi = \phi\,(x, y, z)$$

$$C = \oint d\phi = [\phi]_A^A = \phi_A - \phi_A = 0$$

[$\because$ in closed curve the upper and lower limits coincide]

Corollary : Circulation of any constant vector along any closed curve is zero.

13.1 Kelvin's Theorem

The acceleration of the circulation (C_{ac}) is equal to the circulation of the acceleration.

Proof : We know circulation C is given by (def)

$$C = \oint \vec{V} \cdot d\vec{r} \qquad .....(13.1)$$

and $$C_{ac} = \oint \frac{d\vec{V}}{dt} \cdot d\vec{r} \quad .....(13.2)$$

Differentiating (13.1) w.r.t 't' time, we have

$$\frac{dC}{dt} = \frac{d}{dt} \oint \vec{V} \cdot d\vec{r} = \oint \frac{d}{dt} (\vec{V} \cdot d\vec{r})$$

[since the curve always consists of the same particles]

$$= \oint \left[\frac{d\vec{V}}{dt} . d\vec{r} + \vec{V} . \frac{d}{dt}(d\vec{r}) \right]$$

$$\frac{dC}{dt} = \oint \frac{d\vec{V}}{dt} . d\vec{r} + \oint \vec{V} . \frac{d}{dt} (d\vec{r}) \quad .....(13.3)$$

We shall now prove that the second integral is zero. Let the position of the curve be L at time $t = t_0$ and L' at time $t = t_0 + dt$, that is moved to this position in the time interval of dt. Let the points A, B on L be moved to the new positions A_1, B_1 on the curve L' as shown in the Fig. 13.3.

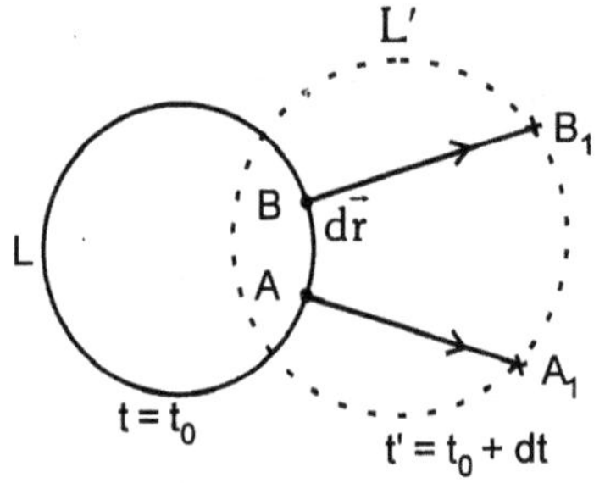

Fig. 13.3

Then we have

$$\overrightarrow{AB} = d\vec{r}, \; \overrightarrow{AA_1} = \vec{V}_A \, dt, \; \overrightarrow{BB_1} = \vec{V}_B \, dt,$$

$$\overrightarrow{A_1B_1} = d\vec{r} + \frac{d}{dt} (d\vec{r}) dt \quad .....(13.4)$$

where $\vec{V}_A$, $\vec{V}_B$ are the velocities at points A and B respectively.

Consider the polygon AA_1B_1B. These points A, A_1, B_1 and B form a closed polygon

$$\therefore \quad \overrightarrow{AA_1} + \overrightarrow{A_1B_1} + \overrightarrow{B_1B} + \overrightarrow{BA} = 0 \quad ......(13.5)$$

using the relations (13.4), eq. (13.5) becomes

$$\vec{V}_A \, dt + \left(d\vec{r} + \frac{d}{dt}(d\vec{r}) dt \right) - \vec{V}_B \, dt - d\vec{r} = 0$$

or $$\frac{d}{dt}(d\vec{r})\,dt = \vec{V}_B dt - \vec{V}_A dt$$

or $$\frac{d}{dt}(d\vec{r}) = \vec{V}_B - \vec{V}_A$$

or $$\frac{d}{dt}(d\vec{r}) = d\vec{V} \qquad .....(13.6)$$

Substituting this value in the second integral, of eq. (13.3) we have

$$\oint \vec{V}.\frac{d}{dt}(d\vec{r}) = \oint \vec{V}.d\vec{V} = \oint d\left(\frac{1}{2}V^2\right)$$

$$= \left[\frac{1}{2}V^2\right]_A^A = 0$$

$$\therefore \oint \vec{V}.\frac{d}{dt}(d\vec{r}) = 0 \qquad .....(13.7)$$

using (13.7), the eq. (13.3) becomes

$$\frac{dC}{dt} = \oint \frac{d\vec{V}}{dt}.d\vec{r}$$

i.e., $$C_{ac} = \frac{dC}{dt} = \oint \frac{d\vec{V}}{dt}.\vec{dr}$$

which is the required result.

13.2 Bjerkne's Theorem

Bjerkne's circulation theorem is a combination of equations of motion and Kelvin's circulation theorem.

Equations of motion neglecting friction $\left(\alpha \frac{d\vec{\tau}}{dz}\right)$ is given by

$$\frac{d\vec{V}}{dt} = -\alpha\nabla p - 2\vec{\Omega} \times \vec{V} + \vec{g} \qquad .....(13.8)$$

Where $2\vec{\Omega} \times \vec{V}$ = Coriolis force, $\vec{g}$ = gravity, $\alpha\nabla p$ = pressure force.

Kelvin's theorem is

$$\frac{dC}{dt} = \oint \left(\frac{d\vec{V}}{dt}.d\vec{r}\right) \qquad .....(13.9)$$

From (13.8) and (13.9) we have

$$\frac{dC}{dt} = \oint[-\alpha\nabla p\,.\,d\vec{r} - 2\,(\vec{\Omega} \times \vec{V}).\,d\vec{r} + \vec{g}\,.\,d\vec{r}]$$

or $$\frac{dc}{dt} = -\oint \alpha\nabla p.\, d\vec{r} - \oint 2\,(\vec{\Omega} \times \vec{V})\,.\, d\vec{r} + \oint \vec{g}\,.\, d\vec{r}$$

$$= -\oint \alpha\nabla p.\, d\vec{r} - \oint 2\,(\vec{\Omega} \times \vec{V})\,.\, d\vec{r} + \oint \vec{g}.\, d\vec{r} \qquad ....(13.10)$$

$\oint \vec{g}\,.\, d\vec{r} = 0$, Since $\vec{g}$ is the (gradient) ascendent of the geopotential $[\vec{g_a} = -\nabla\psi_a]$ and the circulation of an ascendent is zero .....(13.11)

$$\oint \alpha\nabla p\,.\, d\vec{r} = \oint \alpha \left[i\frac{\partial p}{\partial x} + j\frac{\partial p}{\partial y} + k\frac{\partial p}{\partial z} \right].\,[\,idx + jdy + kdz]$$

$$= \oint \alpha \left[\frac{\partial p}{\partial x}dx + \frac{\partial p}{\partial y}dy + \frac{\partial p}{\partial z}dz \right]$$

$$\oint \alpha\nabla p.\, \vec{dr} = \oint \alpha dp \qquad .....(13.12)$$

[Total differential of p(x, y, z) is $dp = \frac{\partial p}{\partial x}dx \frac{\partial p}{\partial y}dy + \frac{\partial p}{\partial z}dz$]

Substituting (13.11) and (13.12) in (13.10) we have

$$\frac{dC}{dt} = -\oint \alpha dp - 2\oint(\vec{\Omega} \times \vec{V})\,.\, d\vec{r} \qquad .....(13.13)$$

eq. (13.13) is called the circulation theorem of Bjerkne's.

Solenoids : A Solenoid is a three dimensional body formed by the intersection of two sets of equi-scalar surfaces.

An equi-scalar surface meaning a certain scalar property (such as pressure, volume, density, temperature etc.,) is constant along that surface.

Isobaric-isosteric solenoids are formed by the intersection of isobaric surfaces and isosters (surfaces of constant specific volume, $\alpha = \frac{1}{\rho}$ reciprocal of density).

Note : Presence of more number of solenoids in a vertical cross-section indicates more available energy to begin circulation. Existence of solenoids represents a setup for direct circulation.

Solenoidal : A vector field $\vec{F}$ (x, y, z) such that Div $\vec{F} = 0 = \nabla\,.\,\vec{F}$, is called solenoidal.

This means that the field has no sources or sinks.

Theorem : $\oint -\alpha dp = N(\alpha, p)$ = Number of solenoids enclosed by the curve in (α, p) diagram.

Proof. Refer to the Fig. (13.4 a) as the point p moves around the physical curve in space we can have one-one correspondence with point p on (α, p) diagram as shown in Fig. (13.4 b).

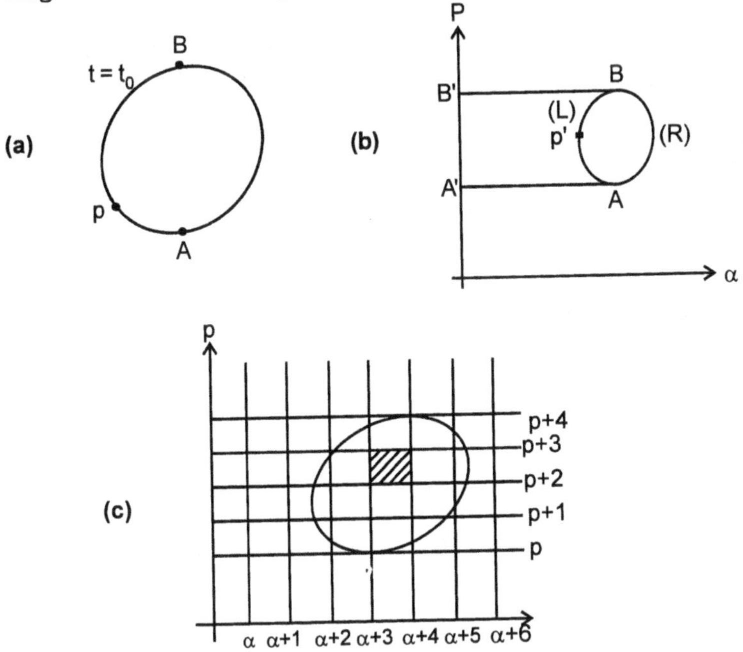

Fig. 13.4 (a,b,c)

$$\oint \alpha dp = \int_{\substack{A \\ \text{Over(R)}}}^{B} \alpha dp + \int_{\substack{B \\ \text{Over(L)}}}^{A} \alpha dp$$

Where first integral is taken over the curve A to B along (R) while the second integral is taken over the curve B to A along (L).

$$\int_{\substack{A \\ \text{Over(R)}}}^{B} \alpha dp = \text{Area A'A(R)BB'A'} \qquad \text{.....(13.14)}$$

$$\int_{\substack{B \\ \text{Over(L)}}}^{A} \alpha dp = -\int_{\substack{A \\ \text{Over (L)}}}^{B} \alpha dp = -\text{ Area B'A'A(L)BB'} \qquad \text{....(13.15)}$$

$$\therefore \oint \alpha dp = \text{Area A'A(R)BB'A'} - \text{Area B'A'A(L)BB'}$$

$$= \text{area of the closed curve A(R)B(L)A}$$

$$\oint \alpha dp = \text{Area enclosed by the curve in } (\alpha, p) \text{ diagram.}$$

Now consider the (α, p) diagram as shown in Fig. 13.4 (C). Let the isobaric lines p, p + 1, p + 2, p + 3, p + 4... and the isosteric lines α, α+ 1, α+ 2, α + 3, α + 4....be drawn. These lines form square net over (α, p) diagram with each square is unity, since the isolines are drawn at unit intervals. Each square in this diagram is an isosteric-isobaric solenoid. Thus the whole area enclosed by curve in (α, p) diagram represents solenoids.

∴ $\oint \alpha dp$ = The number of solenoids enclosed by the curve in (α, p) diagram = N(α, p).

(a) **The Solenoidal vector** : From the given meteorological data of specific volume (α) and pressure (p) to find the number of solenoids.

Let the isobaric surfaces p, p + 1, p + 2, p + 3, p + 4 ... and isosteric surfaces α, α +1, α + 2, α + 3, α + 4... intersect as shown in Fig 13.5.

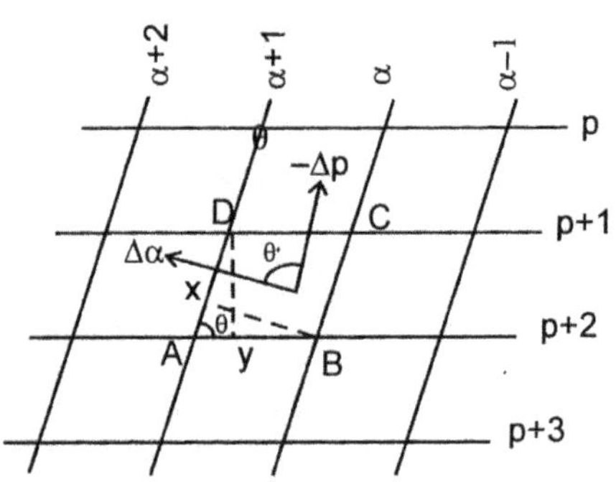

Fig. 13.5

The area of the solenoid ABCD (parallelogram) is

$$= AB \times AD \sin\theta \qquad (\because DY = AD \sin\theta) \qquad \text{......(13.16)}$$

where θ = is the angle between isosters and isobars

$$Bx = \text{magnitude of isosteric gradient} = \frac{1}{|\nabla \alpha|}$$

$$Dy = \text{magnitude of isobaric gradient} = \frac{1}{|\nabla P|}$$

From the Fig. 13.5

$$AB = \frac{Bx}{\sin\theta} = \frac{1}{|\nabla\alpha| \sin\theta}$$

$$AD = \frac{Dy}{\text{Sin } \theta} = \frac{1}{|\nabla p| \sin\theta}$$

∴ Area of the solenoid

$$ABCD = \frac{1}{|\nabla\alpha| \sin\theta} \cdot \frac{1}{|\nabla p| \sin\theta} \cdot \sin\theta = \frac{1}{|\Delta\alpha||\nabla p| \sin\theta}$$

∴ Number of solenoids per unit area

$$S = \frac{1}{\text{area}} = |\nabla\alpha||\nabla p| \sin\theta \qquad .....(13.17)$$

The solenoidal vector $\vec{S}$ is defined as

$$\vec{S} = (\nabla\alpha) \times (-\nabla p) \qquad .....(13.18)$$

whose magnitude is S, given by eq. (13.17) and direction is perpendicular into the paper.

$$-\oint \alpha dp = -\oint \alpha \nabla p.d\vec{r} \qquad (\because d\phi = \nabla\phi.d\vec{r})$$

$$= -\int_A \nabla \times (\alpha \nabla p).d\vec{A} \qquad \text{(By stoke's theorem)}$$

$$= -\int_A (\nabla\alpha \times (\nabla p).d\vec{A}$$

[Stokes theorem $\int_C \vec{V}.dS = \int_S (\nabla \times \vec{V}).\vec{n}\, dS$]

$$\therefore -\oint \alpha dp = \int_A \vec{S}.d\vec{A} \qquad \text{[using eq. (13.18)]}$$

i.e. The number of solenoids in (α-p) diagram = Area integral in the geometrical space of the solenoidal vector.

(b) *The Coriolis term* : Consider the integral $\oint(\vec{\Omega} \times \vec{V}).d\vec{r}$ which can be written as $\oint (\vec{V} \times d\vec{r}) . \vec{\Omega}$ [properties cross product, dot product vectors.]

i.e.,
$$\oint (\vec{\Omega} \times \vec{V}) . d\vec{r} = \oint (\vec{V} \times d\vec{r}) . \vec{\Omega}$$

Since under integral

$$\text{LHS} = \begin{vmatrix} dx & dy & dz \\ \Omega_x & \Omega_y & \Omega_z \\ u & v & w \end{vmatrix}$$

is equal to

$$\text{RHS} = \begin{vmatrix} \Omega_x & \Omega_y & \Omega_z \\ u & v & w \\ dx & dy & dz \end{vmatrix}$$

But $\vec{V} \times d\vec{r}$ = magnitude of the area covered in unit time by the particles located on the vector in $d\vec{r}$

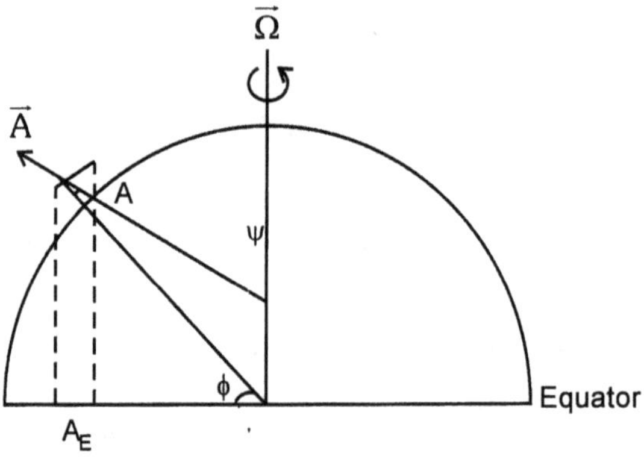

Fig. 13.6

$\therefore \oint \vec{V} \times d\vec{r}$ = Change in area enclosed by the curve per unit time

$$\therefore \oint (\vec{V} \times d\vec{r}).\vec{\Omega} = \vec{\Omega}.\frac{d\vec{A}}{dt} \qquad .....(13.19)$$

where

$\vec{A}$ = area vector which is normal to the plane of the physical curve.

and $\vec{\Omega}$ = points along the axis of rotation of the earth (Fig. 13.6).

From eq. (13.19) and Fig. 13.6 we have

$$\oint(\vec{\Omega} \times \vec{V}).d\vec{r} = \oint(\vec{V} \times d\vec{r}).\vec{\Omega}$$

$$= \vec{\Omega}.\frac{d\vec{A}}{dt} = \Omega\frac{dA_E}{dt} \qquad .....(13.19)$$

where A_E = Area of the projection of A on equatorial plane. $|\vec{\Omega}| = \Omega$

Substituting solenoidal theorem viz $-\oint \alpha dp = N(\alpha, -p)$

and eq. (13.19) (the above coriolis term) in Bjerkne's circulation theorem

$$[\text{Viz } \frac{dC}{dt} = -\oint \alpha dp - 2\oint(\vec{\Omega} \times \vec{V}).d\vec{r}]$$

We have

$$\frac{dC}{dt} = N(\alpha - p) - 2\Omega \frac{dA_E}{dt} \quad .....(13.20)$$

The angle ψ is the colatitude if the plane of the curve is horizontal. In that case $A_E = A \sin \phi$, where ϕ is latitude as shown in Fig. (13.7)

then $$\frac{dA_E}{dt} = \frac{d}{dt}(A \sin \phi)$$

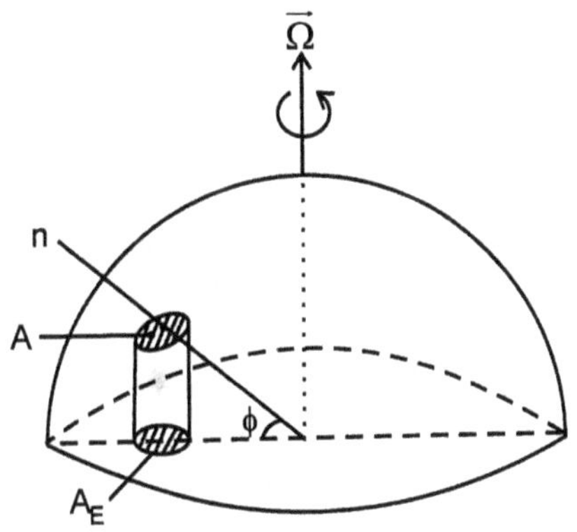

Fig. 13.7

13.3 Application of Bjerkne's Circulation Theorem

$$\frac{dC}{dt} = N(\alpha, -p) - 2\Omega \frac{dA_E}{dt} \text{ Bjerkne's Theorem [eq.(13.20)]}.$$

The two terms on the right side of this equation play different roles and hence they are not equally important simultaneously. This will be illustrated in Land and Sea breeze.

When we consider initial circulation from a state of rest, the coriolis term $2\Omega\frac{dA_E}{dt}$ is not important. In that case increase in circulation is equal to the number of solenoids $N(\alpha, -p)$.

In practical application it is convenient to use α and p instead of $\nabla\alpha$, ∇p. We deduce Solenoidal vector with temperature and potential temperature.

$$\vec{S} = -\nabla\alpha \times \nabla p \quad \text{(eq. 13.18)}$$

$$\vec{S} = -\frac{R}{P}\nabla T \times \nabla p \quad \text{[using gas equation } p\alpha = RT]$$

$$\vec{S} = \frac{R}{P}\nabla p \times \nabla T \quad .....(13.21)$$

$$[\because \vec{A} \times \vec{B} = -\vec{B} \times \vec{A}]$$

Also $$\theta = T\left(\frac{p}{p_0}\right)^{\frac{-R}{C_p}} \quad \text{(poisson's equation)}$$

Taking logarithms and differentiating poison's equation

$$\ln\theta - \ln T = -\frac{R}{C_p}[\ln p - \ln p_0]$$

$$\frac{1}{\theta}d\theta - \frac{1}{T}dT = -\frac{R}{C_p}\frac{1}{p}dp \quad (\because p_0 = \text{constant})$$

or $$\frac{1}{\theta}\nabla\theta - \frac{1}{T}\nabla T = -\frac{R}{C_p}\frac{1}{p}\nabla p$$

$$\therefore -\frac{R}{p}\nabla p = \frac{C_p}{\theta}\nabla\theta - \frac{C_p}{T}\nabla T \qquad .....(13.22)$$

Substituting (13.22) in (13.21) we have

$$\vec{S} = -\left[\frac{C_p}{\theta}\nabla\theta - \frac{C_p}{T}\nabla T\right] \times \nabla T$$

or $$\vec{S} = -\frac{C_p}{\theta}\nabla\theta \times \nabla T \qquad [\because \nabla T \times \nabla T = 0]$$

$$\vec{S} = \frac{C_p}{\theta}\nabla T \times \nabla\theta$$

This form of solenoidal vector is useful in vertical cross-section where we use Temperature (T) and potential temperature θ

13.4 Sea Breeze

Let at a certain instant air is in equilibrium over an underlying surface partly on sea and partly on land as shown in Fig. (13.8a). In this state isosteric and isobaric surfaces will be horizontal. During the day land is heated up quickly as compared to sea. Consequently (closer to the land) lower layers of the atmosphere over land gets heated up and hence specific volume will be larger. Isosteric surfaces will get tilted as shown in Fig. (13.8a). In this situation circulation develops from $\nabla\alpha$ to $-\nabla p$ i.e., from sea to land in lower layer and from land to sea in the upper layer as shown by dotted lines with arrows. In this initial state the coriolis term is negligible. But as the circulation increases, the mass transport of cold air below and warm air above takes isobaric surfaces begin to tilt and the coriolis effect shown is the wind. On the south-north oriented coast (like the west coast of peninsular India) initially sea-breeze will be from west (in the lower levels) slowly becomes from north westerly direction.

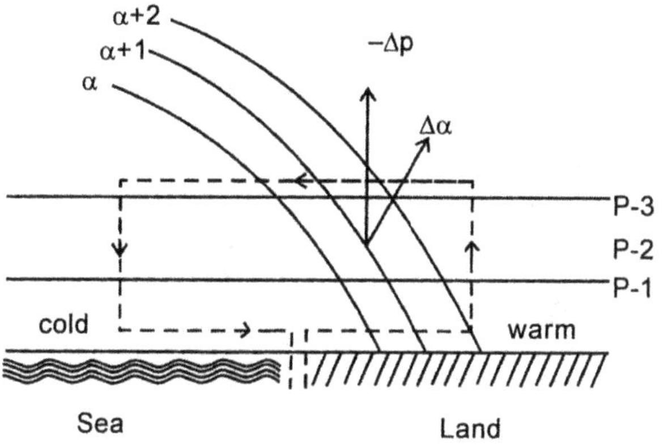

Fig. 13.8(a) Sea breeze.

13.4.1 Land Breeze

During the night land is comparatively cooler while sea is still warmer. The lower layers of the atmosphere over land becomes cooler quickly as compared to over the sea. As a result isosteric surfaces will tilt in opposite direction as against the day time (as shown in Fig (13.8b). Circulation develops from ∇p to $\nabla \alpha$ i.e., from land to sea in the lower layers and from sea to land in the upper layers as shown by dotted lines with arrows in Fig. (13.8b).

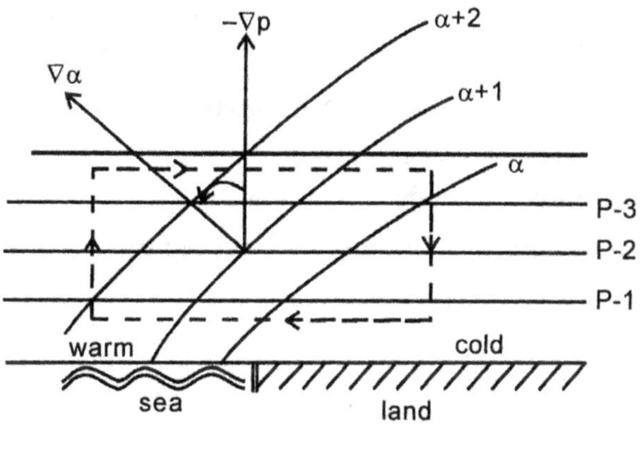

Fig. 13.8(b) Land breeze.

13.5 Vorticity

If $\vec{V}$ be the velocity vector of a fluid particle then the quantity

$\vec{W} = \frac{1}{2} \nabla \times \vec{V}\ \frac{1}{2}$; curl $\vec{V}$ is called the vorticity vector or simply the vorticity. In fact the velocity of rotation (angular velocity) of an infinitesimal element is the vorticity.

If $\vec{W} = i\xi + j\eta + k\zeta$ and $\vec{V} = iu + jv + kw$ Then

$$i\xi + j\eta + k\zeta = \begin{vmatrix} i & j & k \\ \frac{\partial}{\partial x} & \frac{\partial}{\partial y} & \frac{\partial}{\partial z} \\ u & v & w \end{vmatrix}$$

i.e., $$\xi = \frac{\partial w}{\partial y} - \frac{\partial v}{\partial z},\ \eta = \frac{\partial u}{\partial z} - \frac{\partial w}{\partial x},\ \zeta = \frac{\partial v}{\partial x} - \frac{\partial u}{\partial y}$$

13.5.1 Vortex lines

A vertex line is a curve drawn in the fluid such that the tangent to the curve at each point is in the direction of vorticity vector $\vec{W}$ at that point. The differential equations of the system of vortex lines are given by

$$\frac{dx}{\xi} = \frac{dy}{\eta} = \frac{dz}{\zeta}$$

The motion of a fluid is called irrotational if the vorticity $\vec{W}$ of every fluid particle is zero. When the vorticity is not zero, the motion is called rotational. In case of irrotational motion there are no vortex lines of the fluid.

13.5.2 The vertical component of vorticity ζ

Generally atmospheric flow is horizontal. The vertical component of vorticity ζ is of most significant in describing large scale flow.

In meteorology the term vorticity refers to the vertical component ζ.

$$\zeta = k \cdot \nabla \times \vec{V} = \frac{\partial v}{\partial x} - \frac{\partial u}{\partial y}$$

13.5.3 Definition

The microscopic measure of fluid rotation is called vorticity. It is a vector field and defined as the curl of velocity.

13.6 Stoke's Theorem

The circulation C along a closed curve is equal to the area A enclosed by the curve multiplied by the mean vorticity component $\bar{q}_n$, normal to the area.

$$\oint \vec{V}\,.\,\delta\vec{r} = \bar{q}_n A \text{ or } \oint (udx + vdy) = \iint \left(\frac{\partial v}{\partial x} - \frac{\partial u}{\partial y}\right) dxdy$$

Proof: For simplicity consider a plane rectangular elemental area with sides δx, δy, $-\delta x$, $-\delta y$ in order. The circulation [C = $\bar{V}_t$ L] is of the flows along these sides AB, BC,CD, DA respectively as shown in the Fig (13.9a).

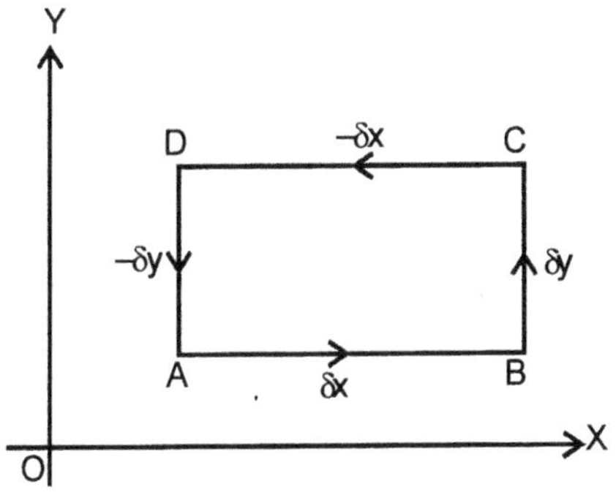

Fig. 13.9(a)

Let u, v be the velocity components at A. Then by Taylors expansion the velocity components (up to first order) at B are u + $\frac{\partial u}{\partial x}$ δx, v + $\frac{\partial v}{\partial x}$ δx.

The velocity components at C are

$$\left[u + \frac{\partial u}{\partial x}\delta x + \frac{\partial}{\partial y}(u + \frac{\partial u}{\partial x}\delta x)\delta y\right], \left[v + \frac{\partial v}{\delta x}\delta x + \frac{\partial}{\partial y}\left(v + \frac{\partial v}{\partial x}\delta x\right)\delta y\right]$$

or

$$\left[u + \frac{\partial u}{\partial x}\delta x + \frac{\partial u}{\partial y}\delta y\right], \left[v + \frac{\partial v}{\partial x}\delta x + \frac{\partial v}{\partial y}\delta y\right]$$

The velocity components at D are

$$u + \frac{\partial u}{\partial y}\,\delta y,\ v + \frac{\partial v}{\partial y}\,\delta y$$

[C = Tangential mean velocity × length of the curve = V_tL,

C = circulation along AB, BC, CD, DA]

$\therefore$ Elemental circulation $\delta c = \frac{1}{2}\left[u+u+\frac{\partial u}{\partial x}\delta x\right]\delta x$

$$+\frac{1}{2}\left[v+\frac{\partial v}{\partial x}\delta x+v+\frac{\partial v}{\partial x}\delta x+\frac{\partial v}{\partial y}\delta y\right]\delta y +$$

$$\frac{1}{2}\left[\left(u+\frac{\partial u}{\partial x}\delta x+\frac{\partial u}{\partial y}\delta y\right)+u+\frac{\partial u}{\partial y}\delta y\right](-\delta x)+$$

$$\frac{1}{2}\left[v+\frac{\partial v}{\partial y}\delta y+v\right](-\delta y),$$

$$\delta c = \left[\frac{\partial v}{\partial x}-\frac{\partial u}{\partial y}\right]\delta x\delta y, \qquad .....(13.23)$$

or $\delta c = q_n\ \delta A$ or $\zeta\ \delta A$

where $\left|\bar{q}_n\right| = q_n = \zeta$ and $A = \delta x\delta y$

For finite area we obtain by integrating eq. (13.23).

$$C = \int_A q_n\delta A = \overline{q_n}\ A \qquad .....(13.24)$$

where $\overline{q_n}$ = mean of q_n over the area A

i.e. $\oint(udx + vdy) = \iint\left(\frac{\partial v}{\partial x}-\frac{\partial u}{\partial y}\right)dxdy$

[since $C = \oint \vec{V}\,.\,d\vec{r}\ \oint(iu + jv)\,.\,(i\delta x + j\delta y)]$

From eq. (13.23) we have $q_n = \frac{\partial c}{\partial A}$, which means that vorticity is circulation along the border of unit area.

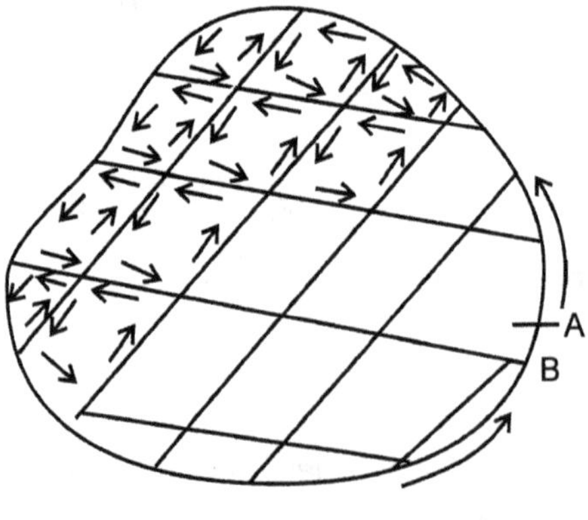

Fig. 13.9(b)

Form (13.24) we have $\bar{q}_n = \frac{C}{A}$. This means that vorticity over the finite area A is the circulation along the border divided by the area $\left(\zeta = \frac{C}{A}\right)$

Now consider a closed curve divided into a number of unit areas as shown in Fig. (13.9b). As pointed out above vorticity is equal to the circulation along the border of a unit area, the sum of the vorticities within the closed curve = The circulations along the borders of unit areas. The flows along the borders cancel except on the outer boundary which is equal to the circulation along the closed curve AB, which generalizes the proof for the theorem.

Consider the circulation around a circle of radius r in a velocity field corresponding to a rotation with constant angular velocity ω. Fig. (13.9C)

$$C = \oint r\omega \,.\, r d\theta = \int_0^{2\pi} r^2 \omega d\theta$$

$$= r^2\omega \left[\theta\right]_0^{2\pi}$$

$$C = 2\pi r^2 \omega.$$

∴ Circulation per unit area $= \frac{C}{\pi r^2} = 2\omega$

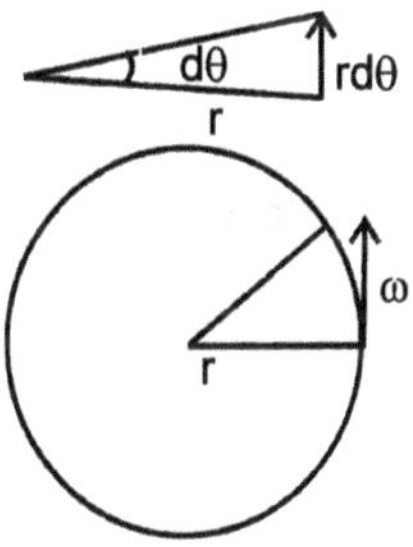

Fig. 13.9(c)

This shows that the vorticity of motion of solid rotation is equal to twice the angular velocity $\left[\because \delta C = \zeta A \text{ or } \zeta = \frac{\delta C}{A}\right]$

Interpretation : If we suppose that an elemental area suddenly solidified, but otherwise followed the motion of the fluid in contact with it. The vorticity of the elemental area is equal to twice the observed angular velocity of the solid element. Vorticity may be regarded as a measure of local angular velocity of the fluid.

13.7 Divergence

We defined divergence as the dot product of 'del' (∇) and velocity vector $\vec{V}$. We shall now illustrate the concept of divergence.

Consider an elemental rectangular area (ABCD) with sides δx, δy as shown in the Fig. 13.10a.

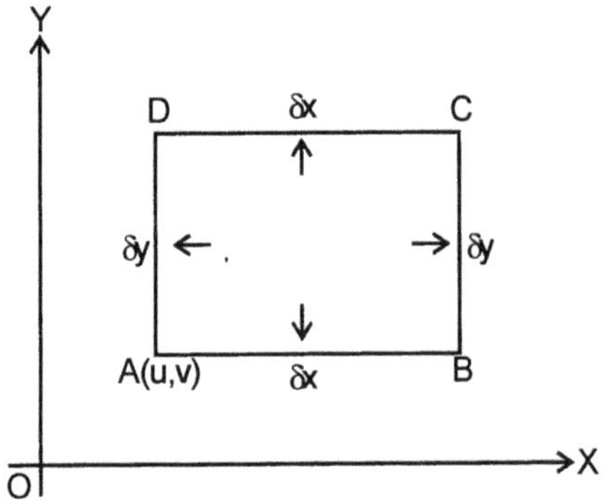

Fig. 13.10(a)

Let u, v be the components of velocity at A in x, y directions. Then the components of velocity at B (by Taylors expansion) are

$$u + \frac{\partial u}{\partial x}\delta x, \; v + \frac{\partial v}{\partial x}\delta x.$$

at C are $\left(u + \frac{\partial u}{\partial x}\delta x + \frac{\partial u}{\partial y}\delta y\right), \; \left(v + \frac{\partial v}{\partial x}\delta x + \frac{\partial v}{\partial y}\delta y\right),$

and at D are $u + \frac{\partial u}{\partial y}\delta y, \; v + \frac{\partial v}{\partial y}\delta y,$

The outflow across AB

$$= -\frac{1}{2}\left[v + v + \frac{\partial v}{\partial x}\delta x\right]\delta x,$$

[(Average of velocities at A, B in y - direction) $\times \delta x$] .....(i)

The outflow across CD

$$= \frac{1}{2}\left[v + \frac{\partial v}{\partial y}\delta y + v + \frac{\partial v}{\partial x}\delta x + \frac{\partial v}{\partial y}\delta y\right]\delta x$$

[(Average of velocity at D and C in y - direction) × δx] .....(ii)

Net outflow across AB and CD in y-direction (i + ii)

$$= \frac{\partial v}{\partial y}\delta x \delta y \qquad .....(13.25)$$

Again the outflow across AD in x-direction.

$$AD = -\frac{1}{2}\left[u + u + \frac{\partial u}{\partial y}\delta y\right]\delta y \qquad .....(iii)$$

The outflow across BC in x-direction.

$$= \frac{1}{2}\left[u + \frac{\partial u}{\partial x}\delta x + u + \frac{\partial u}{\partial x}\delta x + \frac{\partial u}{\partial y}\delta y\right]\delta y \qquad .....(iv)$$

∴ Net outflow across AD and BC in x direction (iii + iv)

$$= \frac{\partial u}{\partial x}\delta x \delta y \qquad .....(13.26)$$

From (13.25) and (13.26) it follows that the net outflow across ABCD

$$\delta G = \left(\frac{\partial u}{\partial x} + \frac{\partial v}{\partial y}\right)\delta x \delta y.$$

or

$$\frac{\delta G}{\delta A} = \frac{\partial u}{\partial x} + \frac{\partial v}{\partial y} = D \qquad ......(13.27)$$

where $\delta A = \delta x \delta y$ = Area

$D = \dfrac{\partial u}{\partial x} + \dfrac{\partial v}{\partial y}$ = Divergence in 2-dimension

It follows from eq. (13.27), net outflow per unit area is divergence.

We can write eq. (13.27) as

$$\delta G = D\,\delta A \qquad ......(i)$$

Integrating

$$\oint dG = \int_A D\,dA$$

$$G = \oint V_n dl = \int_A D\,dA \qquad .....(13.28)$$

where V_n = normal component

dl = element of the curve

eq. (13.28) is Gauss's theorem. RHS of (13.28) is area integral of divergence. Consider Fig. (13.10b). Each small element in this Fig. (13.10b), we have

$$\delta G = D\,\delta A$$

$$\therefore \Sigma\,\delta G = \Sigma\,D\,\delta A \qquad \text{.....(ii)}$$

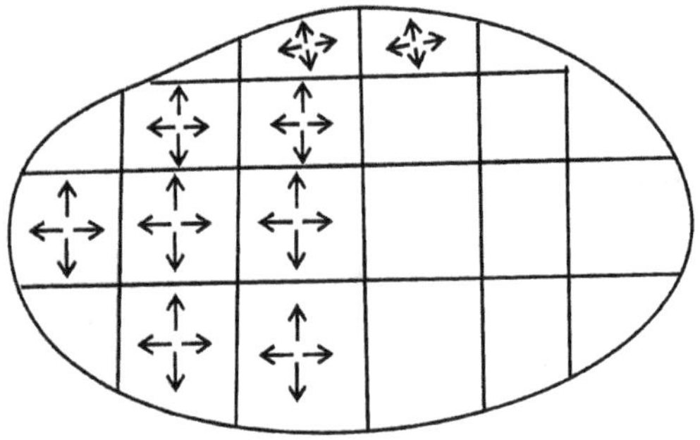

Fig. 13.10(b)

In this summation over all small elements, the RHS of (ii) gives area integral of divergence in the limit, while LHS of (ii) gives total outflow across the closed curve, because the contribution of interior elements get canceled. This gives the Gauss's theorem

viz $$G = \oint V_n dl = \int_A D\,dA$$

13.7.1 Another Interpretation of Divergence

Consider the Fig (13.10(c)).

$$\delta A = \delta x \delta y$$

$$\ln \delta A = \ln \delta x + \ln \delta y$$

Differentiating w.r.t time 't' we have

$$\frac{1}{\delta A}\frac{d(\delta A)}{dt} = \frac{1}{\delta x}\frac{d(\delta x)}{dt} + \frac{1}{\delta y}\frac{d(\delta y)}{dt} \qquad \text{.....(13.29)}$$

For interpretation of the term $\frac{1}{\delta x}\frac{d(\delta x)}{dt}$ consider the Fig (13.10c).

Let at time $t = t_0$ the element δx is in the position AB and let at $t = t_0 + dt$ is in the position of A_1B_1. Let the projections of AB and A_1B_1 on x-axis be $A_0B_0 = \delta x$ and $A_2B_2 = \delta x_1$ respectively. Let the coordinates of A_0, B_0, A_2, B_2 be X_{A0}, X_{B0}, X_{A2}, X_{B2}.

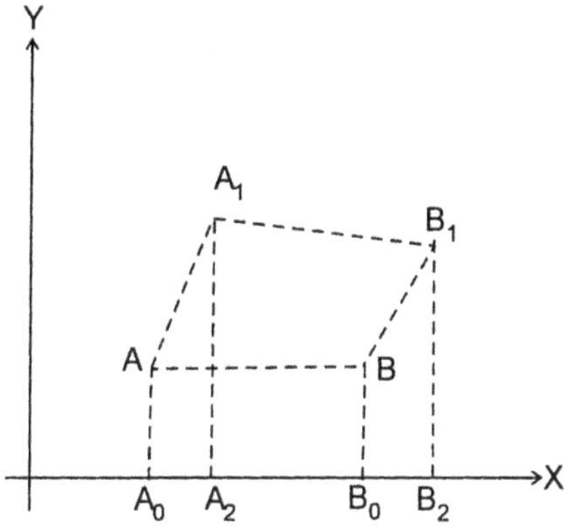

Fig. 13.10 (c)

Now $\frac{1}{\delta x}\frac{d}{dt}(\delta x) = \lim_{\delta t \to 0} \frac{1}{\delta x}.\frac{\delta x_1 - \delta x}{\delta t}$

$$= \lim_{\delta t \to 0} \frac{1}{\delta x}\frac{(A_2B_2 - A_0B_0)}{\delta t}, \ [\delta x_1 = X_{B2} - X_{A2}, \delta x = X_{B0} - X_{A0}]$$

$$= \lim_{\delta t \to 0} \frac{1}{\delta x}\frac{[(X_{B2} - X_{A2}) - (X_{B0} - X_{A0})]}{\delta t}$$

$$= \lim_{\delta t \to 0} \frac{1}{\delta x}\left[\frac{(X_{B2} - X_{B0})}{\delta t} - \frac{(X_{A2} - X_{A0})}{\delta t}\right]$$

$$\frac{1}{\delta x}\frac{d}{dt}(\delta x) = \frac{1}{\delta x}[U_B - U_A] = \frac{\partial u}{\partial x} \quad .....(13.30)$$

similarly we can write

$$\frac{1}{\delta y}\frac{d(\delta y)}{dt} = \frac{\partial v}{\partial y} \quad .....(13.31)$$

substituting eq. (13.30) and eq. (13.31) in eq. (13.29) we have.

$$\frac{1}{\delta A}\frac{d(\delta A)}{dt} = \frac{\partial u}{\partial x} + \frac{\partial v}{\partial y} = D$$

∴ We can interpret divergence is equal to the change in the area of the fluid element per unit of time and per unit area.

13.8 Deformation

$F1 = \left(\frac{\partial u}{\partial x} - \frac{\partial v}{\partial y}\right)$ is called deformation stretching

$$F2 = \left(\frac{\partial v}{\partial x} + \frac{\partial u}{\partial y}\right) \text{ is called deformation shearing}$$

F_1, F_2 are component of deformation.

Note : ζ, D, $F^2_1 + F^2_2$ are invariant under rotation of coordinate axes.

Consider the rate of change per unit time of the ratio $\frac{\delta x}{\delta y}$

Let $$D_f = \frac{\delta x}{\delta y}$$

$$\therefore \ln D_f = \ln \delta x - \ln \delta y$$

Differentiating w.r.t to (time) we have

$$\frac{1}{D_f}\frac{d}{dt}(D_f) = \frac{1}{\delta x}\frac{d\delta x}{dt} - \frac{1}{\delta y}\frac{d(\delta y)}{dt} \qquad .....(13.32)$$

Using eq.(13.30) and (13.31) in (13.32) we have

$$\frac{1}{D_f}\frac{dD_f}{dt} = \frac{\delta u}{\delta x} - \frac{\delta v}{\delta y} = F_1 \qquad .....(13.33)$$

eq. (13.33) shows that deformation component F_1 is equal to the stretching in x-direction.

Note: Taylors series for functions of two variables

$$f(x, y) = f(a, b) + (x - a) f_x(a, b) + (y - b) f_y(a, b) +$$

$$\frac{1}{\underline{|2}}[(x - a)^2 f_{xx}(a, b) + 2(x - a)(y - b) f_{xy}(a,b)$$

$$+ (y - b)^2 f_{yy}(a, b)]$$

where $f_x(a, b)$, $f_y(a, b)$ denote partial derivatives w.r.t x, y at x = a, y = b, etc.

13.9 Differential Properties of the Wind Field

Consider a point (x_0, y_0) where the velocity components are u_0, v_0. In the small neighborhood of x_0, y_0 say $x_0 \pm \delta x$, $y_0 \pm \delta y$ (by Taylor's expansion) the velocity field u, v may be written as.

$$u = u_0 + \left(\frac{\partial u}{\partial x}\right)_0 \delta x + \left(\frac{\partial u}{\partial y}\right)_0 \delta y +$$

$$\frac{1}{\underline{|2}}\left[\frac{\partial^2 u}{\partial x^2}\delta^2 x + 2\frac{\partial^2 u}{\partial x \partial y}\delta x \delta y + \frac{\partial^2 v}{\partial y^2}\delta^2 y\right]_0 + \ldots$$

$$v = v_0 + \left(\frac{\partial v}{\partial x}\right)_0 \delta x + \left(\frac{\partial v}{\partial y}\right)_0 \delta y +$$

$$\frac{1}{\underline{|2}}\left[\frac{\partial^2 v}{\partial x^2}\delta^2 x + 2\frac{\partial^2 v}{\partial x \partial y}\delta x \delta y + \frac{\partial^2 v}{\partial y^2}\delta^2 y\right]_0 + \ldots$$

Neglecting second and higher order terms, we have

$$u = u_0 + \left(\frac{\partial u}{\partial x}\right)_0 \delta x + \left(\frac{\partial u}{\partial y}\right)_0 \delta y$$

$$v = v_0 + \left(\frac{\partial v}{\partial x}\right)_0 \delta x + \left(\frac{\partial v}{\partial y}\right)_0 \delta y \qquad \ldots\ldots(13.34)$$

where $\delta x = x - x_0$

$\delta y = y - y_0$

The above equation can be written as

$$u = u_0 + \frac{1}{2}\left(\frac{\partial u}{\partial x} + \frac{\partial v}{\partial y}\right)_0 \delta x + \frac{1}{2}\left(\frac{\partial u}{\partial x} - \frac{\partial v}{\partial y}\right)_0 \delta x$$

$$-\frac{1}{2}\left(\frac{\partial v}{\partial x} - \frac{\partial u}{\partial y}\right)_0 \delta y + \frac{1}{2}\left(\frac{\partial v}{\partial x} + \frac{\partial u}{\partial y}\right)_0 \delta y$$

$$v = v_0 + \frac{1}{2}\left(\frac{\partial u}{\partial x} + \frac{\partial v}{\partial y}\right) + \delta y - \frac{1}{2}\left(\frac{\partial u}{\partial x} - \frac{\partial v}{\partial y}\right)\delta y +$$

$$\frac{1}{2}\left(\frac{\partial v}{\partial x} - \frac{\partial u}{\partial y}\right)\delta x + \frac{1}{2}\left(\frac{\partial v}{\partial x} - \frac{\partial u}{\partial y}\right)\delta y$$

or

$$\left.\begin{aligned} u &= u_0 + \frac{1}{2} D\delta x - \frac{1}{2}\zeta\delta y + \frac{1}{2} F_1\delta x + \frac{1}{2} F_2\delta y \\ v &= v_0 + \frac{1}{2} D\delta y + \frac{1}{2}\zeta\delta x - \frac{1}{2} F_1\delta y + \frac{1}{2} F_2\delta x \end{aligned}\right\} \qquad \ldots\ldots(13.35)$$

where $D = \left(\frac{\partial u}{\partial x} + \frac{\partial v}{\partial y}\right)_0$ = Divergence in two dimension

$\zeta = \left(\frac{\partial v}{\partial x} - \frac{\partial u}{\partial y}\right)_0$ = vorticity

$$F_1 = \left(\frac{\partial u}{\partial x} - \frac{\partial v}{\partial y}\right)_0 \quad = \text{Deformation stretching}$$

$$F_2 = \left(\frac{\partial v}{\partial x} + \frac{\partial u}{\partial y}\right)_0 \quad = \text{Deformation shearing}$$

F_1, F_2 are components of deformation.

eq. (13.35) shows that wind motion is a combination of translation (u_0, v_0), divergence (D), Vorticity (ζ) and deformation (F_1, F_2)

13.10 Some Flow Patterns

At any instant wind field may be characterised by a family of stream-lines.

In the Fig. 13.11, L is a streamline, at the point p on the streamline tangent makes an angle θ with the x-axis and u, v are the components of velocity $\vec{V}$.

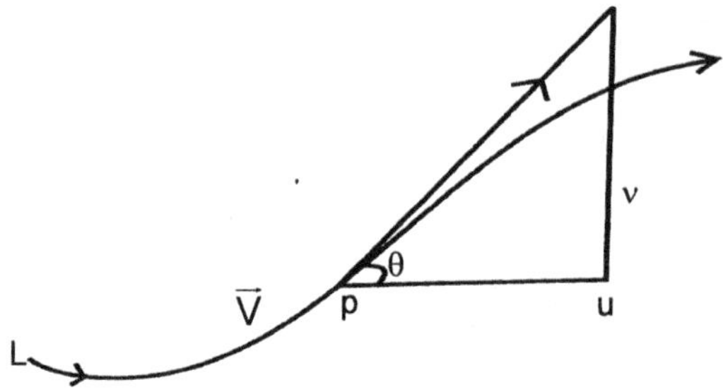

Fig. 13.11

$$\text{then} \quad \tan\theta = \frac{dy}{dx} = \frac{v}{u} \qquad .....(13.36)$$

$$\text{or} \quad \frac{dx}{u} = \frac{dy}{v} \quad \text{Equations of streamline at p.}$$

From the differential properties of wind field described above eq. (13.35)we have

$$\left.\begin{aligned} u &= u_0 + \frac{1}{2} D.\delta x - \frac{1}{2}\zeta\delta y + \frac{1}{2} F_1\delta x + \frac{1}{2} F_2\delta y \\ v &= v_0 + \frac{1}{2} D\delta y + \frac{1}{2}\zeta\delta x - \frac{1}{2} F_1\delta y + \frac{1}{2} F_2\delta x, \end{aligned}\right\} \qquad ....(13.35)$$

where $\quad \delta x = x - x_0$, $\delta y = y - y_0$

Case 1

Suppose D = constant and $\zeta = F_1 = F_2 = 0$, $u_0 = v_0 = 0$.

$\therefore \quad \delta x = x$, $\delta y = y$

eq (13.35) reduces to

$$\frac{dx}{dt} = u = \frac{1}{2} Dx \qquad .....(i)$$

$$\frac{dy}{dt} = v = \frac{1}{2} Dy \qquad .....(ii)$$

From (i) and (ii) $\frac{dy}{dx} = \frac{y}{x}$ or $\frac{dy}{y} = \frac{dx}{x}$ .....(13.37)

Integrating we have ln y – ln x = constant.

or $\frac{y}{x}$ = const. = c i.e. y = cx

Case 2

If ζ = const.

$$D = F_1 = F_2 = 0, u_0 = v_0 = 0 \text{ implies } \delta x = x, \delta y = y$$

eq. (13.35) reduces to

$$\frac{dx}{dt} = u = -\frac{1}{2} \zeta y \qquad (iii)$$

$$\frac{dy}{dt} = v = +\frac{1}{2} \zeta x \qquad (iv)$$

From (iii) and (iv) $\frac{dy}{dx} = -\frac{x}{y}$, or xdx + ydy = 0.

$x^2 + y^2 = C^2$ (const) i.e., streamlines are concentric circles with centre (0,0). If $\zeta > 0$, we have anti-clock wise rotation and $\zeta < 0$ clock wise rotation.

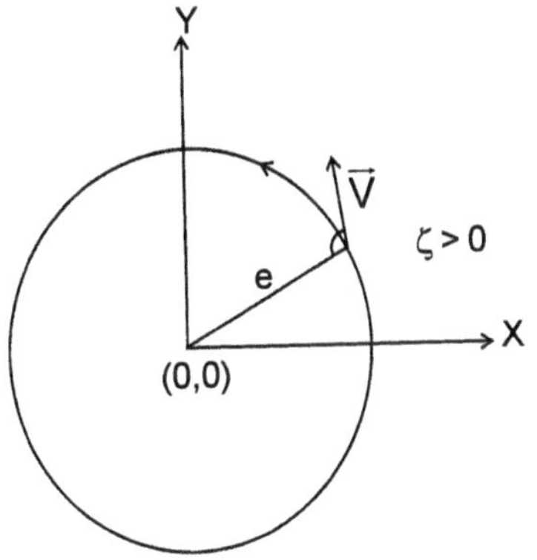

Fig. 13.12

Case 3

If F_1 = const. and

$$\zeta = D = F_2 = 0,\ u_0 = v_0 = 0 \text{ implies } \delta x = x,\ \delta y = y$$

eq. (13.35) reduces to

$$u = \frac{1}{2} F_1 x \text{ and } v = -\frac{1}{2} F_1 y$$

i.e.,

$$\frac{dx}{dt} = \frac{1}{2} F_1 x, \qquad \frac{dy}{dt} = -\frac{1}{2} F_1 y$$

$$\therefore \frac{dy}{dx} = \frac{-y}{x} \text{ or } \frac{dx}{x} + \frac{dy}{y} = 0.$$

Integrating we have ln x + ln y = const. or xy = const.

∴ streamlines are hyperbolas with coordinate axes as asymptotes [Fig. 13.13.]

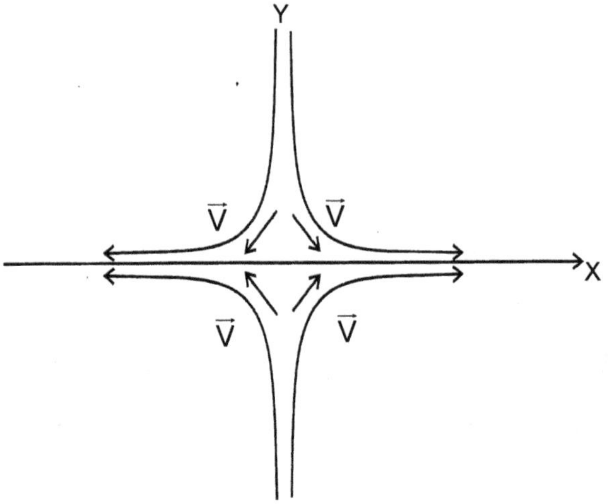

Fig. 13.13

Case 4

If F_2 = const and $D = \zeta = F_1 = 0$, $u_0 = v_0 = 0$, implies $\delta x = x$, $\delta y = y$ eq. (13.35) reduces to

$$u = \frac{dx}{dt} = \frac{1}{2} F_2 y, \quad v = \frac{dy}{dt} = \frac{1}{2} F_2 x$$

$$\therefore \frac{dy}{dx} = \frac{x}{y} \qquad \text{or } xdx - ydy = 0$$

Integrating we have $x^2 - y^2$ =const.

$\therefore$ streamlines are hyperbolas with asymptotes $x = y$ and $x = -y$ (Fig.13.14.) In case (3) and ease (4) streamlines are hyperbolas , and in these two cases we have deformation fields.

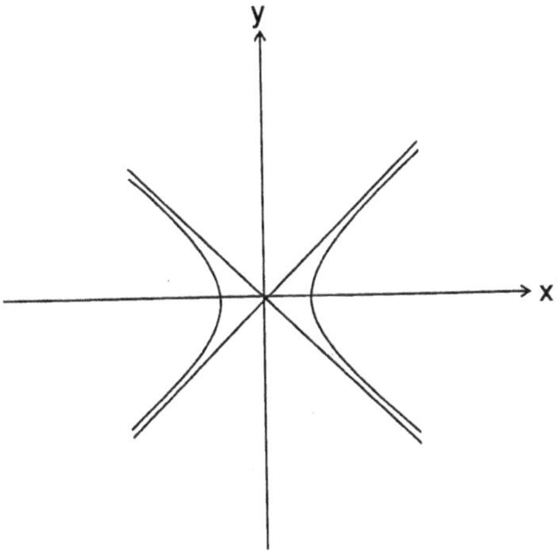

Fig. 13.14

Questions

1. Define Flow and circulation. Show that the circulation of an ascendent along any closed curve is zero.
2. State and prove Kelvin's theorem.
3. Prove Bjerknes circulation theorem.
4. Define solenoid and solenoidal. Show that $\oint \alpha \, dp = N(\alpha,p)$,

   where $N(\alpha,p)$ = no. of solenoids enclosed by the curve in α-p diagram.
5. From a given meteorological data of specific volume (α) and pressure (p) defining solenoidal vector $\vec{S}$, find the number of solenoids or show

   that $\oint \alpha \, dp = \int_A \vec{S}.\vec{dA}$

6. Using the concept of solenoids, show that the Bjerkne's circulation theorem can be written as $\frac{dC}{dt} = N(\alpha - p) - 2\Omega\frac{dA_E}{dt}$

Where $\vec{A}$ = area vector which is normal to the plane of physical curve,
A_E = Area of projection of A on equatorial plane

$\vec{\Omega}$ = Angular velocity of the earth

7. If $\vec{S} = \nabla\alpha \times (-\nabla p)$ is a solenoidal vector, show that $\vec{S} = \frac{C_p}{\theta}\nabla T \times \nabla\theta$

8. Using the concept of solenoids explain the phenomena of sea breeze and land breeze.

Vorticity

9. Define vorticity vector and vorticity. Write the differential equations of vortex lines.

10. State and prove Stokes theorem of circulation. Show that the vorticity of an elemental area is equal to twice the observed angular velocity of the solid element.

Divergence

11. Define divergence. Explain the concept of divergence and show that the net out flow per unit area is divergence. Hence derive Gauss's theorem.

12. (a) Considering an element area ($\delta A = \delta x \delta y$), show that divergence is equal to the change in the area of the fluid element per unit of time and per unit area.

(b) Define deformation stretching (F_1) and deformation shearing (F_2). Show that deformation component F_1 is equal to the stretching in x-direction.

13. Considering a neighbouring of a point (x_0, y_0) in a fluid, using Taylors expansion, write velocity field of u.v (components). Show that wind motion is a combination of translation, divergence, vorticity and deformation.

14. Define Streamlines. Write the differential equations of streamlines. Write the differential properties of wind field. Discuss the following cases.

Let $u_o = v_o = 0$ (initial velocity components) and

$\delta x = x - x_0$, $\delta y = y - y_0$

(i) D = const, and $\zeta = F_1 = F_2 = 0$

(ii) ζ = const, and $D = F_1 = F_2 = 0$

(iii) F_1 = const, and $\zeta = D = F_2 = 0$

(iv) F_2 = const, and $D = \zeta = F_1 = 0$

CHAPTER 14

The Vorticity Equation

The scale approximate equations of motions in local cartesian coordinates are given by

$$\frac{du}{dt} = -\frac{1}{\rho}\frac{\partial p}{\partial x} + fv + \frac{1}{\rho}F_x \qquad .....(14.1)$$

$$\frac{dv}{dt} = -\frac{1}{\rho}\frac{\partial p}{\partial y} - fu + \frac{1}{\rho}F_y \qquad .....(14.2)$$

$$\frac{dw}{dt} = 0 = -\frac{1}{\rho}\frac{\partial p}{\partial z} - g \qquad .....(14.3)$$

Where $f = 2\Omega \sin \phi$ = Coriolis parameter, ϕ latitude

Since

$$\frac{du}{dt} = \frac{\partial u}{\partial t} + \frac{\partial u}{\partial x}\frac{dx}{dt} + \frac{\partial u}{\partial y}\frac{dy}{dt} + \frac{\partial u}{\partial z}\frac{dz}{dt}$$

$$= \frac{\partial u}{\partial t} + u\frac{\partial u}{\partial x} + v\frac{\partial u}{\partial y} + w\frac{\partial u}{\partial z}$$

and

$$\frac{dv}{dt} = \frac{\partial v}{\partial t} + u\frac{\partial v}{\partial x} + v\frac{\partial v}{\partial y} + w\frac{\partial v}{\partial z}$$

For simplicity neglecting friction term and using above relation, the equations (14.1), (14.2) reduces to

$$\frac{\partial u}{\partial t}+u\frac{\partial u}{\partial x}+v\frac{\partial u}{\partial y}+w\frac{\partial u}{\partial z}-fv=-\alpha\frac{\partial p}{\partial x} \quad .....(14.4)$$

$$\frac{\partial v}{\partial t}+u\frac{\partial v}{\partial x}+v\frac{\partial v}{\partial y}+w\frac{\partial v}{\partial z}+fu=-\alpha\frac{\partial p}{\partial y} \quad .....(14.5)$$

Differentiating (14.4) w.r.t y and (14.5) w.r.t x we have

$$\frac{\partial^2 u}{\partial t\partial y}+\frac{\partial u}{\partial y}\frac{\partial u}{\partial x}+u\frac{\partial^2 u}{\partial x\partial y}+\frac{\partial v}{\partial y}\frac{\partial u}{\partial y}+v\frac{\partial^2 u}{\partial y^2}+\frac{\partial w}{\partial y}\frac{\partial u}{\partial z}+w\frac{\partial^2 u}{\partial y\partial z}-v\frac{\partial f}{\partial y}-f\frac{\partial v}{\partial y}$$

$$=-\alpha\frac{\partial^2 p}{\partial y\partial x}-\frac{\partial \alpha}{\partial y}\frac{\partial p}{\partial x} \quad .....(14.6)$$

$$\frac{\partial^2 v}{\partial t\partial x}+\frac{\partial u}{\partial x}\frac{\partial v}{\partial x}+u\frac{\partial^2 v}{\partial x^2}+\frac{\partial v}{\partial x}\frac{\partial v}{\partial y}+v\frac{\partial^2 v}{\partial y\partial x}+\frac{\partial w}{\partial x}\frac{\partial v}{\partial z}+w\frac{\partial^2 v}{\partial z\partial x}+u\frac{\partial f}{\partial x}+f\frac{\partial u}{\partial x}$$

$$=-\alpha\frac{\partial^2 p}{\partial y\partial x}-\frac{\partial \alpha}{\partial x}\frac{\partial p}{\partial y} \quad .....(14.7)$$

Subtracting (14.6) from (14.7) we have

$$\frac{\partial}{\partial t}\left(\frac{\partial v}{\partial x}-\frac{\partial u}{\partial y}\right)+\frac{\partial u}{\partial x}\left(\frac{\partial v}{\partial x}-\frac{\partial u}{\partial y}\right)+u\frac{\partial}{\partial x}\left(\frac{\partial v}{\partial x}-\frac{\partial u}{\partial y}\right)+\frac{\partial v}{\partial y}\left(\frac{\partial v}{\partial x}-\frac{\partial u}{\partial y}\right)$$

$$+v\left[\frac{\partial}{\partial y}\left(\frac{\partial v}{\partial x}-\frac{\partial u}{\partial y}\right)\right]+w\left[\frac{\partial}{\partial z}\left(\frac{\partial v}{\partial x}-\frac{\partial u}{\partial y}\right)\right]+v\frac{\partial f}{\partial y}+\frac{\partial w}{\partial x}\frac{\partial v}{\partial z}-\frac{\partial w}{\partial y}\frac{\partial u}{\partial z}+$$

$$f\left(\frac{\partial u}{\partial x}+\frac{\partial v}{\partial y}\right)=\frac{\partial \alpha}{\partial y}\frac{\partial p}{\partial x}-\frac{\partial \alpha}{\partial x}\frac{\partial p}{\partial y}$$

$$\left(\because \frac{\partial f}{\partial x}=0\right)$$

or $$\frac{\partial \zeta}{\partial t}+u\frac{\partial \zeta}{\partial x}+v\frac{\partial \zeta}{\partial y}+w\frac{\partial \zeta}{\partial z}+(\zeta+f)\left(\frac{\partial u}{\partial x}+\frac{\partial v}{\partial y}\right)+\left(\frac{\partial w}{\partial x}\frac{\partial v}{\partial z}-\frac{\partial w}{\partial y}\frac{\partial u}{\partial z}\right)$$

$$+v\frac{\partial f}{\partial y}=\left(\frac{\partial \alpha}{\partial y}\frac{\partial p}{\partial x}-\frac{\partial \alpha}{\partial x}\frac{\partial p}{\partial y}\right)$$

where $$\zeta = \frac{\partial v}{\partial x} - \frac{\partial u}{\partial y}$$

Since f depends on y and not on x, and

$$\frac{dy}{dt}.\frac{df}{dy} = \frac{df}{dt} \text{ i.e., } v\frac{df}{dy} = \frac{df}{dt}$$

and $$\frac{d\zeta}{dt} = \frac{\partial\zeta}{\partial t} + u\frac{\partial\zeta}{\partial x} = v\frac{\partial\zeta}{\partial y} + w\frac{\partial\zeta}{\partial z}$$

The above equation be comes

$$\frac{d\zeta}{dt} + \frac{df}{dt} + (\zeta + f)\left(\frac{\partial u}{\partial x} + \frac{\partial v}{\partial y}\right) + \left(\frac{\partial w}{dx}\frac{\partial v}{\partial z} - \frac{\partial w}{\partial y}\frac{\partial u}{\partial z}\right)$$

$$= \left(\frac{\partial p}{\partial x}\frac{\partial \alpha}{\partial y} - \frac{\partial p}{\partial y}\frac{\partial \alpha}{\partial x}\right)$$

or $$\frac{d}{dt}(\zeta + f) = -(\zeta + f)\left(\frac{\partial u}{\partial x} + \frac{\partial v}{\partial y}\right) - \left(\frac{\partial w}{\partial x}\frac{\partial v}{\partial z} - \frac{\partial w}{\partial y}\frac{\partial u}{\partial z}\right)$$

$$+ \left(\frac{\partial p}{\partial x}\frac{\partial \alpha}{\partial y} - \frac{\partial p}{\partial y}\frac{\partial \alpha}{\partial y}\right) \quad .....(14.8)$$

The eq. (14.8) is the vorticity equation

LHS of (14.8) $= \frac{d}{dt}(\zeta + f)$ = Rate of change of absolute vorticity

RHS of (14.8) $= (\zeta + f)\left(\frac{\partial u}{\partial x} + \frac{\partial v}{\partial y}\right)$ Divergence term or generation of vorticity by horizontal divergence,

$\left(\frac{\partial w}{\partial x}\frac{\partial v}{\partial z} - \frac{\partial w}{\partial y}\frac{\partial u}{\partial z}\right)$ tilting and twisting term or vertical velocity developed by tilting of horizontally oriented components of vorticity,

and $\left(\frac{\partial p}{\partial x}\frac{\partial \alpha}{\partial y} - \frac{\partial p}{\partial y}\frac{\partial \alpha}{\partial x}\right)$ Soliniodal term.

14.1 Vorticity Equation in Local Coordinate System with pressure as z-axis

We know from total differential,

$$\frac{du}{dt} = \frac{\partial u}{dt} + u\frac{\partial u}{\partial x} + v\frac{\partial u}{\partial y} + \omega\frac{\partial u}{\partial p}$$

$$\frac{dv}{dt} = \frac{\partial v}{\partial t} + u\frac{\partial v}{\partial x} + v\frac{\partial v}{\partial y} + \omega\frac{\partial v}{\partial p} \qquad .....(14.9)$$

and $\psi = gz$ and $\frac{\partial \psi}{\partial p} = -\alpha$ or $\delta\psi = -\alpha\delta p$ (above are exact differentials)

The scale approximated equations of motion (eq. 12.26) are

$$\frac{du}{dt} = -\alpha\frac{\partial p}{\partial x} + fv + \alpha F'_x \qquad \frac{dv}{dt} = -\alpha\frac{\partial p}{\partial y} - fu + \alpha F'_y$$

$$0 = -\alpha\frac{\partial p}{\partial z} - g \qquad .....(14.10)$$

Using eq. (14.9) we write eq. (14.10) as

$$\frac{\partial u}{\partial t} + u\frac{\partial v}{\partial x} + v\frac{\partial u}{\partial y} + \omega\frac{\partial u}{\partial p} + = -\frac{\partial \psi}{\partial x} + fv + F_x \qquad .....(14.11)$$

$$\frac{\partial v}{\partial t} + u\frac{\partial v}{\partial x} + v\frac{\partial v}{\partial y} + \omega\frac{\partial v}{\partial p} = -\frac{\partial \psi}{\partial y} - fu + F_y \qquad .....(14.12)$$

Differentiating (14.11) w.r.t y and (14.12) w.r.t x we have

$$\frac{\partial}{\partial t}\left(\frac{\partial u}{\partial y}\right) + \frac{\partial u}{\partial y}\frac{\partial u}{\partial x} + u\frac{\partial^2 u}{\partial x\partial y} + \frac{\partial v}{\partial y}\frac{\partial u}{\partial y} + v\frac{\partial^2 u}{\partial y^2} + \frac{\partial w}{\partial y}\frac{\partial u}{\partial p} + \omega\frac{\partial^2 u}{\partial p\partial y}$$

$$= -\frac{\partial^2 \psi}{\partial x\partial y} + v\frac{\partial f}{\partial y} + f\frac{\partial v}{\partial y} + \frac{\partial}{\partial y}F_x \qquad .....(14.13)$$

$$\frac{\partial}{\partial t}\left(\frac{\partial v}{\partial x}\right) + \frac{\partial u}{\partial x}\frac{\partial v}{\partial x} + u\frac{\partial^2 v}{\partial x^2} + \frac{\partial v}{\partial x}\frac{\partial v}{\partial y} + v\frac{\partial^2 v}{\partial y\partial x} + \frac{\partial \omega}{\partial x}\frac{\partial v}{\partial p} + \omega\frac{\partial^2 v}{\partial p\partial x}$$

$$= -\frac{\partial^2 \psi}{\partial y\partial x} - f\frac{\partial u}{\partial x} - u\frac{\partial f}{\partial x} + \frac{\partial}{\partial x}F_y \qquad .....(14.14)$$

Subtracting (14.13) from (14.14) we have

$$\frac{\partial}{\partial t}\left(\frac{\partial v}{\partial x}-\frac{\partial u}{\partial y}\right)+\frac{\partial u}{\partial x}\left(\frac{\partial v}{\partial x}-\frac{\partial u}{\partial y}\right)+u\left[\frac{\partial}{\partial x}\left(\frac{\partial v}{\partial x}-\frac{\partial u}{\partial y}\right)\right]+\frac{\partial v}{\partial y}\left(\frac{\partial v}{\partial x}-\frac{\partial u}{\partial y}\right)+$$

$$v\left[\frac{\partial}{\partial y}\left(\frac{\partial v}{\partial x}-\frac{\partial u}{\partial y}\right)\right]+\frac{\partial \omega}{\partial x}\frac{\partial v}{\partial p}-\frac{\partial \omega}{\partial y}\frac{\partial u}{\partial p}+\omega\left[\frac{\partial}{\partial p}\left(\frac{\partial v}{\partial x}-\frac{\partial u}{\partial y}\right)\right]$$

$$= -u\frac{\partial f}{\partial x}-v\frac{\partial f}{\partial y}-f\left(\frac{\partial u}{\partial x}+\frac{\partial v}{\partial y}\right)+\frac{\partial F_y}{\partial x}-\frac{\partial F_x}{\partial y} \qquad .....(14.15)$$

It may be noted that

$$\frac{\partial f}{\partial t}=0, \frac{\partial f}{\partial p}=0, \text{ and } \zeta D = \left(\frac{\partial v}{\partial x}-\frac{\partial u}{\partial y}\right)\left(\frac{\partial u}{\partial x}+\frac{\partial v}{\partial y}\right)$$

eq. (14.15) becomes

$$\frac{\partial}{\partial t}(\zeta+f)+u\frac{\partial(\zeta+f)}{\partial x}+v\frac{\partial(\zeta+f)}{\partial y}+\omega\frac{\partial(\zeta+f)}{\partial p}$$

$$= -(\zeta+f)D+\left(\frac{\partial u}{\partial p}\frac{\partial \omega}{\partial y}-\frac{\partial v}{\partial p}\frac{\partial \omega}{\partial x}\right)+\frac{\partial F_y}{\partial x}-\frac{\partial F_x}{\partial y}$$

putting $\zeta + f = \eta$ we have

$$\frac{\partial \eta}{\partial t}+u\frac{\partial \eta}{\partial x}+v\frac{\partial \eta}{\partial y}+\omega\frac{\partial \eta}{\partial p} = -\eta D +$$

$$\left(\frac{\partial u}{\partial p}\frac{\partial \omega}{\partial y}-\frac{\partial v}{\partial p}\frac{\partial \omega}{\partial x}\right)+\left(\frac{\partial F_y}{\partial x}-\frac{\partial F_x}{\partial y}\right) \qquad .....(14.16)$$

we can write

$$\frac{\partial u}{\partial p}\frac{\partial \omega}{\partial y}-\frac{\partial v}{\partial p}\frac{\partial \omega}{\partial x} = k\,.\left(\frac{\partial \vec{v}}{\partial p}\times\nabla\omega\right)$$

$$= k.\begin{vmatrix} i & j & k \\ \frac{\partial u}{\partial p} & \frac{\partial v}{\partial p} & \frac{\partial \omega}{\partial p} \\ \frac{\partial \omega}{\partial x} & \frac{\partial \omega}{\partial y} & \frac{\partial \omega}{\partial z} \end{vmatrix} \qquad .....(14.17)$$

and $$\frac{\partial F_y}{\partial x}-\frac{\partial F_x}{\partial y} = k.(\nabla \times \vec{F})$$

$$= k. \begin{vmatrix} i & j & k \\ \frac{\partial}{\partial x} & \frac{\partial}{\partial y} & \frac{\partial}{\partial z} \\ F_x & F_y & F_z \end{vmatrix} \qquad .....(14.18)$$

using (14.17) and (14.18) eq. (14.16) becomes

$$\frac{d\eta}{dt} = -\eta D + k.\left(\frac{\partial \vec{v}}{\partial p} \times \nabla\omega\right) + k.(\nabla \times \vec{F}) \qquad .....(14.19)$$

Equation (14.19) is called vorticity equation and the quantity $\eta = \zeta + f$ is called absolute vorticity, (since $\vec{V}_a = \vec{V}_r + \vec{V}_E$ where $\vec{V}_r$ = relative velocity, of the earth). It is the vorticity of absolute wind.

where

$\vec{V}_E$ = Earth's velocity

$\vec{V}_a$ = Absolute wind velocity

$-\eta D$ = Divergence term

ζ = Vertical component of vorticity of $\vec{V}_r$

$\left(\frac{\partial v}{\partial p} \times \nabla\omega\right)$ = Twisting term

$k.(\nabla \times \vec{F})$ = Friction term.

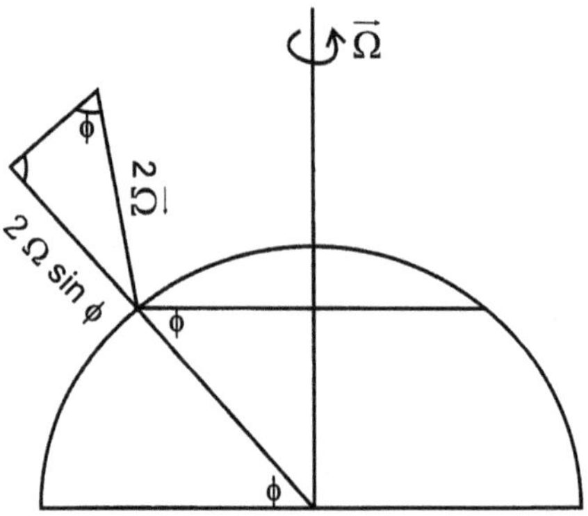

Fig. 14.1

Note : In general when $D > 0$ (divergence) there is tendency to develop anticyclonic vorticity but when $D < 0$ (convergence) it will develop cyclonic vorticity.

Questions

1. Derive vorticity equation in local Cartesian coordinates, hence show that the rate of change of absolute vorticity is equal to the sum of divergence term, tilt and twisting term and solenoidal term.
2. Derive the vorticity equation in local coordinate system with pressure as z-axis.

CHAPTER 15

The Divergence Equation

The equations of motion in local cartesian coordinates with pressure as z-axis are given

$$\frac{\partial u}{\partial t}+u\frac{\partial u}{\partial x}+v\frac{\partial u}{\partial y}+\omega\frac{\partial u}{\partial z} = -\frac{\partial \psi}{\partial x} + fv + F_x \qquad .....(15.1)$$

$$\frac{\partial v}{\partial t}+u\frac{\partial v}{\partial x}+v\frac{\partial v}{\partial y}+\omega\frac{\partial v}{\partial p} = -\frac{\partial \psi}{\partial y} - fu + F_y \qquad .....(15.2)$$

Differentiating (15.1) w.r.t to x and (15.2) w.r.t y- we have

$$\frac{\partial}{\partial t}\left(\frac{\partial u}{\partial x}\right)+\left(\frac{\partial u}{\partial x}\right)^2+u\frac{\partial^2 u}{\partial x^2}+\frac{\partial v}{\partial x}\frac{\partial u}{\partial y}+v\frac{\partial^2 u}{\partial y\partial x}+\frac{\partial \omega}{\partial x}\frac{\partial u}{\partial p}+\omega\frac{\partial^2 u}{\partial p\partial x}$$

$$= -\frac{\partial^2 \psi}{\partial x^2}+f\frac{\partial v}{\partial x}+v\frac{\partial f}{\partial x}+\frac{\partial F_x}{\partial x} \qquad .....(15.3)$$

$$\frac{\partial}{\partial t}\left(\frac{\partial v}{\partial y}\right)+\frac{\partial u}{\partial y}\frac{\partial v}{\partial x}+u\frac{\partial^2 v}{\partial x\partial y}+\left(\frac{\partial v}{\partial y}\right)^2+v\frac{\partial^2 v}{\partial y^2}+\frac{\partial \omega}{\partial y}\frac{\partial v}{\partial p}+\omega\frac{\partial^2 v}{\partial p\partial y}$$

$$= -\frac{\partial^2 \psi}{\partial y^2}-u\frac{\partial f}{\partial y}-f\frac{\partial u}{\partial y}+\frac{\partial F_y}{\partial y} \qquad .....(15.4)$$

Adding these equations (15.3) and (15.4) we have

$$\frac{\partial}{\partial t}\left(\frac{\partial u}{\partial x}+\frac{\partial v}{\partial y}\right)+u\frac{\partial}{\partial x}\left(\frac{\partial u}{\partial x}+\frac{\partial v}{\partial y}\right)+v\frac{\partial}{\partial y}\left(\frac{\partial u}{\partial x}+\frac{\partial v}{\partial y}\right)+\omega\frac{\partial}{\partial p}\left(\frac{\partial u}{\partial x}+\frac{\partial v}{\partial y}\right)$$

$$+\left(\frac{\partial u}{\partial x}\right)^2+2\frac{\partial u}{\partial y}.\frac{\partial v}{\partial x}+\left(\frac{\partial v}{\partial y}\right)^2+\frac{\partial \omega}{\partial x}.\frac{\partial u}{\partial p}+\frac{\partial \omega}{\partial y}.\frac{\partial v}{\partial p}\ + = -\frac{\partial^2\Psi}{\partial x^2}-\frac{\partial^2\Psi}{\partial y^2}+$$

$$f\left(\frac{\partial v}{\partial x}-\frac{\partial u}{\partial y}\right)+v\left(\frac{\partial f}{\partial x}\right)-u\frac{\partial f}{\partial y}+\left(\frac{\partial F_x}{\partial x}+\frac{\partial F_y}{\partial y}\right)$$

or

$$\frac{\partial D}{\partial t}+u\frac{\partial D}{\partial x}+v\frac{\partial D}{\partial y}+\omega\frac{\partial D}{\partial p}+\left(\frac{\partial u}{\partial x}\right)^2+2\frac{\partial u}{\partial y}\frac{\partial v}{\partial x}+\left(\frac{\partial v}{\partial y}\right)^2+$$

$$\frac{\partial \omega}{\partial x}\frac{\partial u}{\partial p}+\frac{\partial \omega}{\partial y}\frac{\partial v}{\partial p} = -\nabla^2\Psi + f\zeta + v\frac{\partial f}{\partial y} - u\frac{\partial f}{\partial y} + \nabla.\vec{F} \quad .....(15.5)$$

we know total differential

$$\frac{dD}{Dt}=\frac{\partial D}{\partial t}+u\frac{\partial D}{\partial x}+v\frac{\partial D}{\partial y}+\omega\frac{\partial D}{\partial p}$$

[Also Jacobian $J = \dfrac{\partial(x,y,z)}{\partial(a,b,c)} = \begin{vmatrix} \frac{\partial x}{\partial a} & \frac{\partial y}{\partial a} & \frac{\partial z}{\partial a} \\ \frac{\partial x}{\partial b} & \frac{\partial y}{\partial b} & \frac{\partial z}{\partial b} \\ \frac{\partial x}{\partial c} & \frac{\partial y}{\partial c} & \frac{\partial z}{\partial c} \end{vmatrix}$

where x, y, z are functions of a,b,c

and $\dfrac{dJ}{dt}=\dfrac{\partial(u,y,z)}{\partial(a,b,c)}+\dfrac{\partial(x,v,z)}{\partial(a,b,c)}+\dfrac{\partial(x,y,\omega)}{\partial(a,b,c)}$

where $u = \dfrac{dx}{dt}$ etc.]

$$\therefore\ J\,(v,u) = \frac{\partial(v,u)}{\partial(x,y)} = \begin{vmatrix} \frac{\partial v}{\partial x} & \frac{\partial u}{\partial x} \\ \frac{\partial v}{\partial y} & \frac{\partial u}{\partial y} \end{vmatrix} = \frac{\partial v}{\partial x}\frac{\partial u}{\partial y}-\frac{\partial v}{\partial y}\frac{\partial u}{\partial x}$$

With this notation the expression

$$\left(\frac{\partial u}{\partial x}\right)^2 + 2\frac{\partial u}{\partial y}\frac{\partial v}{\partial x} + \left(\frac{\partial v}{\partial y}\right)^2 = \left(\frac{\partial u}{\partial x} + \frac{\partial v}{\partial y}\right)^2 + 2\left[\frac{\partial u}{\partial y}\frac{\partial v}{\partial x} - \frac{\partial u}{\partial x}\frac{\partial v}{\partial y}\right]$$

$$= D^2 + 2J\,(v,u) \qquad \text{.....(15.6)}$$

and

$$\frac{\partial \omega}{\partial x}\frac{\partial u}{\partial p} + \frac{\partial \omega}{\partial y}\frac{\partial v}{\partial p} = \left(i\frac{\partial \omega}{\partial x} + j\frac{\partial \omega}{\partial y}\right)\cdot\left(i\frac{\partial u}{\partial p} + j\frac{\partial v}{\partial p}\right)$$

$$= \nabla\,\omega . \frac{\partial \vec{v}}{\partial p} \qquad \text{[in horizontal motion]} \qquad \text{.....(15.7)}$$

substituting (15.6) and (15.7) in the equation (15.5) we have

$$\frac{dD}{dt} + D^2 + 2J\,(v,u) + \nabla\omega . \frac{\partial \vec{v}}{\partial p}$$

$$= -\nabla^2 \psi + f\zeta + v\frac{\partial f}{\partial x} - u\frac{\partial f}{\partial y} + \nabla \vec{F} \qquad \text{.....(15.8)}$$

Equation (15.8) is the divergence equation.

On scale analysis, neglecting second and higher order terms, and friction term the above equation further reduces to

$$2J(v,\,u) = -\nabla^2 \psi + f\,\zeta + v\frac{\partial f}{\partial x} - u\frac{\partial f}{\partial y} \qquad \text{.....(15.9)}$$

eq. (15.9) is independent of time derivatives.

Question

1. Derive the divergence equation in local cartesian coordinates with pressure as z-axis.

CHAPTER 16

Balanced Motion

16.1 Definition

A motion is said to be balanced if the sum of the forces acting on the body is zero, or the sum of the forces acting on the body has no component along the velocity. In such a situation the K.E of the individual particle of the air remains constant. In atmospheric motions, the pressure and velocity distributions of weather disturbances are governed by empirical balance of forces. To get this empirical relations we shall consider steady flows (that are independent of time) and vertical components of forces are negligible (that is, the flow is horizontal).

In meteorological applications the fictitious forces such as coriolis force, centrifugal force are treated as real forces. Coriolis force as we know is not a force caused by some real field of forces, but it results from the rotating frame of reference.

Let us consider a balanced motion of a rotating cylindrical vessel having water and rotating about its vertical axis as shown in the Fig.16.1 (like kitchen grinder). Assume that the vessel has water of constant density ρ. Let the angular velocity of the rotating cylinder be $\vec{\Omega}$. Each particle of the fluid describes a circular motion with the axis of rotation as centre of circle. The balancing forces acting on the particle are the horizontal pressure force $\vec{P}$ and the centrifugal force $\vec{G}$. The balance can be utilised to find the shape of the free surface of rotating water.

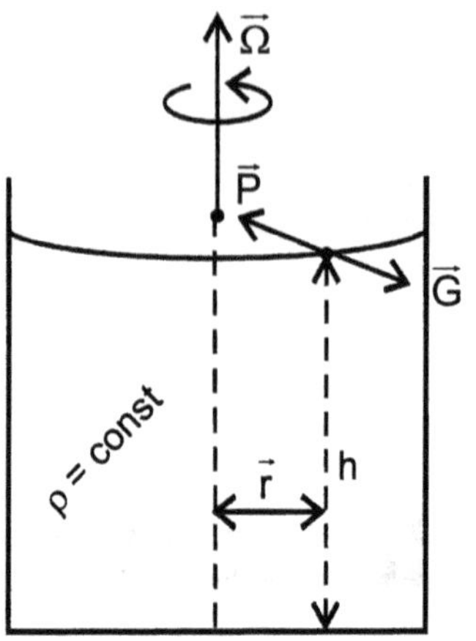

Fig. 16.1 Balanced motion of a rotating cylindrical vessel, with water.

We have

$$-\frac{1}{\rho}\frac{dp}{dt}-\vec{\Omega}\times\left(\vec{\Omega}\times\vec{r}\right)=0$$

or $$-\alpha\frac{dp}{dr}=\vec{\Omega}\times\left(\vec{\Omega}\times\vec{r}\right) \quad .....(16.1)$$

At the surface of the water p = 0.

From the hydrostatic equation we have

$$[p - p_0 = g\rho\,(h - z)]$$

$$p = g\rho(h - z) \quad .....(16.2)$$

where h (is a function of radius)=h(r), is the depth of the water in the vessel.

Differentiating eq. (16.2) w.r.t r we have

$$\alpha\,\frac{dp}{dr}=g\frac{dh}{dr} \quad .....(16.3)$$

since $\frac{dz}{dr} = 0$, z being constant

substituting (16.3) in (16.1) we have

$$-g\,\frac{dh}{dr} = -\,\Omega^2\,r$$

or $$\frac{dh}{dr} = \frac{\Omega^2 r}{g}$$

Integrating we have

$$\int_{h_0}^{h} \frac{dh}{dr} dr = \frac{1}{g}\int_{0}^{r} \Omega^2 r dr$$

$$h - h_0 = \frac{\Omega^2}{2g} r^2$$

or $$h = h_0 + \frac{\Omega^2}{2g} r^2$$

This equation represents a parabola.

∴ The free surface of water in the vessel will have the shape of a parabola, as shown in Fig. 16.1.

Ex : A liquid is kept in a cylindrical vessel and rotated about its axis. The liquid rises at the sides. If the radius of the vessel is 5 cm and the speed of the rotation is 2 rps, find the difference in the height of the liquid at the centre of vessel and its sides.

Sol:

We know

$$h - h_0 = \frac{\Omega^2 r^2}{2g}$$

given $$r = \frac{5}{100} \text{ m } \Omega = 2 \times 2\pi \text{ radions, } g = 9.8 \text{ ms}^{-2}$$

∴ $$h - h_0 = \frac{(4\pi)^2 \times (0.05)^2}{2 \times 9.8} \simeq 0.022 \text{ m}$$

Ex : An oil drop falls through air with a terminal velocity $v = 5 \times 10^{-4}\, ms^{-1}$. Calculate the radius of the drop and the terminal velocity of a drops of half of this radius. η = viscosity of air = 1.8×10^{-5} N m^{-2}s, density of oil ρ= 900 kg m^{-3}. Neglect the density of air compared to that of oil.

Sol :

(i) we know

$$v = \frac{2}{9}\frac{(\rho - \sigma) r^2 g}{\eta} \simeq \frac{2}{9}\frac{\rho r^2 g}{\eta} \qquad (\because \sigma \ll \rho)$$

∴ $$r = \sqrt{\frac{9\eta v}{2\rho g}} = \sqrt{\frac{9 \times 1.8 \times 10^{-5} \times 5 \times 10^{-4}}{2 \times 900 \times 9.8}}$$

$$\simeq 2.14 \times 10^{-6} \text{m} = 2.14\ \mu$$

(ii) we know $\frac{v_1}{v_2} = \frac{r_1^2}{r_2^2}$, given $r_2 = \frac{r_1}{2}$, $v_1 = 5 \times 10^{-4}\,ms^{-1}$

$$\therefore \quad v_2 = v_1 \times \left(\frac{r_2}{r_1}\right)^2 = 5 \times 10^{-4} \left(\frac{\frac{r_1}{2}}{r_1}\right)^2 \simeq 1.24 \times 10^{-4}\ ms^{-1}$$

Question

1. Define balanced motion. Show that the balanced motion of a rotating cylinder having water rotating about its vertical axis causes liquid rise at the sides and that the surface of the water in the cylinder assumes the shape of a parabola.

CHAPTER 17

Natural Coordinates and Equations of Motion

In this system of coordinates one axis is taken in the tangential direction of streamline i.e., oriented parallel to the direction of flow (s),second coordinate is taken perpendicular to the flow (n) or normal to the streamline and the third coordinate (z) is taken perpendicular to the plane containing the first two coordinates. The unit vectors in these directions are denoted by $\hat{t}$, $\hat{n}$ and k respectively. Let P, Q be the positions of the particle at time t and t + dt, as shown in the Fig. 17.1.

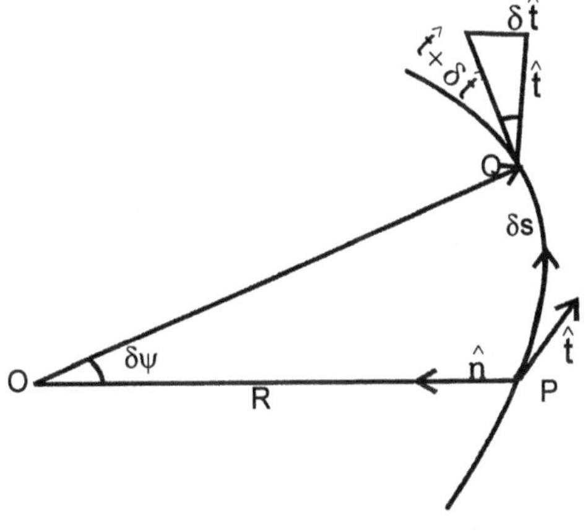

Fig. 17.1

The horizontal velocity $\vec{V} = V\hat{t}$ .....(17.1)

where V = magnitude of velocity $|\vec{V}| = \frac{ds}{dt}$ = speed, and $|\hat{t}| = 1$

Differentiating (17.1) w.r.t (time)

$$\frac{d\vec{V}}{dt} = \hat{t}\frac{dv}{dt} + v\frac{d\hat{t}}{dt} \quad .....(17.2)$$

The derivative of unit vector $\hat{t}$ w.r.t time t = $\left(\frac{d\hat{t}}{dt}\right)$ can be found from the Fig. 17.1 as follows.

$$\delta\psi = \frac{\delta s}{R} \qquad \left(\frac{\text{arc}}{\text{radius}} = \text{radian}\right)$$

where R = radius of curvature, taken positive when $\hat{n}$ is positive (cyclonic curvature)

From the congruent triangles.

$$\frac{\delta s}{R} = \frac{|\delta\hat{t}|}{|\hat{t}|} = |\delta\hat{t}| \qquad \because |\hat{t}| = 1$$

Again from Fig. 17.1 as $\delta s \to 0$

$$\frac{d\hat{t}}{ds} = \frac{\hat{n}}{R} \qquad \text{Since } \delta\hat{t} \text{ is parallel to } \hat{n} \quad .....(17.3)$$

Thus

$$\frac{d\hat{t}}{dt} = \frac{d\hat{t}}{ds}\cdot\frac{ds}{dt} = V\frac{\hat{n}}{R} \quad .....(17.4)$$

[since $\frac{ds}{dt}$ = v and $\frac{d\hat{t}}{ds} = \frac{\hat{n}}{R}$ from (17.3)]

Substituting (17.4) in (17.2), we have

$$\frac{d\vec{V}}{dt} = \hat{t}\frac{dv}{dt} + V\frac{\hat{n}}{R}V$$

$$\frac{d\vec{V}}{dt} = \hat{t}\frac{dv}{dt} + \hat{n}\frac{V^2}{R} \quad .....(17.5)$$

Equation (17.5) expresses that acceleration is equal to rate of change of wind speed plus centripetal acceleration created by the curvature of the trajectory.

Since coriolis force acts perpendicular to the direction of motion, we have

$$f k \times \vec{V} = fV \hat{n} \qquad .....(17.6)$$

[Note : The horizontal momentum equatins (on scale approximation) are

$$\frac{du}{dt} - fv = -\alpha\frac{\partial p}{\partial x} \text{ and } \frac{dv}{dt} + fu = -\alpha\frac{\partial p}{\partial y}$$

In vector form $\frac{d\vec{V}_h}{dt} - f\vec{V}_h \times k = -\alpha\ \nabla_{hp} = i\ \frac{\partial}{\partial x} + j\ \frac{\partial}{\partial y}$

Where $\vec{V_h} = iu + jv$, $\nabla_h = i\ \frac{\partial}{\partial x} + j\ \frac{\partial}{\partial y}$

Here horizontal velocity $\vec{V_h}$ is denoted as $\vec{V}$.]

∴ The horizontal equations of motion in natural coordinates may be written as

$$\frac{dv}{dt} = -\alpha\frac{\partial p}{\partial s} \qquad .....(17.7)$$

and $$\frac{V^2}{R} + fV = -\alpha\frac{\partial p}{\partial n} \qquad .....(17.8)$$

eq. (17.7) and (17.8) expresses balance of forces in parallel to the flow and perpendicular to it (flow).

For motion parallel to the isobars $\frac{\partial p}{\partial s} = 0$, and the speed is constant following the motion ($\because \frac{dv}{dt} = 0$)

17.1 Baric and Antibaric Flow

Horizontal equations of motion in natural coordinates are given by

$$\frac{dv}{dt} = -\alpha\frac{\partial p}{\partial s} \text{ and } \frac{v^2}{R} + fv = -\alpha\frac{\partial p}{\partial n}$$

or $$K_H v^2 + fv = -\alpha \frac{\partial p}{\partial n} \qquad .....(17.9)$$

as shown in Fig 17.2

where s = tangential direction of streamline

or s = direction of flow

n = perpendicular to streamline.

R = Radius of curvature. $K_H = \dfrac{1}{R}$

b = pressure (gradient) force

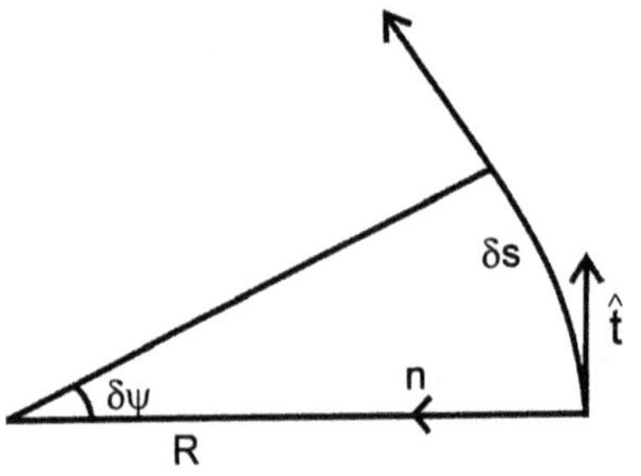

Fig. 17.2

Let N be the coordinate in the direction of horizontal pressure gradient. Let θ = Cross isobar wind direction as shown in following figure, + ve towards the lower pressure.

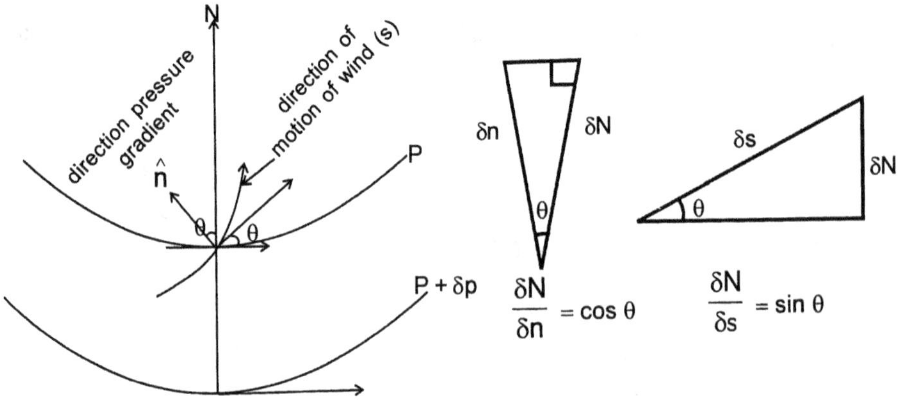

$$\frac{\partial p}{\partial s} = \frac{\partial p}{\partial N} \cdot \frac{\partial N}{\partial s} = \frac{\partial p}{\partial N} \text{ Sin } \theta$$

$$\frac{\partial p}{\partial n} = \frac{\partial p}{\partial N} \cdot \frac{\partial N}{\partial n} = \frac{\partial p}{\partial N} \text{ Cos } \theta \quad .....(17.10)$$

The components horizonal pressure gradient force in natural coordinates are given by

$$bs = -\alpha \frac{\partial p}{\partial s} = -\alpha \frac{\partial p}{\partial N} \sin \theta \qquad \text{[using (17.10)]}$$

$$bn = -\alpha \frac{\partial p}{\partial n} = -\alpha \frac{\partial p}{\partial N} \cos \theta \qquad \text{[using (17.10)]} \quad .....(17.11)$$

The horizontal coriolis force $-f\,v\,\hat{n}$ has only one component

$$c_n = -f\,v \quad .....(17.12)$$

$$\therefore \quad \frac{dv}{dt} = -\alpha \frac{\partial p}{\partial s} = -\alpha \frac{\partial p}{\partial N} \sin \theta = bs \quad \text{[from 17.11]} \quad ......(17.13)$$

$$k_H v^2 = -fv - \alpha \frac{\partial p}{\partial n}$$

$$= -f\,v - \alpha \frac{\partial p}{\partial N} \cos \theta = c_n + b_n$$

[from (17.11) and (17.12)] ....(17.14)

Equation (17.14) is the basis for the classification depending on the horizontal pressure gradient (b) and coriolis force (c_n).

Baric flow: If the normal pressure force (b) is in opposite direction to the coriolis force (c_n) the flow is termed baric [Fig. 17.3 (a).]

Antibaric flow: If the normal pressure force is along the coriolis force, the flow is termed anti-baric Fig. 17.3 (b).

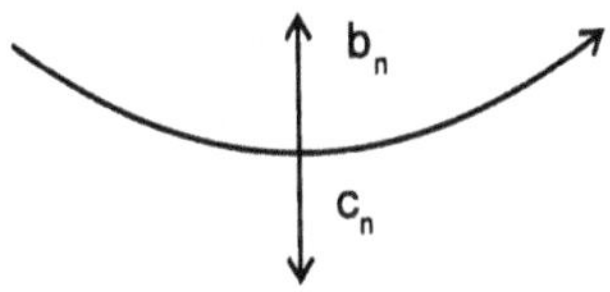

Baric flow

Fig. 17.3 (a)

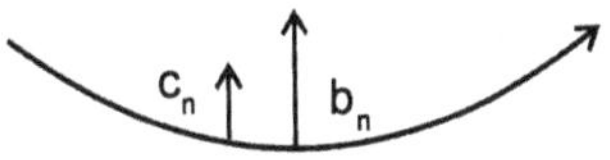

Anti-baric flow

Fig. 17.3 (b)

For baric flow θ lies between $-\frac{\pi}{2}$ to $\frac{\pi}{2}$ or $-\frac{\pi}{2} \leq \theta \leq \frac{\pi}{2}$

For anti baric flow $|\theta| > \frac{\pi}{2}$

Note: Large-scale atmospheric flow is nearly baric.

Another classification of wind flow is defined as follows.

Cyclonic flow: If the centripetal acceleration ($K_H V^2$) is opposite to the coriolis force it is called cyclonic flow. (Fig. 17.3(c), 17.3(f))

Anti cyclonic flow: If the centripetal acceleration ($K_H V^2$) is directed along the coriolis force, then it is called anti cyclonic flow. (Fig. 17.3 (d), 17.3 (g)).

The above two types of flows in northern hemisphere and southern hemisphere are given below Fig. 17.3(c), 17.3(d), 17.3(e) for northern hemisphere and Fig. 17.3 (f) Fig. 17.3 (g) and Fig 17.3 (h) for southern hemisphere.

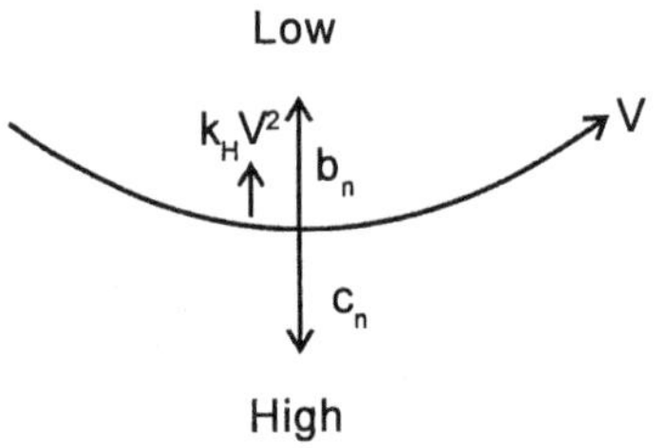

Fig. 17.3 (c) Baric cyclonic flow (in Northern hemisphere).

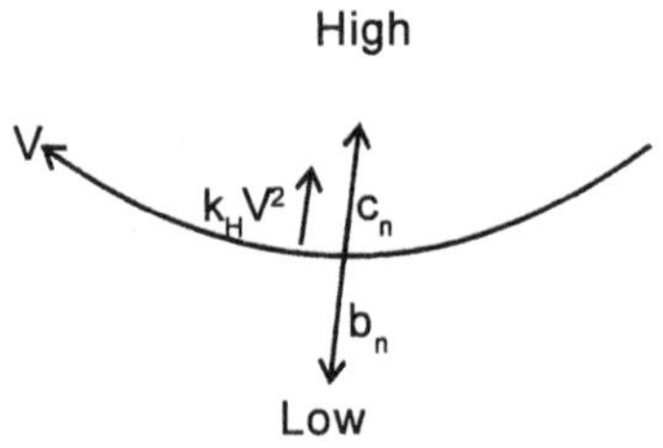

Fig. 17.3 (d) Baric anti cyclonic flow (in Northern hemisphere).

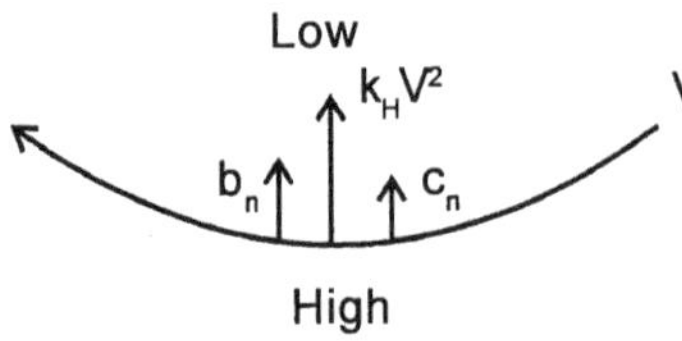

Fig. 17.3 (e) Antibaric anticyclonic flow. (in Northern hemisphere).

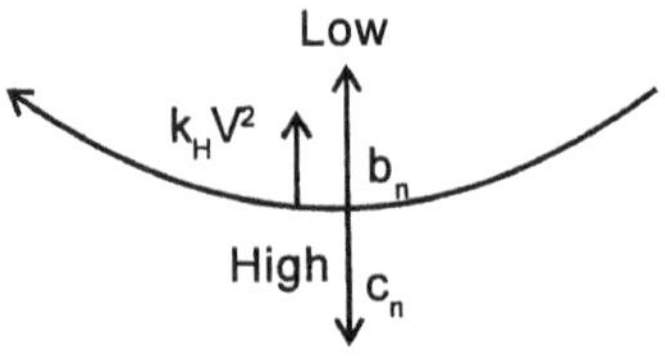

Fig. 17.3 (f) Baric cyclonic flow. (in Southern hemisphere).

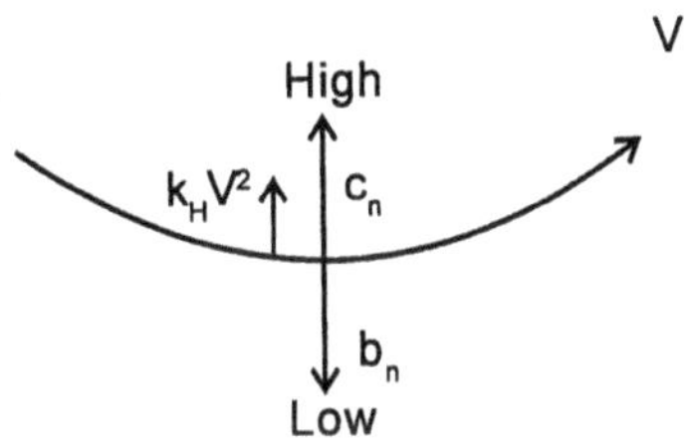

Fig. 17.3 (g) Baric anticyclonic flow (in Southern hemisphere).

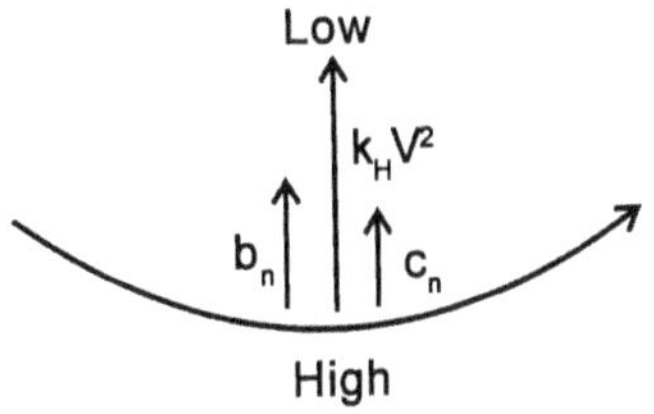

Fig. 17.3 (h) Antibaric anticyclonic flow. (in Southern hemisphere).

17.2 Vorticity in Natural Coordinates

Let AD = BC = δn and < AOB = $\delta\psi$ as shown in the Fig. 17.4.

Let OA = R_s = Radius of curvature of the streamline.

Let V be the velocity along the streamline AB, and $-\left(V + \frac{\partial V}{\partial n}\delta n\right)$ be the velocity along CD, while the velocity along AD and BC is zero.

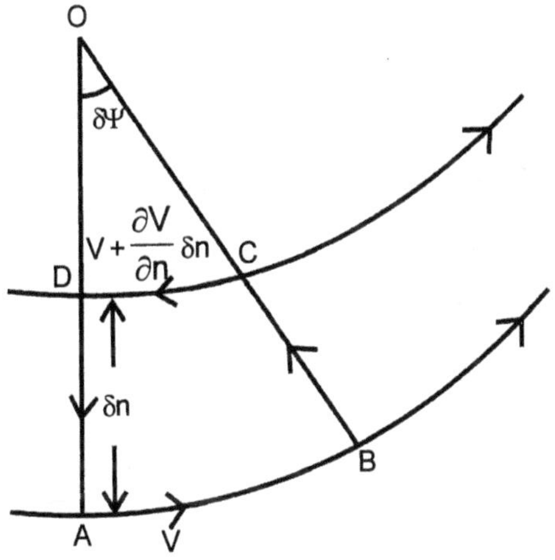

Fig. 17.4 Circulation for an infinitesimal loop in natural coordinates.

Infinitesimal circulation

$\delta C =$ Flow along AB + Flow along BC + Flow along CD + Flow along DA

[We know, flow = Average velocity × length of the arc,

length of arc = radius × radian]

$$\delta C = V R_s \delta\Psi - \left(V + \frac{\partial V}{\partial n}\delta n\right)(R_s - \delta_n)\,\delta\psi$$

$$= VR_s\delta\Psi - (VR_s\delta\Psi - V\delta n\delta\Psi + \frac{\partial V}{\partial n}\,\delta n R_s\delta\Psi - \frac{\partial V}{\partial n}\,\delta n^2\,\delta\Psi)$$

$$\delta C = V\delta n\delta\Psi - R_s\frac{\partial V}{\partial n}\,\delta n\delta\psi + \frac{\partial V}{\partial n}\,\delta n^2\,\delta\psi$$

Infinitesimal area $\delta A \simeq R_s\delta\psi\delta n$

we know vorticity $\zeta = \dfrac{\delta C}{\delta A}$

$$\therefore \qquad \zeta = \frac{\delta C}{\delta A} = \frac{V\delta n\delta\Psi - R_s\dfrac{\partial V}{\partial n}\delta n\delta\Psi + \dfrac{\partial V}{\partial n}\delta n^2\delta\Psi}{R_s\delta\Psi\delta n}$$

$$= \frac{V}{R_s} - \frac{\partial V}{\partial n} + \frac{\partial V}{\partial n}\frac{\partial n}{R_s}$$

In the limit $\zeta = \dfrac{V}{R_s} - \dfrac{\partial V}{\partial n}$

[∵ second order infintesional $\dfrac{\partial V}{\partial n}\cdot\dfrac{\delta n}{R_s}$ is negligible]

It follows from this equation that vorticity as a function of curvature $\left(\dfrac{V}{R_s}\right)$ and shear $\dfrac{\partial V}{\partial n}$.

Questions

1. What are the natural coordinates, Show that acceleration is equal to rate of change of wind speed plus centripetal acceleration.

   Show that the horizontal equations of motion in natural coordinates are expressed as

$$\frac{dV}{dt} = -\alpha \frac{\partial p}{\partial s} \text{ and } \frac{V^2}{R} + fV = -\alpha \frac{\partial p}{\partial n}$$

   Where R is radius of curvature.

Baric and Antibaric Flow

2. Define natural coordinates. Derive components of pressure gradient force (b) in natural coordinates, horizontal coriolis force (c_n). Hence show that $bs = \frac{dV}{dt}$ and $K_H V^2 = c_n + b_n$.

3. Based on c_n, b_n and $K_H V^2$, define diagrammatically baric flow, antibaric flow, cyclonic flow, anticyclonic flow, baric cyclonic, baric anticyclonic flow and antibaric anticyclonic flow both in northern and southern hemispheres.

Vorticity in Natural Coordinates

4. Derive vorticity equation in natural coordinates and show that ζ is a function of curvature $\left(\frac{V}{R_s}\right)$ and shear $\left(\frac{\partial V}{\partial n}\right)$.

CHAPTER 18

Geostrophic Wind

18.1 Definition

The horizontal wind occurring in a frictionless flow when the pressure gradient force is in balance with the coriolis force is called geostrophic wind. That is, the wind that occurs in atmosphere which is hydrostatic and horizontal in the absence of acceleration and friction.

The general equations of motion is given by

$$\frac{d\vec{V}}{dt} = -\alpha\nabla p - 2\vec{\Omega} \times \vec{V} + \vec{g} + \frac{d\vec{\tau}}{dz}$$

In case of geostrophic balance

$$\alpha\nabla p = -2\vec{\Omega} \times \vec{V_g} \qquad .....(18.1)$$

where $\vec{V_g} = iu_g + jv_g$ geostrophic wind vector, u_g, v_g are geostrophic wind components in x, y directions.

$f = 2\Omega \sin\phi$ = coriolis parameter, ϕ = latitude

or

$$\alpha\frac{\partial p}{\partial x} = fv_g \qquad .....(18.2)$$

$$\alpha\frac{\partial p}{\partial y} = -fu_g \qquad .....(18.3)$$

Geostrophic wind as a finite difference

$$V_g = \frac{\alpha}{f}\frac{\delta p}{\delta n} \quad \text{or} \quad \frac{\alpha}{f}\frac{\Delta p}{\Delta n} \qquad \text{[as in Fig 18.1]}$$

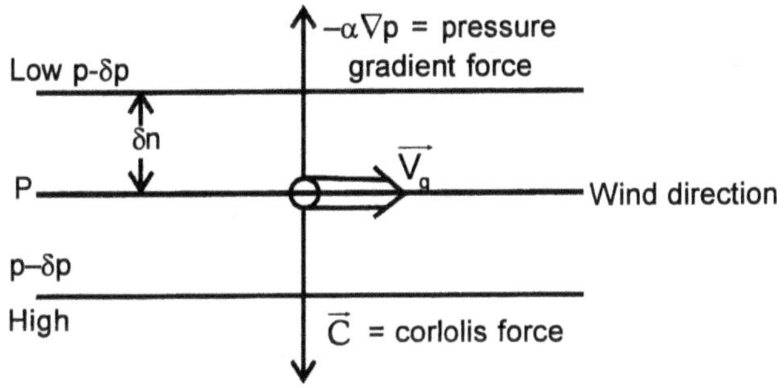

Fig. 18.1 Geostrophic flow in Northern hemisphere.

In geostrophic flow the horizontal components of coriolis force is in balance with the pressure gradient force.

When acceleration is zero $\left(\frac{du}{dt} = 0, \frac{dv}{dt} = 0\right)$ the geostrophic motion is along straight line (since a curved trajectory would require a normal acceleration).

Pressure force $-\alpha\nabla p$ being normal to the isobars (direction from H to L) implies that coriolis force ($\vec{C}$) is also normal to the isobars but in the opposite direction (from L toH) as shown in Fig 18.1. Geostrophic wind vector follows Buys Ballot's Law, namely, looking down stream low pressure will be on the left in the northern hemisphere (and right in the southern hemisphere).

From eq. (18.2) and (18.3) the geostrophic wind vector $\vec{V_g}$ may be written as

$$j\left(\alpha\frac{\partial p}{\partial x} = fv_g\right)$$

$$i\left(-\alpha\frac{\partial p}{\partial y} = fu_g\right)$$

adding

$$f\,(iu_g + jv_g) = j\alpha\frac{\partial p}{\partial x} - i\alpha\frac{\partial p}{\partial y}$$

$$= \alpha k \times \left(i\frac{\partial p}{\partial x} + j\frac{\partial p}{\partial y}\right)$$

or

$$f\vec{V_g} = \alpha k \times \nabla p \qquad .....(18.4)$$

the magnitude of $\vec{V_g}$ is

$$V_g = \frac{\alpha}{f}|\nabla p| = \frac{\alpha}{f}\frac{\partial p}{\partial n}$$

or $$V_g \simeq \frac{\alpha}{f}\frac{\nabla p}{\nabla n} \qquad .....(18.5)$$

The above formulae are in z-system of coordinates.

Let us consider p-system (where pressure is z coordinate). The equations of motion in p-system are

$$\left(\frac{\partial u}{\partial t}\right)_p + u\left(\frac{\partial u}{\partial x}\right)_p + v\left(\frac{\partial u}{\partial y}\right)_p + \omega\frac{\partial u}{\partial p} = -\frac{\partial \Psi}{\partial x} + fv$$

$$\left(\frac{\partial v}{\partial t}\right)_p + u\left(\frac{\partial v}{\partial x}\right)_p + v\left(\frac{\partial v}{\partial y}\right)_p + \omega\frac{\partial v}{\partial p} = -\frac{\partial \Psi}{\partial y} - fu$$

and $$\frac{\partial \Psi}{\partial p} = -\alpha \quad [\text{where } \Psi \text{ is geopotential, } \Psi = gz]$$

Under geostrophic conditions these equations reduce to

$$0 = -\frac{\partial \Psi}{\partial x} + f\,v_g \qquad .....(18.6)$$

$$0 = -\frac{\partial \Psi}{\partial y} - f\,u_g \qquad .....(18.7)$$

The relation between the geostrophic forces are shown in Fig.(18.2).

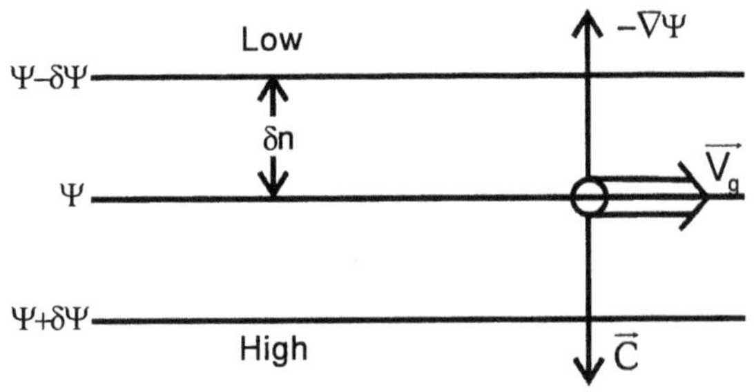

Fig. 18.2

From eq. (18.6) and (18.7) we find that

$$u_g = -\frac{1}{f}\frac{\partial \Psi}{\partial y} \quad (\text{or } -\frac{g}{f}\frac{\partial z}{\partial y})$$

$$v_g = \frac{1}{f}\frac{\partial \Psi}{\partial x} \quad (\text{or } +\frac{g}{f}\frac{\partial z}{\partial x})$$

or

$$\vec{V}_g = (iu_g + jv_g) = \frac{1}{f}\, k \times \nabla\Psi \qquad .....(18.8)$$

[Note : $k \times \left(i\frac{\partial \Psi}{\partial x} + j\frac{\partial \Psi}{\partial y}\right) = j\frac{\partial \Psi}{\partial x} - i\frac{\partial \Psi}{\partial y}$]

The magnitude of $\vec{V}_g$ is given by

$$V_g = \frac{1}{f}|\nabla\Psi| = \frac{1}{f}\frac{\partial \Psi}{\partial n} \qquad .....(18.9)$$

In practical analysis of weather charts the above relation (18.9) in isobaric units is very useful. In z-system the magnitude of geostrophic wind given by eq. (18.5) contains density (= ρ). From the above formulae it follows that geostrophic wind is proportional to the pressure gradient or inversely proportional to the distance between the isobars. That is, closer the isobars the stronger the wind and vice versa. Geostrophic wind eq. (18.5) or (18.8) are independent of time and hence cannot be used for prognosis.

Since $\left(V_g \propto \frac{1}{f}\right)$ geostrophic wind is inversely proportional to the coriolis parameter ($f = 2\Omega \sin \phi$), the geostrophic concept breaks down at equator ($f = 0, \phi = 0$), while the geostrophic speeds become smaller at higher latitudes. Consequently the geostrophic wind formulae are generally used in middle latitudes.

In the vicinity of jet stream the geostrophic wind approximation is poorer because this region has large values of acceleration.

The geostrophic approximation applies only when the isobars are straight and parallel and the pressure distribution is steady and not varying rapidly with time. Geostrophic wind approximation is not valid in case of local winds, such as land and sea breeze, where the balance between pressure gradient and coriolis force is not reached.

Geostrophic wind and actual winds differ because geostrophic wind is a balance motion without acceleration. However it is a good estimate of actual wind above friction layer. It is really difficult to answer how the atmospheric large-scale flow remains in almost in geostrophic balance. Large systematic deviations are observed between actual and geostrophic wind in a strongly curved flow. Geostrophic wind is much larger than actual wind in case of strong cyclonic curvature. However such strong systematic difference is not observed in an anti-cyclonic flow.

Scale analysis of equations of motion shows that:

magnitude of $\frac{du}{dt}$ and $\frac{dv}{dt}$ $\simeq 10^{-4}$ m/s

Horizontal velocity scale V $\simeq$ 10 m/s, vertical velocity w ~ 0.01 m/s

$\frac{df}{dy} = \beta = 10^{-15}$ m/s = Beta parameter

$f \simeq 10^{-4}$/s, = Coriolis parameter, $\alpha\frac{\partial p}{\partial x}$ or $\alpha\frac{\partial p}{\partial y} \simeq 10^{-3}$/s

$2\Omega\, u \sin\phi$ and $2\,\Omega\, v \sin\phi \simeq 10^{-3}$ m/s, $2\Omega\, w \cos\phi \simeq 10^{-6}$ m/s

The largest terms in horizontal flow are pressure force ($\simeq 10^4$dynes/cm^2) and Coriolis force ($\simeq 10^{-4}$/s).

While the remaining terms, namely acceleration, friction force are at least one order of magnitude less than the pressure force. Thus the balance of pressure force and coriolis force is approximated to real situation in the undisturbed atmosphere. Moreover the geostrophic statement is based on meteorological data for large-scale synoptic motions and hence it is an empirical relation.

18.2 Buys Ballot's Law

If an observer stands with his back to the wind, the lower pressure will be to his left in the northern hemisphere and to his right in the southern hemisphere.

Coriolis force: The earth rotates from west to east on its axis with angular velocity of 15° per hour or once in a day. Any object in motion on the surface of the earth is deflected every stage in its motion relative to the moving earth. To visualize the way in which this deflecting force works is illustrated by a simple experiment (Fig. 18.3). Start rotating the turntable of a record player.

With the use of a scale and a piece of chalk draw a straight line lightly from the centre to a point to the end of the disc. To the person drawing the line, the chalk piece travels in a straight line. However if the turntable is stopped from rotation we find the record of the chalk is a curved line which deviated from the straight path in the direction opposite to that of turntables rotation. The deflection is caused by an apparent force. The coriolis force is the apparent force which deflects winds in the atmosphere in a similar fashion due to the rotation of the earth. The coriolis force deflects wind to the right in the northern hemisphere and to the left in the southern hemisphere.

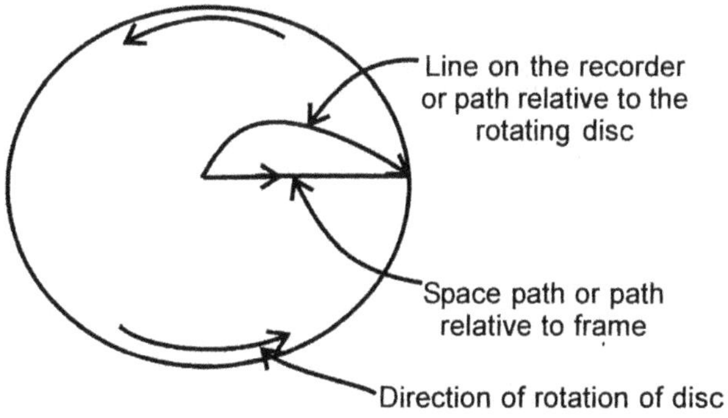

Fig. 18.3 The Coriolis deflecting force working on a body moving outward from the centre of a rotating turntable.

The magnitude of the coriolis force or deflecting force is given by

$$2\Omega V \sin \phi$$

where Ω = angular velocity of the earth.

V = horizontal wind speed

ϕ = latitude.

$f = 2\Omega \sin \phi$ is called coriolis parameter

f is small at the equator ($\phi = 0$, $\sin \phi = 0$) and large at the poles ($\phi = 90°$, $\sin \phi = 1$)

In 1835 the French Physicist Gaspard Coriolis explained mathematically the influence of the earth's rotation on objects moving over its surface. For this reason it is called the coriolis effect. The Fig. (18.4) shows the coriolis effect on winds in both hemispheres.

There is no deflection at the equator but increases slowly, attains maximum at the poles. The coriolis force generates an apparent acceleration and is called coriolis acceleration.

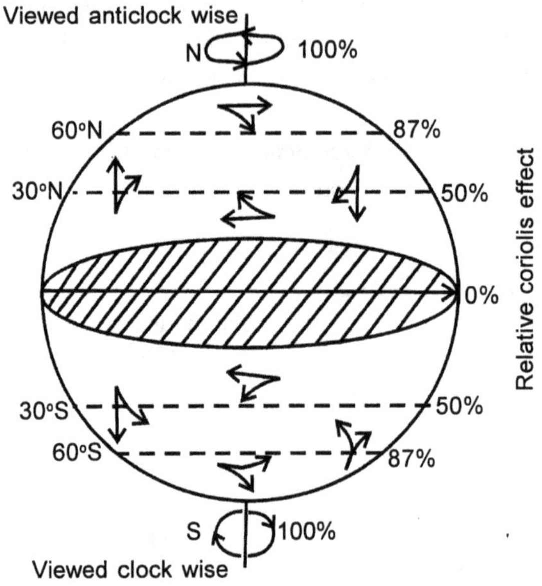

Fig. 18.4 Deflection of wind by coriolis effect.

Note: Since the coriolis force is small in low latitudes and hence geostrophic flow (seldom) rarely occurs between Lat 15° N and 15° S

The friction force is largest near the earth's surface. Air up to 1 km height above ground level is assumed to be in the planetary boundary level (PBL) or friction layer and above 1 km region is called free atmosphere (where the frictional force is negligible).

The values of the coriolis parameter (f) at various latitudes are given below.

Table 18.1

Latitude	0	10	20	43	90
values approximately $f \times 10^{-4}$/sec	0	0.25	0.5	1.0	1.46

f always acts at right angles to wind direction and deflects to the right in northern hemisphere and to the left in the southern hemisphere.

Questions

1. Define geostrophic wind. Derive the general equations of motion as geostrophic approximation. Express geostrophic wind as finite difference. Show that

$$f\,\vec{V_g} = \alpha\, k \times \nabla p$$

2. Assuming the equations of motion in pressure coordinate system deduce the geostrophic approximation as $u_g = -\frac{1}{f}\frac{\partial \Psi}{\partial y}$ and $V_g = \frac{1}{f}\frac{\partial \Psi}{\partial x}$

or $$\vec{V_g} = \frac{1}{f}\, k \times \nabla \Psi$$

Also show that $\left|\vec{V_g}\right| = \frac{1}{f}\frac{\partial \psi}{\partial n}$

3. Write the properties of geostrophic wind and show that geostrophic wind equation is not useful in prognosis and that the geostrophic concept breaksdown at equator and is a poor estimate in the vicinity of jetstream.

4. What is Coriolis force. Write the magnitude formula of Coriolis force and Coriolis parameter as used in meteorology.

   Describe an experiment to illustrate the Coriolis effect. Draw a diagram, how it effects the wind on the globe, give values of Coriolis parameter at latitudes 0, 10, 20, 43 and 90.

CHAPTER 19

The Gradient Wind

19.1 Definition

The wind occurring in a frictionless flow when there is a balance between the pressure force ($\alpha\nabla p$), the coriolis force ($2\vec{\Omega} \times \vec{V}$) and the centrifugal force ($\Omega^2\vec{R}$).

The wind which occurs when the trajectory of the particle is circular and there exists a balance between the pressure force, the coriolis force and the centrifugal force.

$$\alpha\nabla p = -2\vec{\Omega} \times \vec{V} - \Omega^2\vec{R} \quad .....(19.1)$$

In simple terms, frictionless, horizontal motion at constant speed is called gradient flow. The flow is tangential to the isobars. In this case the magnitude of the pressure gradient force and the coriolis force are not equal (either greater or smaller) and hence the motion will be curved either to the left or to the right and not parallel to the isobars. It follows from the definition that the geostrophic wind is a particular case of gradient wind when the centrifugal force is negligible compared to pressure force and coriolis force.

For convenience we shall consider the cases anticlockwise (Fig. 19.1(a)) and clock wise (Fig.19.1(b)) circulations separately.

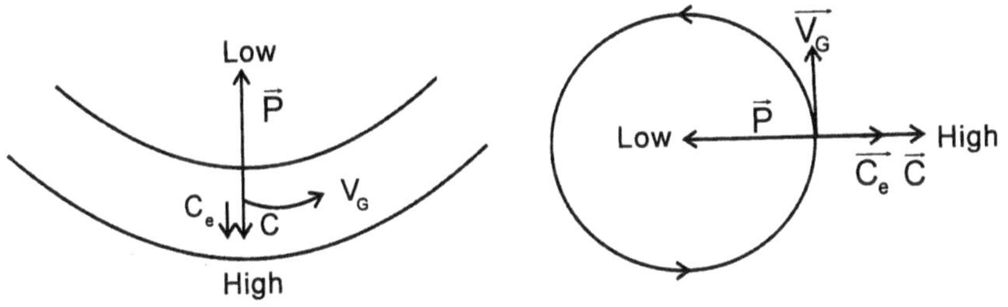

Fig. 19.1

Case I Anticlockwise Circulation

In this case coriolis force $\vec{C}$, centrifugal force ($\vec{C_e}$) act in the same direction while the pressure force ($\vec{P}$) in the opposite direction as shown in Fig (19.1)

The numerical value of the, pressure force ($\vec{P}$) is $\alpha \frac{dp}{dR}$, where R is the radius of circular motion.

The geostrophic wind speed is

$$V_g = \frac{1}{\rho f}|\nabla p| = \frac{1}{\rho f}\frac{\partial p}{\partial n}$$

or

$$\left|\alpha \frac{dP}{dR}\right| \sim fV_g \qquad .....(19.2)$$

$$|\vec{C}| = f\,V_G$$

where V_G = grdient wind speed

$$|\vec{C_e}| = \frac{V_G^2}{R} \qquad .....(19.3)$$

From (Fig. 19.1) we have

$$\vec{C} + \vec{C_e} = \vec{P}$$

$$fV_G + \frac{V_G^2}{R} = fV_g$$

or

$$fV_G + \frac{V_G^2}{R} - fV_g = 0$$

Dividing by $V^2{}_G$ and rearranging terms, we have

$$\frac{fV_g}{V_G^2} - \frac{f}{V_G} - \frac{1}{R} = 0$$

This is a quadratic equation. [We know that, if $ax^2 + bx + c = 0$ then that $x = \frac{-b \pm \sqrt{b^2 - 4ac}}{2a}$.]

Its roots are given by

$$\frac{1}{V_G} = \frac{f \pm \sqrt{f^2 - 4\frac{fV_g}{-R}}}{2fv_g}$$

or $$V_G = \frac{2fV_g}{f \pm \sqrt{f^2 + \frac{4fV_g}{R}}} = \frac{V_g}{\frac{1}{2} \pm \sqrt{\frac{1}{4} + \frac{V_g}{fR}}}$$

It is obvious, in the denominator only plus sign is possible for the radical.

i.e., $$V_G = \frac{V_g}{\frac{1}{2} + \sqrt{\frac{1}{4} + \frac{V_g}{fR}}} \qquad .....(19.4)$$

It follows from (eq. 19.4) that $V_G < V_g$. i.e., Cyclonic gradient wind is sub-geostrophic and $V_G \to V_g$ as $R \to \infty$

when $V_g = 10 \text{ m s}^{-1}$

$f = 10^{-4} \text{s}^{-1}$ (Lat 45 °N) and

$R = 200$ km

we get. $v_G \simeq 7.3$

$$V_G \simeq \frac{V_g}{2} \quad \left(\because \frac{V_g}{fR} \to 0\right)$$

Case II Clockwise Circulation

In this case pressure force ($\vec{P}$) and centrifugal force $\vec{C}_e$ act in the same direction while the coriolis force ($\vec{C}$) acts in the opposite direction as shown. From the Fig. (19.2) we have

$$\vec{C} = \vec{C}_e + \vec{P}$$

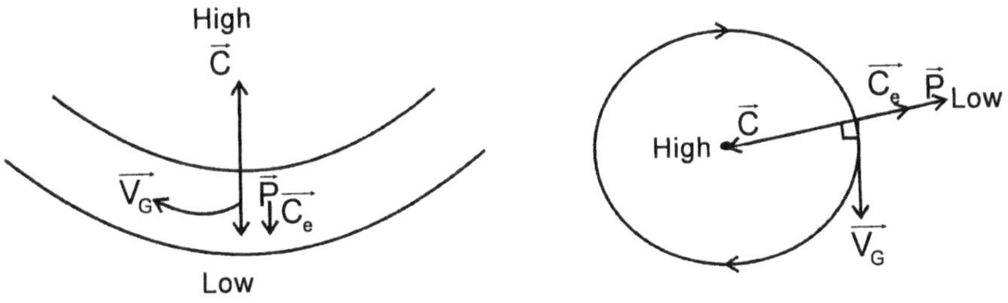

Fig. 19.2

using eq. (19.2) and (19.3) we have

$$fV_G = \frac{V_G^2}{R} + fV_g$$

or
$$fV_G - \frac{V_G^2}{R} - fV_g = 0$$

or
$$\frac{f}{V_G} - \frac{fV_g}{V_G^2} - \frac{1}{R} = 0$$

$$\frac{fV_g}{V_G^2} - \frac{f}{V_G} + \frac{1}{R} = 0$$

This is a quadratic equation in $\frac{1}{V_G}$

∴ its roots are given by

$$\frac{1}{V_G} = \frac{f \pm \sqrt{f^2 - \frac{4fV_g}{R}}}{2fV_g}$$

$$= \frac{1 \pm \sqrt{1 - \frac{4V_g}{Rf}}}{2V_g}$$

$$V_G = \frac{2V_g}{1 \pm \sqrt{1 - \frac{4V_g}{Rf}}}$$

In the denominator only positive sign is possible in front of the radical.

i.e., $$V_G = \frac{2V_g}{1+\sqrt{1-\frac{4V_g}{Rf}}} = \frac{V_g}{\frac{1}{2}+\sqrt{\frac{1}{4}-\frac{V_g}{Rf}}}$$

It follows from above that $V_G > V_g$, i.e., Anti-cyclonic gradient wind is super-geostrophic.

$$\text{As } R \to \infty \quad V_G \to V_g$$

and $$V_G = 2V_g \quad \text{when } \frac{V_g}{Rf} = \frac{1}{4}$$

i.e., the maximum anti-cyclonic gradient wind is twice the geostrophic wind.

Ex : In a cyclonic flow find the difference between geostrophic wind (V_g) and gradient wind V_G at latitude 45 °N

when $R = 1000$ km

$R = 100$ km

Asumme $Vg = 10 \text{ m s}^{-1}$

Sol :

(i) Given $Vg = 10 \text{ m s}^{-1}$

$\phi = 45^\circ$

$R = 1000 \text{ km} = 10^6 \text{ m}$

we know $\Omega = 7.3 \times 10^{-5} \text{s}^{-1}$

$$\therefore \quad f = 2\Omega \sin\phi = 2 \times 7.3 \times 10^{-5} \text{s}^{-1} \times \frac{1}{\sqrt{2}} \simeq 10^{-4} \text{s}^{-1}$$

From eq. (19.4) we have

$$V_G = \frac{V_g}{\frac{1}{2}+\sqrt{\frac{1}{4}+\frac{V_g}{fR}}} = \frac{10}{0.5+\sqrt{0.25+\frac{10}{10^{-4}\text{s}^{-1}\times 10^6}}}$$

$$= \frac{10}{0.5+\sqrt{0.25+0.1}} = \frac{10}{0.5+0.591} \simeq 9.16 \text{ m/s}$$

$$\therefore \quad V_g - V_G = 10 - 9.16 = 0.84 \text{ mps.}$$

(ii) When $R = 100 \text{ Km} = 10^5$ m

we have $$V_G = \frac{10}{0.5+\sqrt{0.25+\frac{10}{10^{-4}\text{s}^{-1}\times 10^5\text{m}}}} = \frac{10}{0.5+\sqrt{0.25+1}}$$

$$= \frac{10}{0.5+1.12} \simeq 6.17$$

$$\therefore \quad V_g - V_G = 10 - 6.17 = 3.83 \text{ mps.}$$

Ex : In an anti cyclonic flow find the difference between gradient wind and geostrophic wind at latitude 45°, V_g = 10 ms^{-1} and radius of curvature is (i) R = 400 km (ii) R = 1000 km.

Sol :

(i) We know, from eq. 19.5

$$V_G = \frac{V_g}{\frac{1}{2}+\sqrt{\frac{1}{4}-\frac{V_g}{Rf}}}$$

when R = 400km = 4×10^5 m

$$V_G = \frac{10}{0.5+\sqrt{0.25-\frac{10}{4\times10^5\times10^{-4}}}} = \frac{10}{0.5+0} = 20.$$

$\therefore$ $V_G - V_g$ = 20 – 10 = 10 mps.

when R = 1000 km = 10^6 m

$$V_G = \frac{10}{0.5+\sqrt{0.25-\frac{10}{10^6\times10^{-4}}}} = \frac{10}{0.5+\sqrt{0.25-0.1}} \simeq 11.27$$

$\therefore$ $V_G - V_g$ = 11.27 – 10 = 1.27 mps.

Questions

1. Define gradient wind. Show that cyclonic gradient wind is sub-geostrophic.
2. Show that anticyclonic gradient wind is supergeostrophic.

   In an anticyclonic flow find the difference between gradient wind and geostrophic wind at lat 45° when V_g = (geostrophic wind) is 10 m/s and radius of curvature R = 400 Km, $\Omega = 7.3 \times 10^{-5}\ s^{-1}$.

CHAPTER 20

Cyclostrophic Flow

Definition

The wind occurring in a frictionless flow when the pressure gradient force is in balance with the centrifugal force is called cyclostrophic wind.

In case of small horizontal motions, such as dust devils, water spouts, Norwesters, Tornadoes, Hurricanes, compared to pressure gradient and centrifugal force the coriolis force is negligible and may be assumed to achieve balance between pressure gradient and centrifugal force. The force balance perpendicular to the direction of flow is given by

$$\frac{V^2}{R} = -\alpha \frac{\partial p}{\partial n}$$

or

$$\Omega^2 \vec{R} = -\alpha \nabla p$$

or

$$V = \left[-R\alpha \frac{\partial p}{\partial n} \right]^{\frac{1}{2}}$$

Where R is radius of curvature of streamline

The cyclostrophic flow may be cyclonic or anticyclonic as shown in Fig. 20.1.

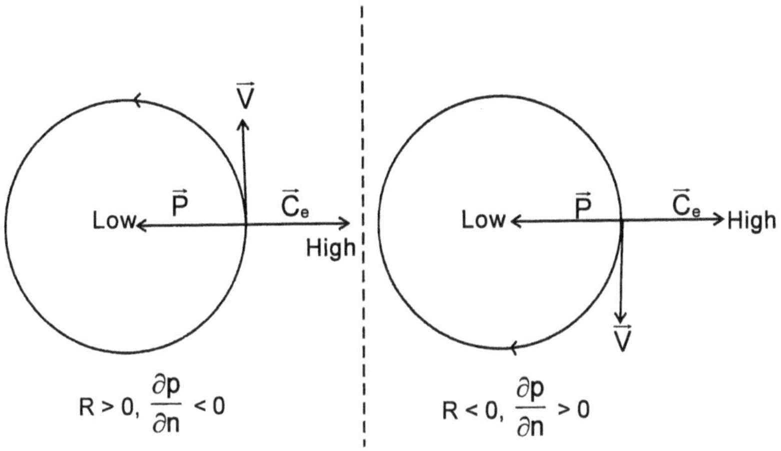

Fig. 20.1 Cyclostrophic flow.

Tornadoes are observed to rotate in cyclonic (anticlockwise) sense in northern hemisphere. However dust devils, water spouts are observed to rotate both cyclonic and anticyclonic sense.

Note : Cyclostrophic wind is valid if the ratio of centrifugal force to coriolis force is large and hence cyclostrophic motion is expected in low latitudes where f is small. Tropical cyclones to some extent obey the cyclostrophic law.

20.1 Inertial Flow

Definition : The wind occurring in a frictionless flow when the centrifugal force is in balance with the coriolis force.

i.e., $$2\vec{\Omega} \times \vec{V} = -\Omega^2\vec{R}$$

or $$fV + \frac{V^2}{R} = 0$$

or $$R = -\frac{V}{f}$$

Inertial law is valid only when the horizontal pressure field is uniform i.e., pressure gradient is approximately zero. Both wind speed and radius of curvature must be constant. Since only coriolis force is present, the inertial flow is necessarily anticyclonic as shown in Fig. 20.2.

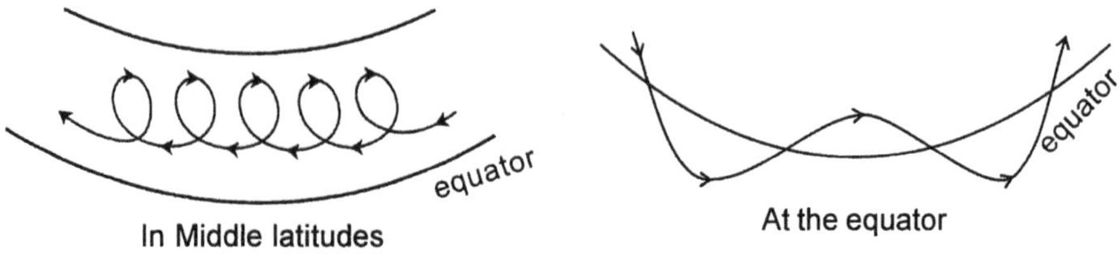

Fig. 20.2 Inertial flow.

20.2 Inference of Divergence, Vertical Motion and Vorticity from Synoptic (Surface and Upper Air) Charts

Analysis of weather charts enables the meteorologist to understand the current weather and to infer the future weather. The three terms Divergence, Vertical motion and Vorticity are the three key terms to understand and unravel the weather on synoptic scale.

20.2.1 Fields of Divergence

The wind field is developed by the variation of wind speed along streamlines and coming together (confluence) or going apart (difluence) of streamlines. Based on wind speed divergence (or convergence) of wind is inferred as follows.

Notation : ⟶ Streamline, ----- isotech

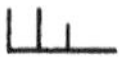 Wind feather 25 knots.

1. When the wind speed increases down stream in a field of parallel streamlines divergence occurs.

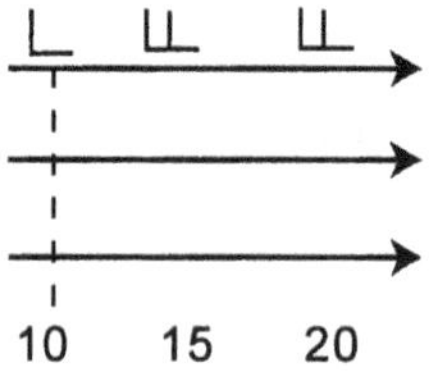

Fig. 20.3.1 Divergence.

2. When the wind speed decreases down stream in a field of parallel streamlines convergence results

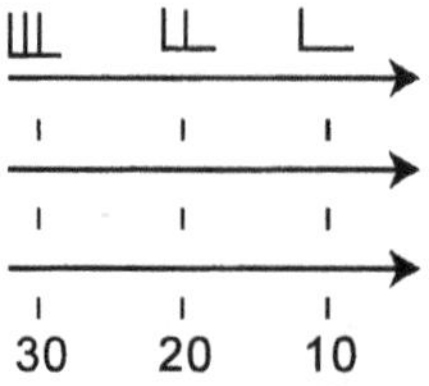

Fig. 20.3.2 Convergence.

3. Fanning out of streamlines is called difluence, converging (coming together) of streamlines is called confluence.

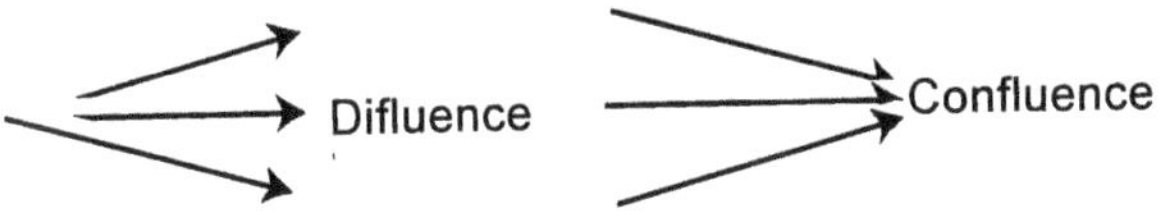

Fig. 20.3.3

4. If the streamlines fan out in an area of uniform wind speed along streamlines divergence occurs.

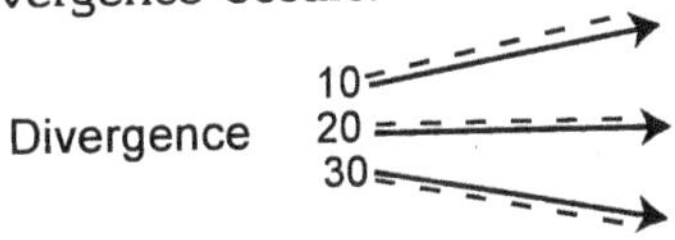

Fig. 20.3.4

5. If the streamlines come together in an area of uniform wind speed along streamlines convergence occurs.

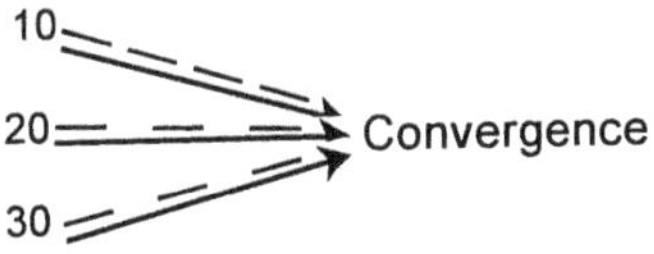

Fig. 20.3.5

Note : Mere fanning out the streamlines or coming together does not indicate the divergence or convergence.

6. If the wind speed increases downstream and simultaneously streamlines fan out indicate divergence.

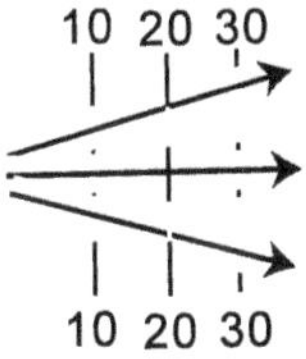

Fig. 20.3.6 Divergence.

7. If the wind speed decreases downstream and simultaneously streamlines come together indicate convergence.

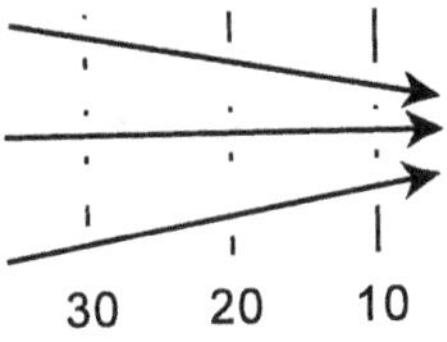

Fig. 20.3.7 Convergence.

8. If the wind speed decreases downstream while streamlines fan out or if the wind speed increases downstream while streamlines come together no inference can be made about convergence or divergence. Such situations arise in middle and higher latitudes.

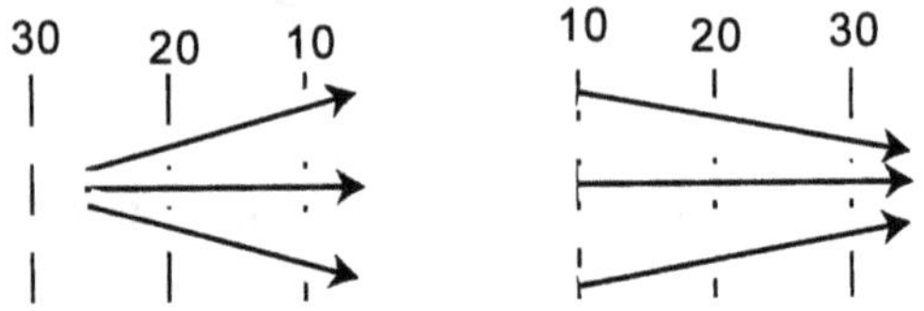

Fig. 20.3.8 Neither convergence nor divergence inferred.

Note: Areas of divergence indicate generation of mechanical Kinetic energy while convergence indicate dissipation of K.E.

Vertical motion of wind: Horizontal convergence (or divergence) of wind near the surface of the earth (in PBL = Planetary Boundary level) causes vertical motion (ascent or decent). This is shown in Fig. 20.4.

In case of low level convergence and simultaneous high level divergence at the same place vertical motion results. Such situations are conducive for triggering convective activity, precipitation and other weather phenomena. Contrary to this, low level divergence and high level convergence leads to subsidence, generally fair weather or smog/fog, inversion of temperature near the earth. A level at which neither convergence nor divergence effectively observed is called level of non-divergence (LND). Below LND either convergence (or divergence) begins and above LND either divergence (or convergence) begins.

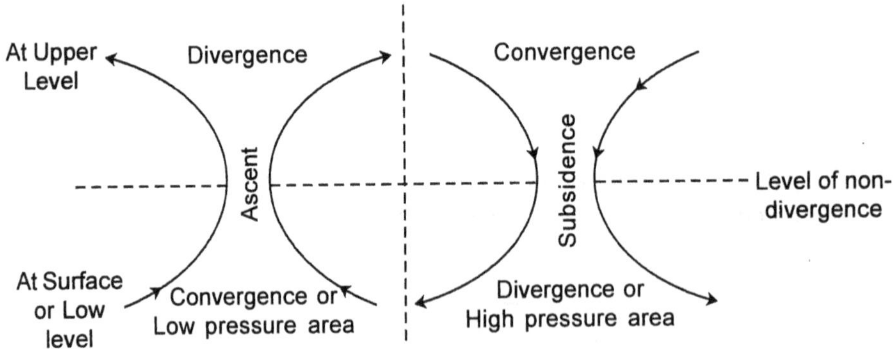

Fig. 20.4

20.2.2 Fields of Vorticity

We know that vorticity $\left(\zeta = \dfrac{V}{R_s} - \dfrac{\partial V}{\partial n}\right)$ has two components.

(i) Curvature $\dfrac{V}{R_s}$ and (ii) another shear $\dfrac{\partial V}{\partial n}$. There are six combinations of these two components which are illustrated for the northern hemisphere.

1. If the flow is of uniform speed or isotachs (equal wind speed lines, in Fig. 20.5(a) indicated by broken lines) are normal to the streamlines then curvature determines the sign of vorticity (positive for cyclonic or negative for anticyclonic).

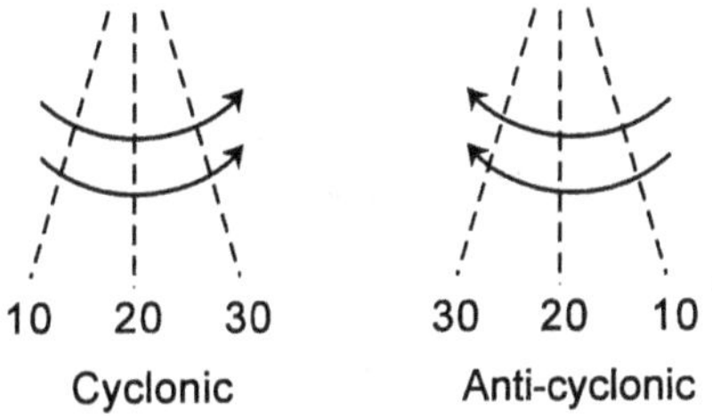

Fig. 20.5(a)

2. In a straight flow, looking down, stream, decreasing of wind speed to the left of the current indicates cyclonic vorticity (Fig. 20.5(b)).

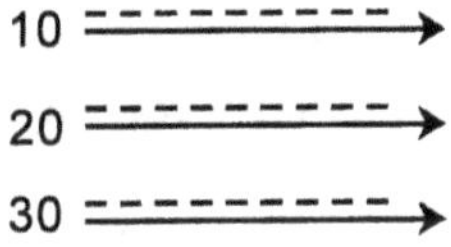

Fig. 20.5(b) Cyclonic vorticity.

3. In a straight flow, looking down stream, increasing of wind speed to left of the current indicates anticyclonic vorticity (Fig. 20.5(c)).

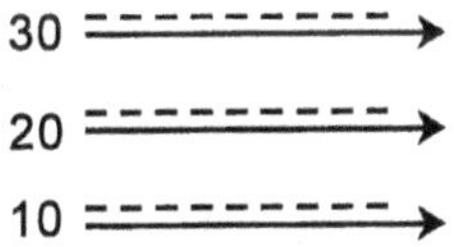

Fig. 20.5(c) Anti-cyclonic vorticity.

4. In streamlines of cyclonic curvature, decreasing of wind speed to the right of the current (looking down stream) indicates cyclonic vorticity (Fig. 20.5(d)).

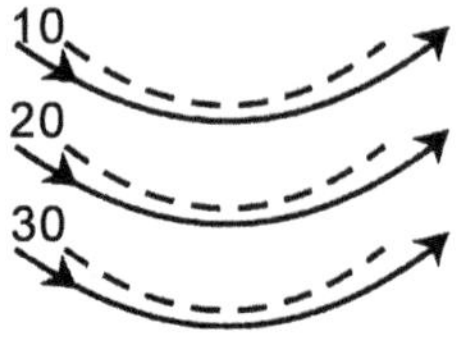

Fig. 20.5(d) Cyclonic vorticity.

5. In streamlines of anticyclonic curvature, increasing of wind speed to the left of the current (looking downstream) indicates anticyclonic vorticity (Fig. 20.5(e)).

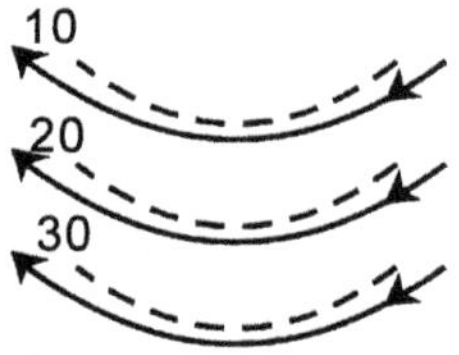

Fig. 20.5(e) Anti-cyclonic vorticity.

6. If the curvature of the streamlines is cyclonic, while the wind speed increases to the left of the current (Fig. 20.5(f)) or if the curvature of the streamlines is anti-cyclonic while the wind speed decreases to the left of the current (Fig. 20.5(g)),in these cases it is difficult to determine the vorticity. In these cases Relative vorticity cannot be determined by inspection.

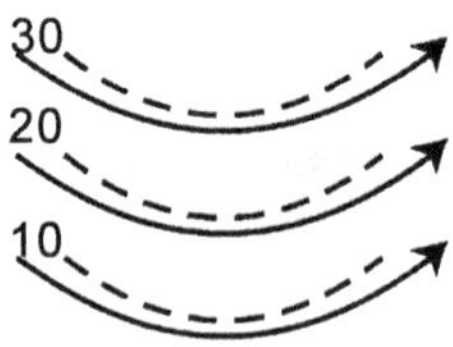

Fig. 20.5(f)

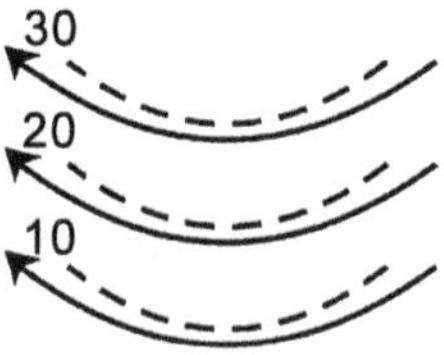

Fig. 20.5(g)

The vorticity of a current that crosses the equator with uniform curvature changes vorticity sign but it is an impediment in understanding the equatorial flow.

In jet-stream zones, a velocity maxima is present. Inspection shows signs of vorticity in some parts of the map only. The following streamline, isotach analysis presents two such models.

In Fig. 20.6, the velocity maxima is located in the trough. North of the axis of jet core, wind has cyclonic vorticity. However South of the jet core it is indeterminate because the curvature is cyclonic while the wind shear is anticyclonic.

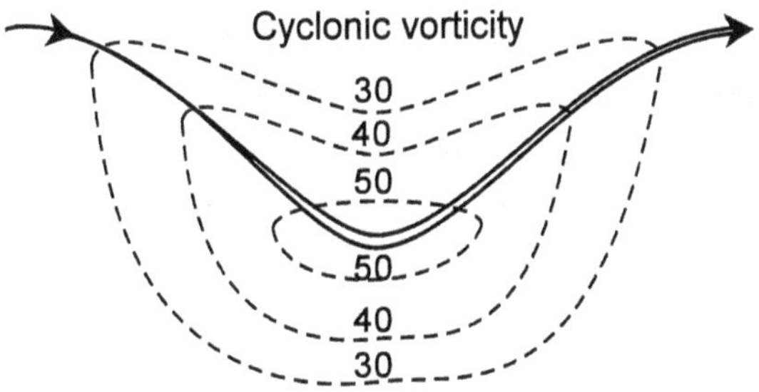

Fig. 20.6

In Fig. 20.7, the velocity maxima is centred in the ridge. To the north of the axis of jet core wind has anticyclonic vorticity. However south of the jet core it is indeterminate, since the curvature is anti cyclonic while the wind shear is cyclonic.

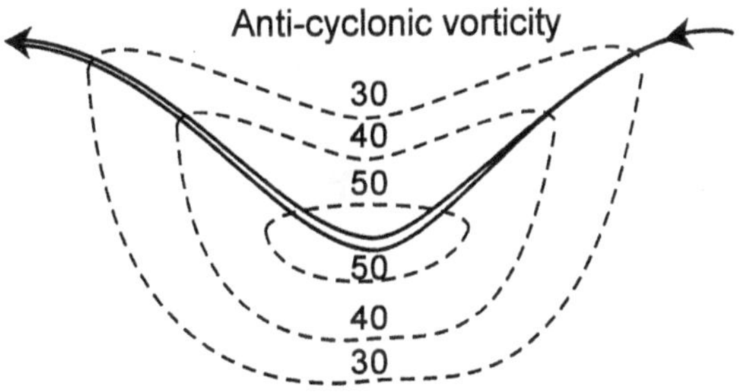

Fig. 20.7

Questions

1. (a) Define cyclostrophic wind and derive a formula.

   (b) Define inertial flow and derive a formula. Represent the inertial flow by a diagram in middle latitudes and at the equator.

2. Draw the field of divergence, convergence with streamlines and isotachs, also confluence and difluence of streamlines. Define level of non-divergence. When do you expect ascent or descent of air at a place.

3. Write the formula of vorticity. What are its components?

   With the help streamlines and isotachs draws the various possible cyclonic and anticyclonic vorticity regions. What happens to the vorticity current when it crosses the equator? Draw cyclonic and anticyclonic vorticity regions in a jet core trough/ridge with the help of isotachs.

CHAPTER 21

Divergence of Geostrophic Wind

By Definition

$$\text{Div } \vec{V}_g = \nabla . \vec{V}_g$$

$$= \left(i\frac{\partial}{\partial x} + j\frac{\partial}{\partial y} \right) . \left(iu_g + jv_g \right) \quad .....(21.1)$$

where x-axis coincides or parallel with the direction of tangent

y-axis is perpendicular to it.

We know

$$u_g = -\frac{\alpha}{f}\frac{\partial p}{\partial y} = -\frac{g\partial z}{f\partial y} \quad (\because \ \alpha dp = -gdz \text{ Hydrostatic equation})$$

$$v_g = \frac{\alpha}{f}\frac{\partial p}{\partial x} = \frac{g\partial z}{f\partial x}$$

$$\therefore \quad \text{Div } \vec{V}_g = \left(i\frac{\partial}{\partial x} + j\frac{\partial}{\partial y} \right) . \left(-i\frac{g}{f}\frac{\partial z}{\partial y} + j\frac{g}{f}\frac{\partial z}{\partial x} \right)$$

$$= -\frac{\partial}{\partial x}\left(\frac{g}{f}\frac{\partial z}{\partial y} \right) + \frac{\partial}{\partial y}\left(\frac{g}{f}\frac{\partial z}{\partial x} \right)$$

$$= -g\frac{\partial f^{-1}}{\partial x}\frac{\partial z}{\partial y} - \frac{g}{f}\frac{\partial^2 z}{\partial x \partial y} + g\frac{\partial f^{-1}}{\partial y}\frac{\partial z}{\partial x} + \frac{g}{f}\frac{\partial^2 z}{\partial x \partial y}$$

$$\text{Div } \vec{V}_g = \frac{-g}{f^2}\frac{\partial f}{\partial y}\frac{\partial z}{\partial x} \qquad \left[\because \frac{\partial f}{\partial x} = 0 \text{ as f is constant along } x\text{-axis}\right]$$

$$= -\frac{g}{f^2}\left(\frac{2\Omega\cos\phi}{R}\right)\frac{\partial z}{\partial x}$$

since $f = 2\Omega \sin \phi$, $\delta y = R\delta\phi$ (see Fig. 21.(a))

$$\text{and } \beta = \frac{\partial f}{\partial y} = 2\Omega \cos \phi \frac{\partial \phi}{\partial y} = \frac{2\Omega\cos\phi}{R}$$

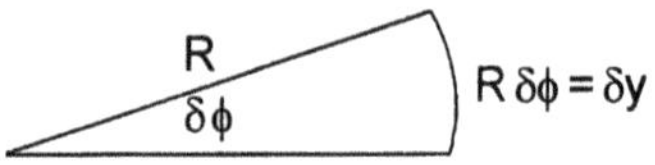

Fig. 21.1(a)

$$\therefore \text{Div } \vec{V}_g = -\frac{g}{f^2}\left(\frac{2\Omega\cos\phi}{R}\right)\left(\frac{fV_g}{g}\right) \quad [\text{using } V_g = \frac{g}{f}\frac{\partial z}{\delta x}]$$

$$= -\frac{2\Omega\cos\phi V_g}{R.2\Omega\sin\phi}$$

i.e., $$\text{Div } \vec{V}_g = \nabla.\vec{V}_g = \frac{-V_g}{R.\tan\phi} \qquad .....(21.2)$$

from eq. (21.2) it follows that geostrophic divergence is found when V_g is negative (between ridge and trough) in the geopotential on constant pressure chart (isobaric surface). Geostrophic convergence is found when V_g is positive (between trough and ridge) on constant pressure chart as showm in Fig. 21.1 (b).

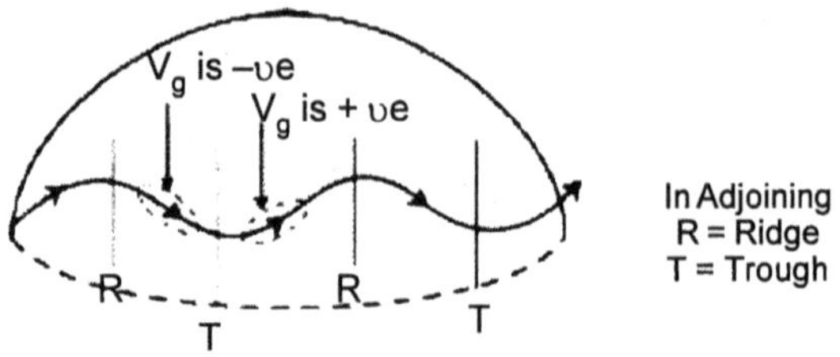

Fig. 21.1 (b)

∴ Div $\vec{V}_g$ is positive (or V_g is negative) between ridge and trough (i.e., on divergence) and Div $\vec{V}_g$ is negative (or V_g is positive) between trough and ridge (i.e., convergence).

This shows that the divergence of geostrophic wind is not a good measure of actual divergence of wind.

21.1 Vorticity of the Geostrophic Wind

The vertical component of the vorticity ζ is defined as $\zeta = k \cdot \nabla \times \vec{V}$

where $\vec{V}$ = velocity of the wind= iu + jv

∴ The vorticity of the geostrophic wind $\vec{V} = iu_g + jv_g$ is given by

$$\zeta_g = k \cdot \nabla \times \vec{V}_g = k \cdot \begin{vmatrix} i & j & k \\ \frac{\partial}{\partial x} & \frac{\partial}{\partial y} & \frac{\partial}{\partial z} \\ u_g & v_g & 0 \end{vmatrix} = \begin{vmatrix} 0 & 0 & 1 \\ \frac{\partial}{\partial x} & \frac{\partial}{\partial y} & \frac{\partial}{\partial z} \\ u_g & v_g & 0 \end{vmatrix}$$

$$\zeta_g = \frac{\partial v_g}{\partial x} - \frac{\partial u_g}{\partial y} \qquad .....(21.3)$$

We know the geostrophic wind components u_g, v_g are given by

$$u_g = -\frac{1}{f}\frac{\partial \Psi}{\partial y} = -\frac{g}{f}\frac{\partial z}{\partial y}$$

(where Ψ is geopotential and given by $\psi = gz$)

$$v_g = \frac{1}{f}\frac{\partial \Psi}{\partial x} = \frac{g}{f}\frac{\partial z}{\partial x} \qquad .....(21.4)$$

Substituting these values in eq. (21.3) we have,

$$\zeta_g = \frac{\partial}{\partial x}\left(\frac{g}{f}\frac{\partial z}{\partial x}\right) - \frac{\partial}{\partial y}\left(-\frac{g}{f}\frac{\partial z}{\partial y}\right)$$

$$= g\frac{\partial f^{-1}}{\partial x}\frac{\partial z}{\partial x} + \frac{g}{f}\frac{\partial^2 z}{\partial x^2} + g\frac{\partial f^{-1}}{\partial y}\frac{\partial z}{\partial y} + \frac{g}{f}\frac{\partial^2 z}{\partial y^2}$$

$$\frac{g}{f}\nabla^2 z - \frac{g}{f^2}\frac{\partial f}{\partial y}\frac{\partial z}{\partial y}$$

$$\zeta_g = \frac{g}{f}\nabla^2 z \quad \frac{g}{f^2}\frac{\partial f}{\partial y}\frac{\partial z}{\partial y}$$

$$\left[\because \frac{df}{dx} = 0 \text{ as f is constant along x-axis and } \nabla^2 z = \frac{\partial^2 z}{\partial x^2} + \frac{\partial^2 z}{\partial y^2}\right]$$

$$\zeta_g = \frac{g}{f}\nabla^2 z - \frac{g}{f^2} 2\Omega\cos\phi \frac{\partial\phi}{\partial y}\left(-u_g \frac{f}{g}\right)$$

$$\left(\text{since } f = 2\Omega \sin\phi,\ \frac{\partial f}{\partial y} = 2\Omega\cos\phi \frac{\partial\phi}{\partial y}\right)$$

$$\zeta_g = \frac{g}{f}\nabla^2 z - \frac{g}{f^2}\left(\frac{2\Omega\cos\phi}{R}\right)\left(-u_g \frac{f}{g}\right) \quad \left(\because \frac{\partial\phi}{\partial y} = \frac{1}{R} \text{ see Fig. 21.2}\right)$$

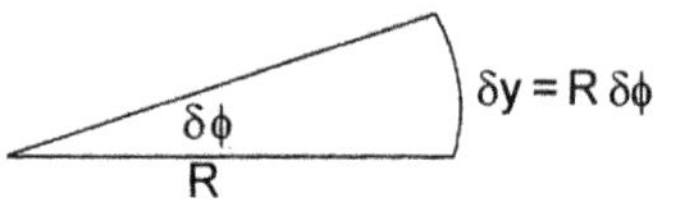

Fig. 21.2

$$\zeta_g = \frac{g}{f}\nabla^2 z + \frac{2\Omega\cos\phi}{R.2\Omega\sin\phi}$$

$$\zeta_g = \frac{g}{f}\nabla^2 z + \frac{u_g}{R\tan\phi} \qquad .....(21.5)$$

[$f = 6.2 \times 10^{-5}\ s^{-1}$ at 25° N and if $u_g = 100\ m^{-1}\ s$,

$\frac{u_g}{R\tan\phi} = 3.4 \times 10^{-5}\ s^{-1}$ at 25° N]

$$\zeta_g \simeq \zeta \simeq \frac{g}{f}\nabla^2 z \qquad .....(21.6)$$

$$\left(\text{since } \frac{u_g}{R\tan\phi} << \frac{g}{f}\nabla^2 z\right.$$

(Note : $\frac{u_g}{R\tan\phi}$ is small in middle and high latitudes)

From eq. (21.6) it follows that the vorticity of the geostrophic wind is a good approximation to the vorticity of the actual wind field. ζ_g is positive in a minimum of the geo-potential of an isobaric map and ζ_g is negative in a maximum of the geo-potential of an isobaric map. The positive value of ζ_g is called cyclonic vorticity and negative value is anticyclonic vorticity.

21.2 The Thermal Wind or Vertical Shear of the Geostrophic Wind

We know the geostrophic wind $\vec{V}_g$ is given by

$$\vec{V}_g = \frac{1}{f} k \times \nabla \Psi \qquad .....(21.7)$$

where $\Psi = gz$ is geo-potential, k= unit vector

f = coriolis parameter, $= 2\Omega \sin \phi$

Differentiating eq. (21.7) partially w.r.t p (pressure)

we have $$\frac{\partial \vec{V}_g}{\partial p} = \frac{1}{f} k \times \nabla \left(\frac{\partial \Psi}{\partial p} \right) \qquad .....(21.8)$$

we can write

$$\frac{\partial \vec{V}_g}{\partial z} = \frac{\partial p}{\partial z} \cdot \frac{\partial \vec{V}_g}{\partial p}$$

$$\frac{\partial \vec{V}_g}{\partial z} = g\rho . \frac{1}{f} k \times \nabla \left(\frac{\partial \Psi}{\partial p} \right)$$

[$\because \delta p = g\rho\delta z$, absolute value $= \rho\delta\Psi$ i.e., $\frac{\partial \psi}{\partial \rho} = \alpha$

$$\therefore \quad \frac{\partial \vec{V}_g}{\partial z} = g\rho \ . \ \frac{1}{f} k \times \nabla\alpha \text{ or } \frac{g}{f} \frac{K \times \nabla\alpha}{\alpha} \qquad .....(21.9)$$

we know $p\alpha = R'T$ (gas equation)

Taking logarithms and Differentiating we have

$$\ln p + \ln \alpha = \ln R' + \ln T$$

$$\frac{1}{p} dp + \frac{1}{\alpha} d\alpha = \frac{1}{T} dT$$

i.e., $$\frac{d\alpha}{\alpha} = \frac{dT}{T}$$ [on isobaric surface p = const. and R' = const.] .....(i)

$$\frac{T}{\theta} = \left(\frac{p}{1000} \right)^k \qquad \text{(Poisson's equation)}$$

$$\ln T - \ln \theta = k [\ln p - \ln 1000]$$

$$\frac{1}{T} dT - \frac{1}{\theta} d\theta = k \left[\frac{1}{p} dp \right]$$

i.e., $\frac{dT}{T} = \frac{d\theta}{\theta}$ ($\because$ dp = 0 on isobaric surface) .....(ii)

From (i) and (ii) we have

$$\frac{d\alpha}{\alpha} = \frac{dT}{T} = \frac{d\theta}{\theta}$$

$$\Rightarrow \quad \frac{\nabla\alpha}{\alpha} = \frac{\nabla T}{T} = \frac{\nabla\theta}{\theta} \qquad .....(21.10)$$

From eq. (21.9) and (21.10) we have

$$\frac{\partial \vec{V}_g}{\partial z} = \frac{g}{f}k \times \frac{\nabla\alpha}{\alpha} = \frac{g}{f}k \times \frac{\nabla T}{T} = \frac{g}{f}k \times \frac{\nabla\theta}{\theta} \qquad ....(21.11)$$

Equation (21.11) is called thermal wind equation. The second relation

viz, $$\frac{\partial \vec{V}_g}{\partial z} = \frac{g}{f}k \times \frac{\nabla T}{T}$$

indicates that the wind increase per unit distance is determined by the horizontal temperature gradient and is directed along the isotherm with cold air to the left as shown in Fig. 21.3, $\frac{\partial \vec{V}_g}{\partial z}$ is called the thermal wind.

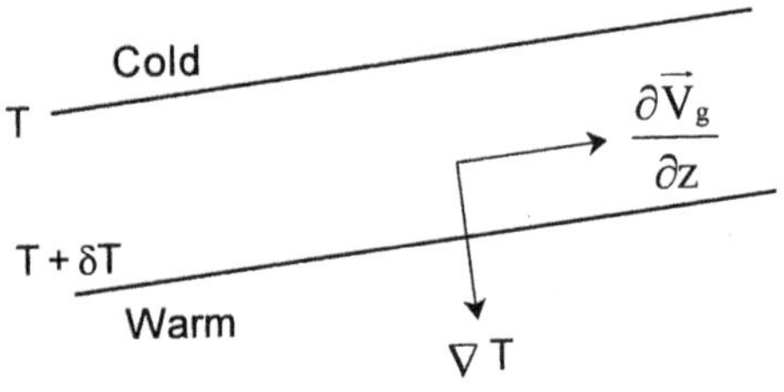

Fig. 21.3

Hodograph : Let the wind observations (pilot Balloon or Rawin) taken during an ascent be plotted with a common origin. The curve through the end points of the wind vectors is called hodograph (See Fig. 21.4). Thermal wind is tangent to the hodograph.

Note : Clock wise turning of wind is called veering. Anti-clockwise turning of wind is called backing Fig. 21.5.

Let $$\frac{\partial \vec{V}}{\partial z} \simeq \frac{\partial \vec{V}_g}{\partial z}$$

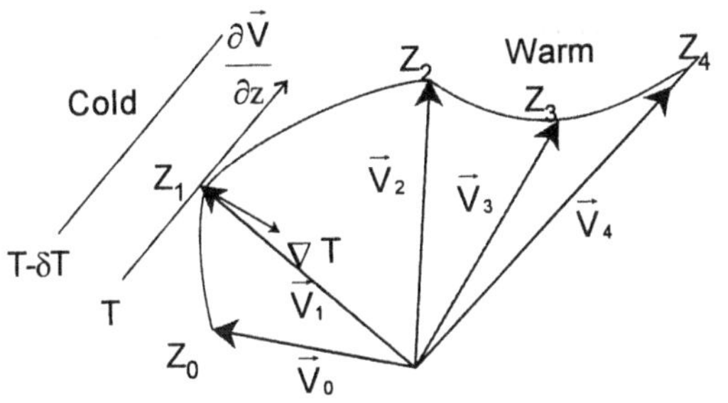

Fig. 21.4 Hydrograph.

Where Z_0, Z_1, Z_2, Z_3 ..... are heights and $\vec{V}_0$ $\vec{V}_1$ $\vec{V}_2$...... are wind velocities at these heights.

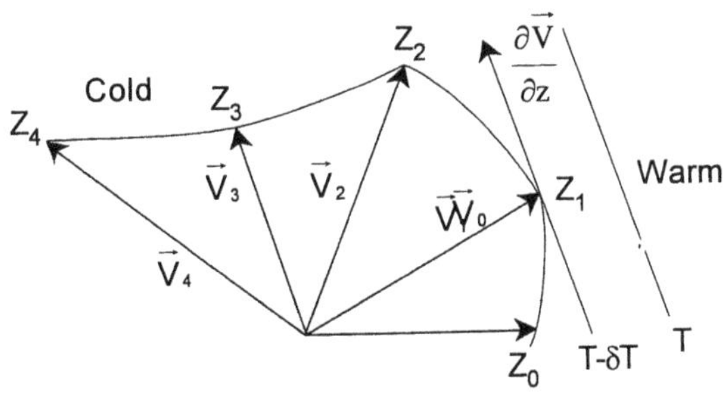

Fig. 21.5

It follows from the Fig. 21.4 that the direction of the isotherms coincide with the direction of $\dfrac{\partial \vec{V}_g}{\partial z}$.

From the direction of the isotherms we can find $\nabla_p T$ as follows :

$$\frac{\partial \vec{V}_g}{\partial z} = -\frac{g}{f} k \times \frac{\nabla_p T}{T} \qquad .....21.12$$

or $$\left|\nabla_p T\right| = \frac{f}{g} T \left|\frac{\partial \vec{V}_g}{\partial z}\right| \qquad ....(21.12')$$

By assuming wind to be geostrophic, we can find the temperature advection with the help of hodograph.

[Note : Horizontal transfer of heat (or a physical quantity) by wind is called advection of heat (or a physical quantity)].

Wind aloft at any level is considered to have two components (i) Geostrophic wind above friction layer (ii) A wind blowing along the mean isotherm with low temperature to its left in the northern hemisphere (or to its right in the southern hemisphere). The second component is the thermal wind Fig. 21.6. Thus thermal wind ($\vec{V}_t$) in a layer is the vector difference of geostrophic wind at upper ($\vec{V}_g U$) and lower levels $\vec{V}_g L$ i.e., $\vec{V}_t = \vec{V}_g U - \vec{V}_g L$.

It is a fictitious wind and it has no actual existence. However the concept is of great practical utility in weather prognosis.

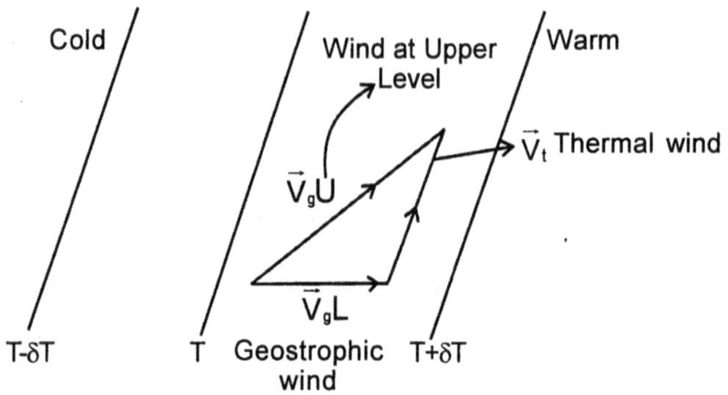

Fig. 21.6

Geostrophic temperature advection is defined as $-\vec{V}_g.\nabla_p T$. Advection may be regarded as the change in temperature in horizontal in adiabatic flow. The thermodynamic equation is

$$\left(\frac{\partial T}{\partial t}\right)_{adv} = -\vec{V}_g.\nabla_p T$$

.....(21.13)

From (eq. 21.11) and Fig. (21.7) we have

$$\nabla_p T = -\frac{f}{g} Tk \times \frac{\partial \vec{V}_g}{\partial z}$$

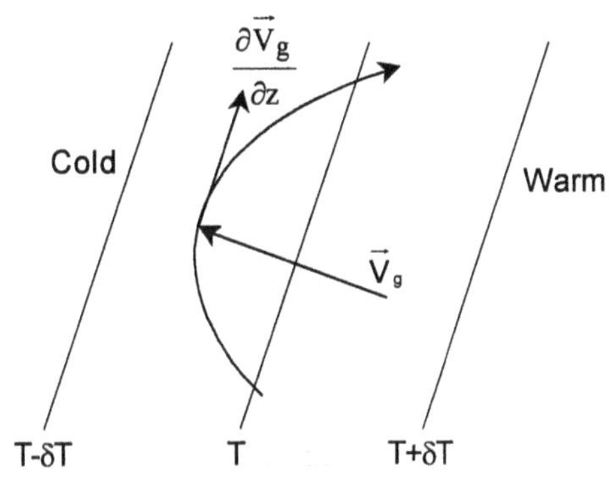

Fig. 21.7

Substituting this in (eq. 21.13), we have

$$\left(\frac{\partial T}{\partial t}\right)_{adv} = \frac{f}{g} T \vec{V}_g \cdot \left(k \times \frac{\partial \vec{V}_g}{\partial z}\right)$$

$$= -\frac{f}{g} Tk \cdot \left(\vec{V}_g \times \frac{\partial \vec{V}_g}{\partial z}\right) \quad \text{(by property of vector product)}$$

We have two typical cases

(i) Consider Fig. 21.7. In this case the wind is veering with height. Horizontal geostrophic wind turns to the right with increasing height. $\left(\frac{\partial T}{\partial t}\right)_{adv}$ is positive hence indicates warm air advection i.e. veering of geostrophic wind indicates warm air advection.

(ii) Consider Fig. 21.8. In this case the wind is backing with height. $\left[\frac{\partial T}{\partial t}\right]_{adv}$ is negative, hence indicates cold air advection. Horizontal geostrophic wind turns to the left with increasing height. i.e., backing of geostropic wind with height indicates cold air advenction.

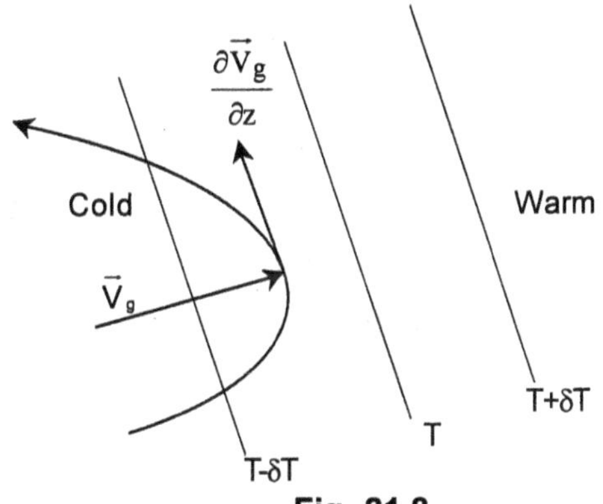

Fig. 21.8

1. The approximate formula for thermal wind $|\vec{V}_T| = V_T$ is given by

$$V_T = \frac{g}{f\bar{T}}\frac{\partial\bar{T}}{\partial n}\delta z$$

where $\frac{\partial\bar{T}}{\partial n}$ = Temperature gradient

$\bar{T}$ = Mean temperature of the layer

δz = Thickness of the layer

2. V_T is also given by

$$V_T = \frac{9.8}{f}\frac{\partial H}{\partial n}$$

where $\frac{\partial H}{\partial n}$ = horizontal gradient of relative geo-potential.

3. Geostrophic wind variation over small height h is given by

$$\frac{\partial\vec{V}_g}{\partial z} \simeq \frac{\vec{V}_g(z+h)-\vec{V}_g z}{h}$$

Barotropic Atmosphere : It was earlier defined as an atmosphere in which isobaric surfaces are also surfaces of constant density i.e., $\rho = \rho(p)$. We know from thermal wind (eq. 21.11)

$$\frac{\partial\vec{V}_g}{\partial z} = \frac{g}{f}k \times \frac{\nabla T}{T} \quad \text{(In a barotropic atmospherer } \nabla_p T = 0\text{)}$$

$$\therefore \frac{\partial\vec{V}_g}{\partial z} = 0 \text{ . i.e., } \vec{V}_g = \text{constant.}$$ = geostrophic wind is constant with height.

Baroclinic Atmosphere : It was earlier defined as an atmosphere which is not barotropic. Consequently in such an atmosphere $\vec{V}_g$ is (not constant) variable with height.

Note :

1. At a given location, thickness between two isobaric surfaces is proportional to mean virtual temperature. $\Delta z \, \alpha \, \bar{T}_v$.
2. Thermal wind is parallel to the isopleths of thickness (Δz) lines, consequently coincide with the mean virtual isotherms.
3. Magnitude of the thermal wind is proportional to the magnitude of the thickness gradient. Closer the thickness lines, stronger the thermal wind and vice versa.

4. In the northern hemisphere, looking downstream the cold area or lower thickness area will be on the left and warm area will be on the right. (The reverse of this in the southern hemisphere).

Questions

1. Define divergence of a geostrophic wind with a formula both in vector and scalar notation. Where do you expect geostrophic convergence and geostrophic divergence in a sinusoidal streamline wave?
2. Define vorticity of a geostrophic wind. Write its components u_g, v_g. How it fits this estimation in actual wind field? Derive the scalar formula for the vorticity of geostrophic wind.
3. Define thermal wind and derive thermal wind equation in the form

   $\frac{\partial \vec{V}_g}{\partial z} = \frac{g}{f} k \times \frac{\nabla T}{T}$ with usual notaion.

4. Define hodograph. How can we find temperature advection with the help of hodograph, assuming wind to be geostrophic?

CHAPTER 22

Circular Vortex Spherical Coordinates and Equation of Motion

In an absolute coordinate system, consider an atmosphere in which all particles rotate in circular orbits around a single axis of rotation. Further assume that the motion is same in all planes w.r.t the same axis of rotation (that is the motion is symmetrical about the axis of rotation).

Let λ be the angle in cylindrical coordinate system. See Fig. 22.1(a). $\overline{\Omega}$ velocity of rotation, $\vec{R}$ = radius vector, z = height.

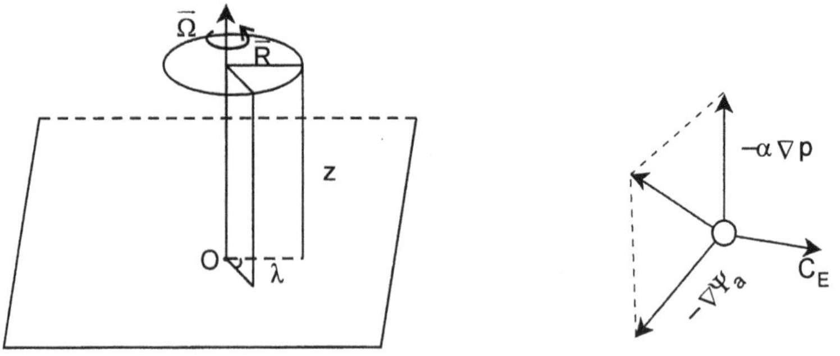

(a) (b)

Fig. 22.1 Circular vortex.

Here angular velocity $\Omega = \Omega\ (R, z)$; $\Psi_a = \Psi_a\ (R, z)$ geopotential :

The linear velocity $V = R\Omega$ and $V = V(R, z)$

$$\text{Spin} = RV = R\,.\,R\Omega = R^2\,\Omega$$

$$[V = \text{arc} = \text{radius} \times \text{radians} = R\Omega]$$

In this system the forces acting on a particle Fig. 22.1(b) are :

(i) Pressure force $(\alpha \nabla p)$

(ii) Gravitational force $\vec{g}_a \simeq \nabla \Psi_a$ and

(iii) Centrifugal force $C_E = \vec{\Omega} \times \left(\vec{\Omega} \times \vec{R}\right)$

Since the motion is stationary $\left(\dfrac{d\vec{V}}{dt} = 0\right)$, in absolute reference frame, we have the equation of motion,

$$-\alpha\,\nabla P - \nabla\Psi_a - \vec{\Omega} \times \left(\vec{\Omega} \times \vec{R}\right) = 0 \qquad \text{.....(22.1)}$$

Let i be the unit vector directed along the radius vector then $i = \nabla R$. Let j be the unit vector directed normal to the meridian, then $j = R \nabla \lambda$ and k the unit vector in the vertical direction, then $k = \nabla z$.

With this notation the centrifugal force C_E, which is directed outwards may be written as $C_E = -\vec{\Omega} \times \left(\vec{\Omega}\right) \times \vec{R} = \Omega^2 R \nabla R$

With this value, the equations of motion (eq. 22.1) becomes

$$-\alpha \nabla p - \nabla \Psi_a + \Omega^2 R \nabla R = 0$$

or $$-\alpha \nabla p - \nabla \Psi_a + \Omega^2 \nabla \left(\frac{1}{2} R^2\right) = 0 \qquad \text{.....(22.2)}$$

By definition, rotation of a vector $\vec{A}$ is given by $\nabla \times \vec{A}$.

Therefore, the rotation of (eq. 22.2), the circular vortex is given by

$$\nabla \times \left[-\alpha \nabla p - \nabla \Psi_a + \Omega^2 \nabla \left(\frac{1}{2} R^2\right)\right] = 0$$

Expanding

$$\nabla \alpha \times \left(-\nabla_p\right) - \nabla \times \nabla \Psi_a + \nabla \Omega^2 \times \nabla \left(\frac{1}{2} R^2\right) = 0$$

$$\therefore \qquad \nabla\alpha \times \left(-\nabla_p\right) = \nabla\Omega^2 \times \nabla\left(-\frac{1}{2}R^2\right) = \vec{S} \text{ say} \qquad .....(22.3)$$

[since $\nabla \times \nabla\Psi_a = 0$], where $\vec{S}$ = Solenoidal vector

equation (22.3) represents the motion of a circular vortex.

22.1 Spherical Coordinates and Equations of Motion

In meteorology the cartesian local coordinates system is sufficient for many purposes. However for the earth, the application of spherical coordinates (natural coordinates) is appropriate as there would be large distortion when we use local coordinates.

In spherical coordinates system, the three coordinates are longitude (λ), Latitude (ϕ) and distance r from the centre of the earth (sphere). Let i, j, k be the unit vectors directed eastward, northward and upward respectively.

As shown in Fig. 22.2(a), λ is measured positive eastward from Greenwich meridian, ϕ is measured from equator to north and south poles. $\phi = 0$ at equator and $\phi = \pi/2$ at north pole, while $\phi = \frac{-\pi}{2}$ at south pole. The radius vector $\vec{r}$ is measured from the centre of the earth and it is related to z (vertical distance from the surface of the earth) by $r = R + z$ where R = 6371 km, radius of the earth. The depth z of the atmosphere is very shallow compared to R and hence r is approximated to R. i.e., $r \sim R$.

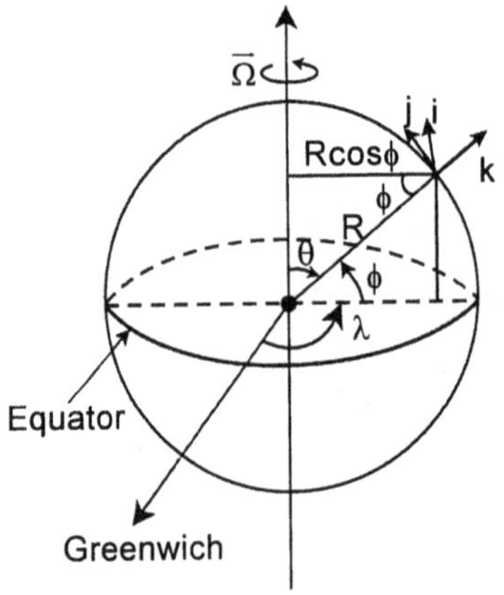

Fig. 22.2 (a)

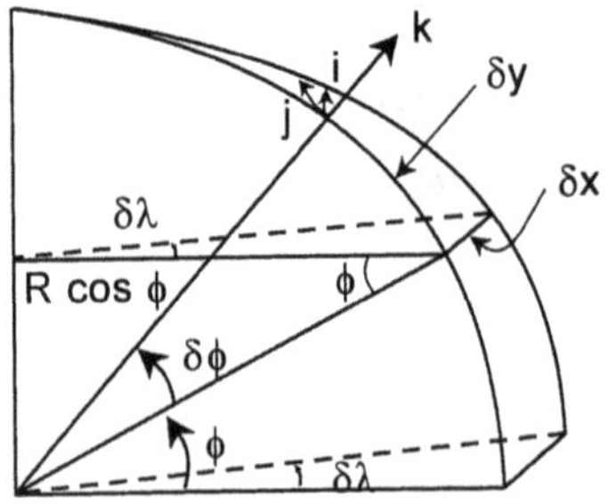

Fig. 22.2 (b)

x-directed eastward, *y*-directed north wards and *z*-directed vertical upwards along *R* [as in Fig. 22.2(b)]

$$dx = R\cos\phi\delta\lambda \qquad (\delta x = R\cos\phi.\delta\lambda)$$

$$dy = R\delta\phi \qquad (\delta y = R\delta\phi)$$

$$dz = dR \qquad .....(22.4)$$

The unit vectors i, j, k are functions of position (and hence not constant in spherical coordinates). They vary with time

Let $\vec{V} = iu + jv + kw$ .....(22.5)

($\vec{V}$ = velocity vector, u, v, w components of velocity in x, y, z directions)

$$\therefore \quad \frac{d\vec{V}}{dt} = i\frac{du}{dt} + j\frac{dv}{dt} + k\frac{dw}{dt} + u\frac{d\vec{i}}{dt} + v\frac{d\vec{j}}{dt} + w\frac{d\vec{k}}{dt} \qquad .....(22.6)$$

To evaluate $\frac{d\vec{i}}{dt}$:

$$\frac{d\vec{i}}{dt} = \frac{\partial\vec{i}}{\partial t} + \frac{\partial\vec{i}}{\partial x}\frac{\partial x}{\partial t} + \frac{\partial\vec{i}}{\partial y}\frac{\partial y}{\partial t} + \frac{\partial\vec{i}}{\partial z}\frac{\partial z}{\partial t}$$

$$= u\frac{\partial\vec{i}}{\partial x} + v\frac{\partial\vec{i}}{\partial y} + w\frac{\partial\vec{i}}{\partial z} \qquad \left(\because \frac{\partial\vec{i}}{\partial t} = 0\right)$$

i is a function of x only (i.e., eastward directed vector is not effected by north-south motion or vertical motion $\frac{\partial\vec{i}}{\partial y} = \frac{\partial\vec{i}}{\partial z} = 0$).

$$\therefore \quad \frac{d\vec{i}}{dt} = u\frac{\partial\vec{i}}{\partial x} \qquad .....(22.7)$$

From Fig. 22.2(c) from similar triangles we have,

$$\lim_{\delta x\to 0}\frac{|\delta\vec{i}|}{\delta x} = \left|\frac{\partial\vec{i}}{\partial x}\right| = \frac{|i|}{R\cos\phi} = \frac{1}{R\cos\phi}$$

The direction of the vector $\frac{\partial\vec{i}}{\partial x}$ is from west to east as shown in Fig. 22.2(c).

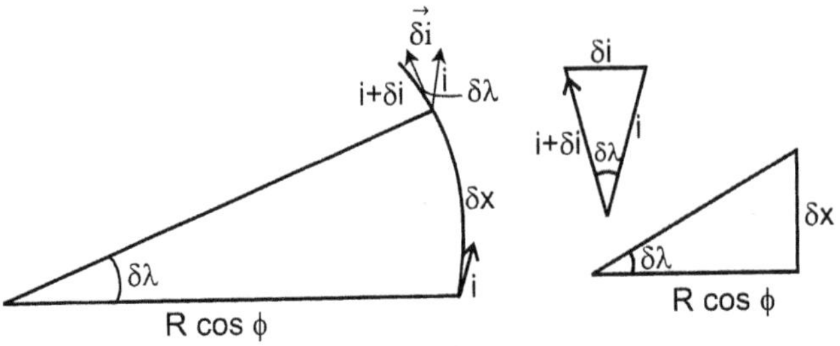

Fig. 22.2(c)

From Fig. 22.2(d). we see that $\vec{\delta i}$ has two components (j sin ϕ, –k cos ϕ)

$$\therefore \qquad \frac{\partial \vec{i}}{\partial x} = \frac{1}{R\cos\phi}\,[j \sin\phi - k \cos\phi]$$

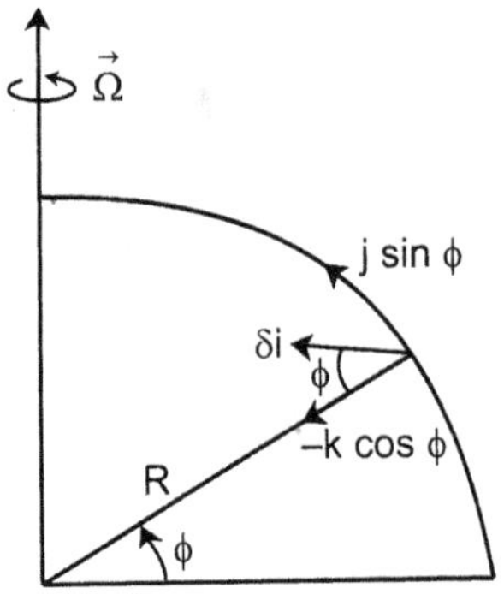

Fig. 22.2(d)

From eq. 22.7,
$$\frac{d\vec{i}}{dt} = u\frac{\partial \vec{i}}{\partial x} = \frac{u(j\sin\phi - k\cos\phi)}{R\cos\phi} \qquad .....(22.8)$$

Note :

$$|\vec{A}| = A = \sqrt{A_x^2 + A_y^2 + A_z^2}$$

$$\left|\frac{\partial \vec{i}}{\partial x}\right| = \frac{1}{R\cos\theta}\sqrt{j^2 \sin^2\phi + k^2 \cos^2\phi} = \frac{1}{R\cos\phi}$$

To evaluate $\frac{d\vec{j}}{dt}$:

we know $\frac{d\vec{j}}{dt} = \frac{\partial \vec{j}}{\partial t} + u\frac{\partial \vec{j}}{\partial x} + v\frac{\partial \vec{j}}{\delta y} + w\frac{\partial \vec{j}}{\partial z}$

j is a function of x and y only, hence $\frac{\partial \vec{j}}{\partial t} = 0,\ \frac{\partial \vec{j}}{\partial z} = 0,$

$$\therefore \qquad \frac{d\vec{j}}{dt} = u\frac{\partial \vec{j}}{\partial x} + v\frac{\partial \vec{j}}{\partial y} \qquad .....(22.9)$$

From similar triangles [Fig. 22.2(e)]

$$\left|\frac{\delta \vec{j}}{\delta x}\right| = \frac{|j|}{h} = \frac{1}{R\cot\phi} = \frac{\tan\phi}{R} \quad \left(\sin\phi = \frac{R\cos\phi}{h} \ \therefore h = R\cot\phi\right)$$

$$\therefore \qquad \frac{\partial \vec{j}}{\partial x} = -\frac{i\tan\phi}{R} \qquad .....(22.10)$$

[since $\vec{\delta j}$ is directed in negative direction of *x* i.e., east to west]

For north ward motion $|\delta\vec{j}| = \delta\phi$ (See Fig. 22.2(f) since $\left|\frac{\delta j}{j}\right| = \delta\phi$

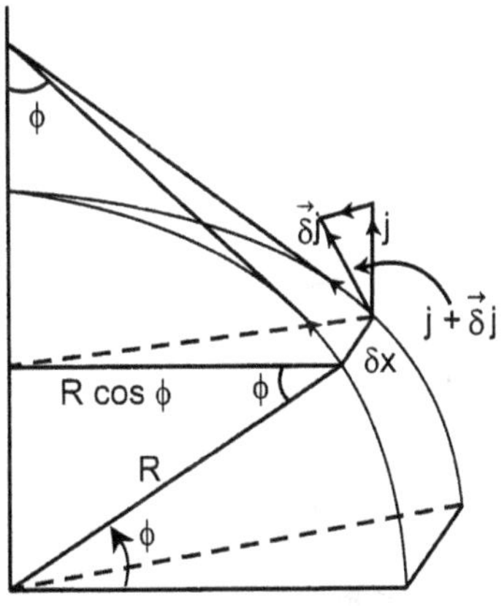

Fig. 22.2(e)

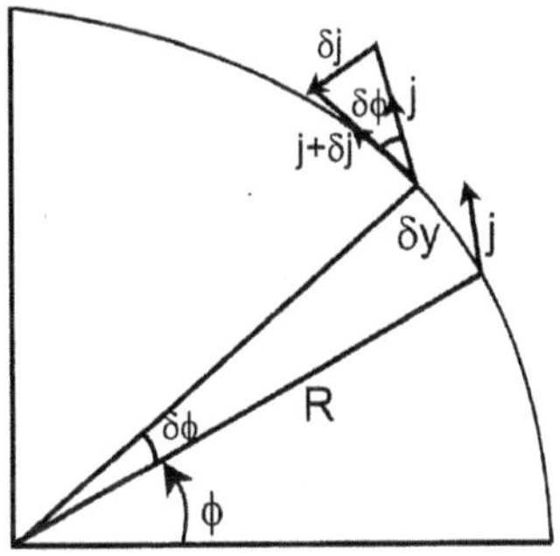

Fig. 22.2(f)

$$\frac{\delta y}{R} = \delta\phi \qquad \left(\frac{\text{Arc}}{\text{radius}} = \text{radian}\right)$$

$$\delta y = R\delta\phi$$

$$\therefore \qquad \frac{|\partial \vec{j}|}{\partial y} = \frac{\delta\phi}{R\delta\phi} = \frac{1}{R}$$

$$\frac{\partial \vec{j}}{\partial y} = \frac{(-k)}{R}, \text{ since } \vec{\delta j} \text{ is directed down wards} \qquad ....(22.11)$$

using eq. (22.10) and (22.11)

$$\frac{d\vec{j}}{dt} = u\frac{\partial \vec{j}}{\partial x} + v\frac{\partial \vec{j}}{\partial y} \quad \text{eq. (22.9) becomes}$$

$$\frac{d\vec{j}}{dt} = u\frac{(-i\tan\phi)}{R} + v\left(\frac{-k}{R}\right) = -i\frac{u\tan\phi}{R} - k\frac{v}{R} \qquad .....(22.12)$$

To evaluate $\frac{d\vec{k}}{dt}$:

we know $\frac{d\vec{k}}{dt} = \frac{\partial \vec{k}}{\partial t} + u\frac{\partial \vec{k}}{\partial x} + v\frac{\partial \vec{k}}{\partial y} + w\frac{\partial \vec{k}}{\partial z}$

Here $\frac{\partial \vec{k}}{\partial t} = 0, \; \frac{\partial \vec{k}}{\partial z} = 0,$

$$\therefore \qquad \frac{d\vec{k}}{dt} = u\frac{\partial \vec{k}}{\partial x} + v\frac{\partial \vec{k}}{\partial y}$$

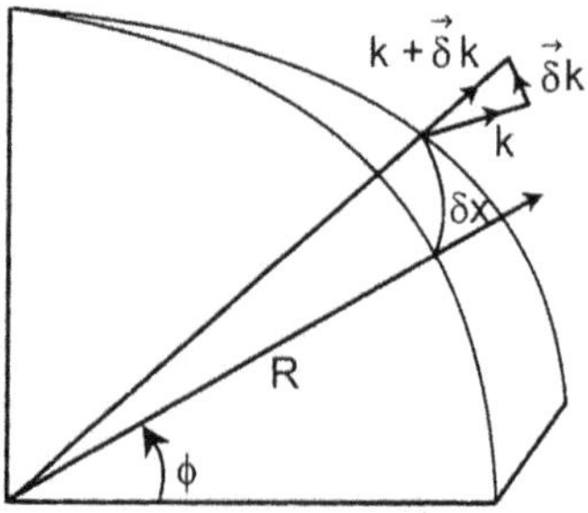

Fig. 22.2(g)

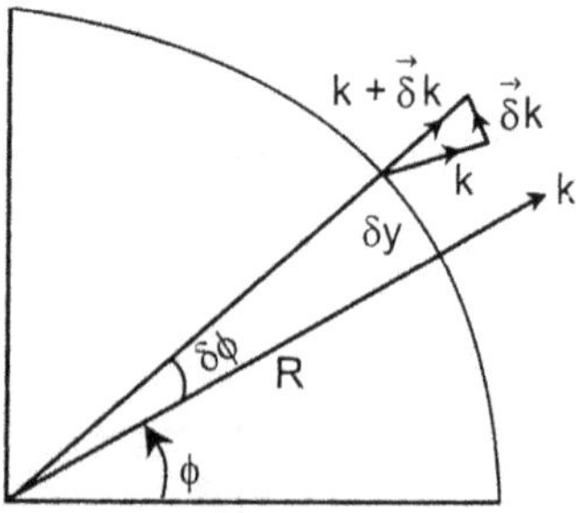

Fig. 22.2(h)

From similar triangles [Fig. 22.2(g) and Fig. 22.2(h)].

we have $\left|\frac{\partial \vec{k}}{\partial x}\right| = \frac{|k|}{R} = \frac{1}{R}$ or $\frac{\partial \vec{k}}{\partial x} = \frac{i}{R}$

$$\left|\frac{\partial \vec{k}}{\partial y}\right| = \frac{|k|}{R} = \frac{1}{R} \quad \text{or} \quad \frac{\partial \vec{k}}{\partial y} = \frac{j}{R}$$

$$\therefore \qquad \frac{d\vec{k}}{dt} = u\frac{\partial \vec{k}}{\partial x} + v\frac{\partial \vec{k}}{\partial y} \text{ becomes}$$

$$\frac{d\vec{k}}{dt} = u.\frac{i}{R} + v\frac{j}{R} = i\frac{u}{R} + j\frac{v}{R} \qquad \text{.....(22.13)}$$

Substituting the values of $\frac{d\vec{i}}{dt}, \frac{d\vec{j}}{dt}, \frac{d\vec{k}}{dt}$ from eq. 22.8, 22.12 and 22.13 in eq. 22.6 we have

$$\frac{d\vec{V}}{dt} = i\frac{du}{dt} + j\frac{dv}{dt} + k\frac{dw}{dt} + u\frac{di}{dt} + v\frac{dj}{dt} + w\frac{dk}{dt} \quad \text{eq. (22.6)}$$

$$= i\frac{du}{dt} + j\frac{dv}{dt} + k\frac{dw}{dt} + u\left[\frac{u(j\sin\phi - k\cos\phi)}{R\cos\phi}\right]$$

$$+v\left[-iu\frac{\tan\phi}{R} - k\frac{v}{R}\right] + w\left[i\frac{u}{R} + j\frac{v}{R}\right]$$

$$\frac{d\vec{V}}{dt} = i\left[\frac{du}{dt} - \frac{uv\tan\phi}{R} + \frac{uw}{R}\right] +$$

$$j\left[\frac{dv}{dt} + u^2\frac{\tan\phi}{R} + \frac{vw}{R}\right] + k\left[\frac{dw}{dt} - \frac{u^2}{R} - \frac{v^2}{R}\right] \qquad \text{.....(A)}$$

In order to find the equation of motion in spherical coordinates we shall find (i) Coriolis force $2\vec{\Omega} \times \vec{V}$, (ii) Pressure gradient force $\alpha\nabla p$ and (iii) Frictional force in spherical coordinates.

1. Coriolis Force

$$-2\vec{\Omega}\times\vec{V} = -2\Omega\begin{vmatrix} i & j & k \\ 0 & \cos\phi & \sin\phi \\ u & v & w \end{vmatrix}$$

(since coriolis force has no component parallel to *i*, while the components along j, k are $2\Omega\cos\phi$, $2\Omega\sin\phi$).

$$= -i\,[2\Omega w\cos\phi - 2\Omega v\sin\phi] - j\,[2\Omega u\sin\phi] + k\,[2\Omega u\cos\phi]$$

$$-2\vec{\Omega}\times\vec{V} = -i\,[we - vf\,] - juf + keu \qquad \text{.....(B)}$$

where $f = 2\Omega\sin\phi$; $e = 2\Omega\cos\phi$

2. Pressure Gradient Force

$$\nabla p = i\frac{\partial p}{\partial x} + j\frac{\partial p}{\partial y} + k\frac{\partial p}{\partial z} = i\frac{\partial p}{\partial\lambda}\frac{\partial\lambda}{\partial x} + j\frac{\partial p}{\partial\phi}\frac{\partial\phi}{\partial y} + k\frac{\partial p}{\partial z}$$

$$= i\frac{1}{R\cos\phi}\frac{\partial p}{\partial\lambda} + j\frac{1}{R}\frac{\partial p}{\partial\phi} + k\frac{\partial p}{\partial z}$$

From eq. (22.4)

$$\left[\because \frac{\partial\lambda}{\partial x} \simeq \frac{1}{R\cos\phi}\;\frac{\partial\phi}{\partial y} = \frac{1}{R}\right]$$

and gravity is represented by $\vec{g} = -gk$ .....(C)

3. Frictional Components are

$$\vec{F} = iF_x + jF_y + kF_z \qquad \text{.....(D)}$$

We know the equation of motion in cartesion coordinates are given by

$$\frac{d\vec{V}}{dt} = -\alpha\nabla p - 2\vec{\Omega}\times\vec{V} + \vec{g} + \alpha\vec{F}$$

Substituting values from (A), (B), (C), (D) in the above equation we have

$$i\left[\frac{du}{dt} - \frac{uv\tan\phi}{R} + \frac{uw}{R}\right] + j\left[\frac{dv}{dt} + \frac{u^2\tan\phi}{R} + \frac{vw}{R}\right] + k\left[\frac{dw}{dt} - \frac{u^2}{R} - \frac{v^2}{R}\right]$$

$$= -i\alpha\left[\frac{1}{R\cos\phi}\cdot\frac{\partial p}{\partial\lambda}\right] - j\alpha\left[\frac{1}{R}\cdot\frac{\partial p}{\partial\phi}\right] - k\alpha\frac{\partial p}{\partial z} +$$

$$[-i\,(we - vf\,) - juf + keu] - gk + i\alpha F_x + j\alpha F_y + k\alpha F_z$$

Equating the like terms (coefficients of i, j, k) we have

$$\frac{du}{dt} - \frac{uv\tan\phi}{R} + \frac{uw}{R} = -\frac{\alpha}{R\cos\phi}\frac{\partial p}{\partial \lambda} - we + vf + \alpha F_x$$

$$\frac{dv}{dt} + \frac{u^2\tan\phi}{R} + \frac{vw}{R} = -\frac{\alpha}{R}\frac{\partial p}{\partial \phi} - uf + \alpha\, F_y$$

$$\frac{dw}{dt} - \frac{u^2}{R} - \frac{v^2}{R} = -\alpha\frac{\partial p}{\partial z} + eu - g + \alpha\, F_y$$

Rearranging, we have

$$\left.\begin{aligned}
\frac{du}{dt} &= -\frac{\alpha}{R\cos\phi}\frac{\partial p}{\partial \lambda} + fv - ew + uv\frac{\tan\phi}{R} - \frac{uw}{R} + \alpha\, F_x && \text{(i)}\\
\frac{dv}{dt} &= -\frac{\alpha}{R}\frac{\partial p}{\partial \phi} - fu - \frac{u^2\tan\phi}{R} - \frac{vw}{R} + \alpha F_y && \text{(ii)}\\
\frac{dw}{dt} &= -\alpha\frac{\partial p}{\partial z} - g + eu + \frac{u^2}{R} + \frac{v^2}{R} + \alpha F_z && \text{(iii)}
\end{aligned}\right\} \quad .....(22.14)$$

Eq. (22.14) (i, ii, iii) are the equations of motion in spherical coordinates.

The individual time derivative may be written as

$$\frac{d}{dt} = \frac{\partial}{\partial t} + u\frac{\delta}{\delta x} + v\frac{\partial}{\partial y} + w\frac{\partial}{\partial z}$$

$$= \frac{\partial}{\partial t} + u\frac{\partial}{\partial \lambda}\frac{\partial \lambda}{\partial x} + v\frac{\partial}{\partial \phi}\frac{\partial \phi}{\partial y} + w\frac{\partial}{\partial z}$$

$$\frac{d}{dt} = \frac{\partial}{\partial t} + u\frac{1}{R\cos\phi}\frac{\partial}{\partial \lambda} + v\frac{1}{R}\frac{\partial}{\partial \phi} + w\frac{\partial}{\partial z} \quad .....(22.15)$$

Note : In the eq. (22.14) (i, ii, iii) if $u \simeq v \approx 10\ ms^{-1}$, $\frac{u^2}{R}, \frac{v^2}{R}$ are of the magnitude $10^{-5}\ ms^{-2}$.

Compared to g ~ 10 ms^{-2}, $\frac{u^2}{R}, \frac{v^2}{R}$ and other second order terms can be neglected. Equations 22.14 (i,ii,iii) reduces to

$$\frac{du}{dt} = -\frac{\alpha}{R\cos\phi}\frac{\partial p}{\partial \lambda} + fv - ew + \alpha F_x$$

$$\frac{dv}{dt} = -\frac{\alpha}{R}\frac{\partial p}{\partial \phi} - fu + \alpha F_y$$

$$\frac{dw}{dt} = -\alpha\frac{\partial p}{\partial z} - g + eu + \alpha F_z \qquad .....(22.16)$$

22.2 Divergence in Horizontal Flow in Spherical Coordinates

Consider an elemental lamina ABCD with velocity components u, v along λ (longitude) and φ (latitude) at the center of the lamina as shown in Fig. 22.3(a). Consider the Figs. 22.3(a), 22.3(b), 22.3(c) and 22.3(d).

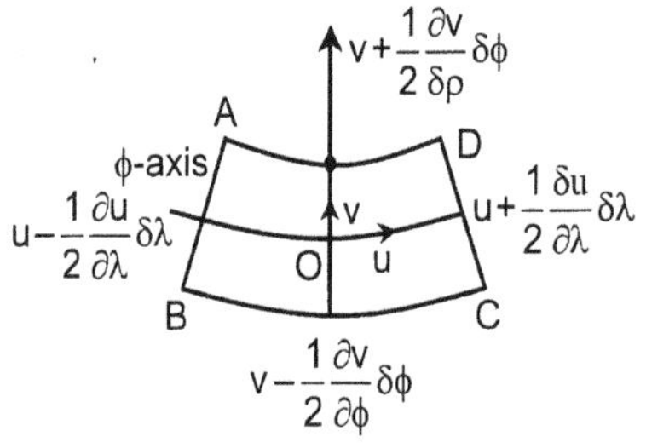

Fig. 22.3(a)

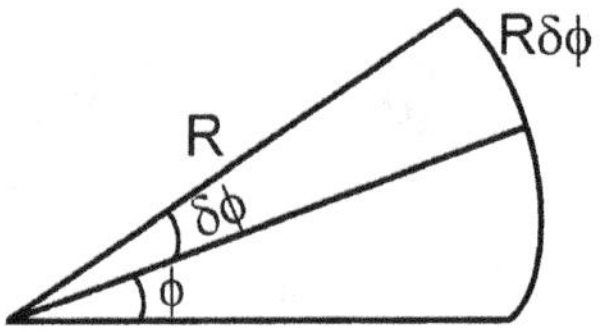

Fig. 22.3(b)

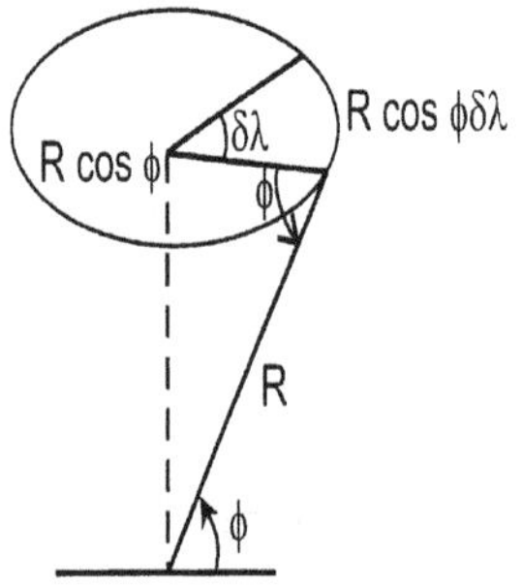

Fig. 22.3(c)

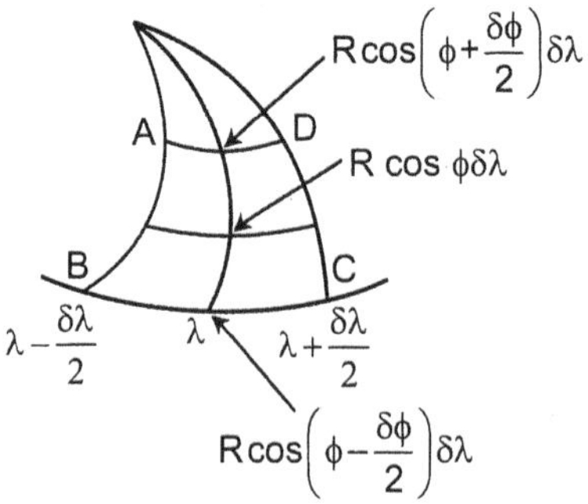

Fig. 22.3(d)

We find velocity of flow across

$$\left.\begin{aligned} BC &= R\cos\left(\phi - \frac{\delta\phi}{2}\right)\delta\lambda \\ AD &= R\cos\left(\phi + \frac{\delta\phi}{2}\right)\delta\lambda \\ CD &= AB = R\delta\phi \end{aligned}\right\} \qquad .....(i)$$

Net flow across the faces CD, AB = [CD × velocity along λ at the centre of CD –AB × velocity along λ at the centre of AB]

$$= \left(u + \frac{1}{2}\frac{\partial u}{\partial \lambda}\delta\lambda\right) R\delta\phi - \left(u - \frac{1}{2}\frac{\partial u}{\partial \lambda}\delta\lambda\right) R\delta\phi$$

$$= R\frac{\partial u}{\partial \lambda}\delta\phi\delta\lambda \qquad .....(ii)$$

Net flow across the faces AD, BC =

$$\left(v + \frac{1}{2}\frac{\partial v}{\partial \phi}\delta\phi\right)\left[R\cos\left(\phi + \frac{\delta\phi}{2}\right)\delta\lambda\right]$$

$$-\left(v - \frac{1}{2}\frac{\partial v}{\partial \phi}\delta\phi\right)\left[R\cos\left(\phi - \frac{\delta\phi}{2}\right)\delta\lambda\right] \qquad .....(iii)$$

∴ Net flow across ABCD = (ii + iii)

$$= \frac{\partial u}{\partial \lambda}\delta\phi\delta\lambda +$$

$$\left\{\left(v + \frac{1}{2}\frac{\partial v}{\partial \phi}\delta\phi\right)\left[R\cos\left(\phi + \frac{\delta\phi}{2}\right)\delta\lambda\right] - \left(v - \frac{1}{2}\frac{\partial v}{\partial \phi}\delta\phi\right)\left[R\cos\left(\phi - \frac{\delta\phi}{2}\right)\delta\lambda\right]\right\}$$

$$= R\frac{\partial u}{\partial \lambda}\delta\phi\delta\lambda + Rv\delta\lambda\left[\cos\left(\phi + \frac{\delta\phi}{2}\right) - \cos\left(\phi - \frac{\delta\phi}{2}\right)\right] +$$

$$+ \frac{\partial v}{\partial \phi}\frac{\delta\phi\delta\lambda}{2}\left[\cos\left(\phi + \frac{\delta\phi}{2}\right) + \cos\left(\phi - \frac{\delta\phi}{2}\right)\right]$$

[Note :

$$\cos C - \cos D = -2\sin\frac{C+D}{2}\sin\frac{C-D}{2},$$

$$\cos C + \cos D = 2\cos\frac{C+D}{2}\cos\frac{C-D}{2}\ ]$$

$$= \frac{\partial u}{\partial y} R\,\delta\phi\,\delta\lambda - 2\,Rv\,\delta\lambda \sin\phi.\sin\frac{\delta\phi}{2} + 2\frac{\partial v}{\partial\phi}\frac{\delta\phi\delta\lambda}{2}.\cos\phi.\cos\frac{\delta\phi}{2}.R \qquad \text{(iv)}$$

Dividing by the area [= (R cos ϕ $\delta\lambda$) (R $\delta\phi$) = R^2 cos ϕ $\delta\lambda\delta\phi$] (v)

the above equation becomes

The Net flow per unit area or divergence= [eq. iv ÷ eq. v]

$$\text{Div } \vec{V} = \frac{\partial u}{\partial\lambda}\frac{1}{R\cos\phi} - \frac{v}{R}\tan\phi\left(\frac{2\sin\frac{\delta\phi}{2}}{\delta\phi}\right) + \frac{\partial v}{R\,\partial\phi}\left(\cos\frac{\delta\phi}{2}\right)$$

In the limit, we have

$$\delta\phi \to 0 \; \frac{2\sin\frac{\delta\phi}{2}}{\delta\phi} = 1,\ \cos\frac{\delta\phi}{2} = 1$$

$$\text{Div } \vec{V} = \nabla_h.\vec{V} = \frac{\partial u}{\partial\lambda}\frac{1}{R\cos\phi} - \frac{v}{R}\tan\phi + \frac{1}{R}\frac{\partial v}{\partial\phi} \qquad \text{.....(vi)}$$

or

$$\text{Div } \vec{V} = \frac{\partial u}{\partial x} + \frac{\partial v}{\partial y} - \frac{v\tan\phi}{R} \qquad \text{.....(22.17)}$$

[∵ $\delta x = R\cos\phi.\delta\lambda$; $\delta y = R\delta\phi$]

eq. (22.17) is the horizontal divergence equation in spherical coordinates.

22.3 Kinematics of the Pressure Field

A meteorological surface chart contains wind direction and velocity, pressure, 24 hour pressure change, temperature, dew point, cloud type amount, rainfall etc. Of these pressure field is a predominant feature, depicted by isobars, which shows low pressure, high pressure centres, troughs, ridges (or wedges) and certain lines along which pressure gradient shows discontinuity (which are called fronts). These features of pressure field analysis is shown in the Fig. 22.5.

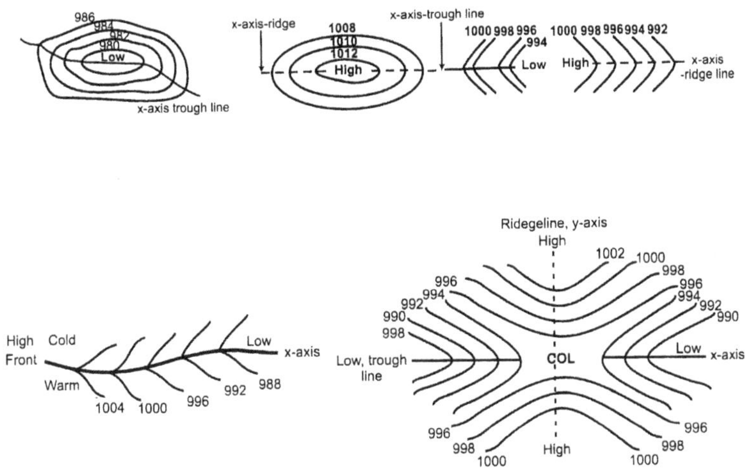

Fig. 22.5 Pressure fields, Showing Low, High, Trough, Ridge, Col, Warm and Cold Fronts.

Let f(x, y, z, t) be any meteorological parameter, which changes both in space and time. As said above the parameters may be scalar (like pressure, rainfall etc.) or a vector (wind direction etc.). We shall consider a scalar field parameter. At any synoptic time, the instantaneous distribution of pressure (p) is represented by isobars (equal pressure lines on sea level). The family of isobars may be denoted by p(x, y, t) = const. The local pressure changes (24 hour pressure change at a place -P_{24}) may be denoted by $\frac{\partial p}{\partial t}$. The isopleths of

$\frac{\partial p}{\partial t}$ = const. are called isallobars. Which are denoted by

$$\frac{\partial p}{\partial t} = b(x, y) = b = \text{const.}$$

For any field parameter f, we can represent by the equation

$$\frac{df}{dt} = \frac{\partial f}{\partial t} + \vec{C}.\nabla f \qquad .....(A)$$

where $\frac{\partial f}{\partial t}$ = local derivative of the parameter f

∇f = gradient of f in the horizontal field

$\vec{C}$ = velocity of the movement

$\frac{df}{dt}$ = total derivative of the parameter f .

We shall consider pressure field p(x, y, t). Put p in place of f in the above equation (A), which gives

$$\frac{dp}{dt} = \frac{\partial p}{\partial t} + \vec{C}.\nabla p \qquad .....(22.17)$$

Along isobar p = const. and hence $\frac{dp}{dt} = 0$. This reduces the above equation to

$$\frac{\partial p}{\partial t} + \vec{C}.\nabla p = 0$$

Equivalently we have

$$\frac{\partial p}{\partial t} + c_n \frac{\partial p}{\partial n} = 0$$

where $\frac{\partial p}{\partial n}$ = magnitude of the horizontal pressure gradient.

or

$$c_n = \frac{-\partial p/\partial t}{\partial p/\partial n} \qquad .....(22.18)$$

It follows from eq. 22.18 that c_n directly varies with pressure tendency $\frac{\partial p}{\partial t}$ and varies inversely with the pressure gradient $\frac{\partial p}{\partial n}$.

22.4 Movement of Troughs and Ridges

Axis of a trough (or a ridge) may be treated as a line along which the pressure attains relative minima (or maxima). If we suppose axis of trough (or ridge) as x-axis, then from calculus the conditions of minima or maxima we have:

For a trough line (condition for Minima) $\frac{\partial p}{\partial x} = 0$ and $\frac{\partial^2 p}{\partial x^2} > 0$

For a ridge line (condition for Maxima) $\frac{\partial p}{\partial x} = 0$ and $\frac{\partial^2 p}{\partial x^2} < 0$

On the axis of trough/ridge $\frac{\partial p}{\partial x} = 0, \quad \therefore \frac{d}{dt}\left(\frac{\partial p}{\partial x}\right) = 0$

From eq. (A) we have (replacing f by $\left(\frac{\partial p}{\partial x}\right)$

$$\frac{d}{dt}\left(\frac{\partial p}{\partial x}\right) = \frac{\partial}{\partial t}\left(\frac{\partial p}{\partial x}\right) + \vec{C}_n.\nabla\left(\frac{\partial p}{\partial x}\right)$$

$$0 = \frac{\partial}{\partial t}\left(\frac{\partial p}{\partial x}\right) + C_n \frac{\partial}{\partial x}\left(\frac{\partial p}{\partial x}\right) \quad \text{(since n concides with x-axis)}$$

$$\therefore \qquad C_n = -\frac{\dfrac{\partial^2 p}{\partial t \partial x}}{\dfrac{\partial^2 p}{\partial x^2}} \quad \text{(which gives the speed of the trough or ridge)}$$

22.5 Movement of Isallobars

For isallobars, put f = $\frac{\partial p}{\partial t}$ (pressure tendency) in eq. (A). which gives

or
$$\frac{d}{dt}\left(\frac{\partial p}{\partial t}\right) = \frac{\partial}{\partial t}\left(\frac{\partial p}{\partial t}\right) + \vec{C}.\nabla\left(\frac{\partial p}{\partial t}\right)$$

$$\frac{d}{dt}\left(\frac{\partial p}{\partial t}\right) = \frac{\partial^2 p}{\partial t^2} + C_n \frac{\partial}{\partial n}\left(\frac{\partial p}{\partial t}\right)$$

where n is perpendicular to the direction of isallobars.

Since $\frac{\partial p}{\partial t}$ = const. along the isallobars, therefore $\frac{d}{dt}\frac{\partial p}{\partial t} = 0$

The above equation reduces to

$$\frac{\partial^2 p}{\partial t^2} + C_n \frac{\partial^2 p}{\partial n \partial t} = 0 \quad \text{or} \quad C_n = -\frac{\dfrac{\partial^2 p}{\partial t^2}}{\dfrac{\partial^2 p}{\partial t \partial n}}$$

which gives the speed of the isallobars

22.6 Movement of a Front

Let C_n denote the movement of a front along x-axis, then

$$C_n = -\frac{\left(\dfrac{\partial p_c}{\partial t} - \dfrac{\partial p_w}{\partial t}\right)}{\left(\dfrac{\partial p_c}{\partial x} - \dfrac{\partial p_w}{\partial x}\right)}$$

where p_c = pressure on the cold side of the front

p_w = Pressure on the warm side of the front.

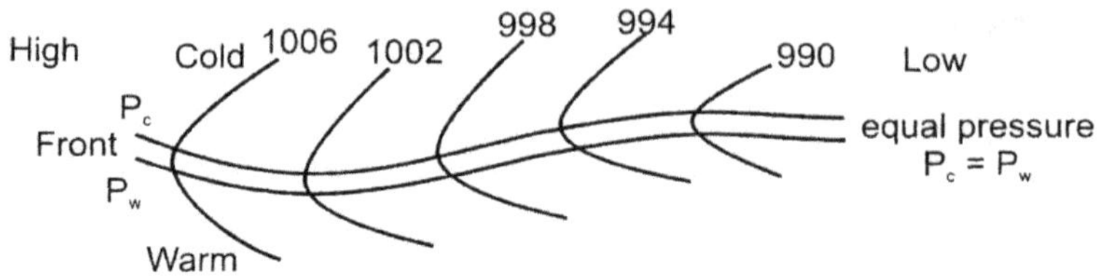

Fig. 22.6

22.7 Movement of a Col

In case of Col, let x-axis denote the trough line joining the two low pressure systems, while y axis denote the ridge line joining the two high pressure systems as shown in Fig. 22.5. In this case the following relations must satisfy simultaneously.

$$\frac{\partial p}{\partial x} = 0, \; \frac{\partial^2 p}{\partial x^2} > 0, \qquad \frac{\partial p}{\partial y} = 0, \; \frac{\partial^2 p}{\partial y^2} < 0$$

$$C_x = \frac{\dfrac{-\partial^2 p}{\partial x \partial t}}{\dfrac{\partial^2 p}{\partial x^2}}, \qquad C_y = -\frac{\dfrac{\partial^2 p}{\partial y \partial t}}{\dfrac{\partial^2 p}{\partial y^2}}$$

where C_x, C_y denote the movement of x-axis and y-axis respectively.

Note :

1. In case of cyclonic centre $\frac{\partial p}{\partial x} = 0, \; \frac{\partial^2 p}{\partial x^2} > 0$

   and anticyclonic centre $\frac{\partial p}{\partial y} = 0, \; \frac{\partial^2 p}{\partial y^2} < 0$.

2. A function f(x) has a maxima or minima at $x = x_1$

   if $f'(x_1) = 0$ and $f''(x_1) < 0$ (condition for maxima),

   and $f'(x_1) = 0$ and $f''(x_1) > 0$ (condition for minima).

Movement of a Cyclonic Centre or an Anticyclonic Centre

1. If the isobars are circular around a cyclonic centre, then

$$\frac{\partial^2 p}{\partial x^2} = \frac{\partial^2 p}{\partial y^2} \quad \text{and} \quad \frac{\partial^2 p}{\partial x^2} > 0, \; \frac{\partial^2 p}{\partial y^2} > 0$$

The centre of the cyclone will move in a negative direction of the isallobaric gradient i.e. $-\nabla\frac{\partial p}{\partial t}$.

2. If the isobars are circular around an Anticyclonic centre, then

$$\frac{\partial^2 p}{\partial x^2} = \frac{\partial^2 p}{\partial y^2} \quad \text{and} \quad \frac{\partial^2 p}{\partial x^2} < 0, \qquad \frac{\partial^2 p}{\partial y^2} < 0$$

The centre of the anticyclone will move in the direction of the isallobaric gradient i.e., $\nabla\frac{\partial p}{\partial t}$. In the above two cases the speed of the motion of the centre is inversely proportional to the curvature of the isobars.

22.8 Intensification and Weakening of a Pressure System

If the pressure gradient ∇p around a centre of a system, increases with time($\nabla p > 0$) then the system is said to be intensifying. On the contrary if the pressure gradient decreases with time ($\nabla p < 0$) around a centre, the system is said to be weakening. In simple terms increasing of number of isobars around the centre with time indicates intensification, while decreasing indicates weakening.

The same terminology is used in case of troughs (or ridges), that is, a trough of low pressure is said to be intensifying or weakening if the pressure gradient around the trough line increases or decreases respectively.

Deepening and Filling of a Pressure System : If the pressure around the centre of a system decreases $\left(\frac{dp}{dt} < 0\right)$ then the system is said to be deepening, but if it increases $\left(\frac{dp}{dt} > 0\right)$ it is said to be filling.

The same terminology is used with regard to troughs & ridges.

Note : Deepening and filling are not synonymous with intensification and weakening. The former terms are related to central pressure while the later terms are related to the pressure gradient.

22.9 The Rate of Intensification of Centre of a Pressure System

At the centre of a pressure system $\nabla p = 0$. The intensity of pressure of system consequently depends on the variation of ∇p with distance from the centre.

Let I denote the intensity of the system then I may be expressed as

$$I = \nabla \cdot \nabla p = \nabla^2 p = \frac{\partial^2 p}{\partial x} + \frac{\partial^2 p}{\partial y^2} = \text{Laplacian of p} \qquad .....(22.19)$$

∇^2 is invariant w.r.t coordinate axes, that is no effect on change in orientation of coordinate axes.

The intensification of a moving pressure centre is given by

$$\frac{dI}{dt} = \frac{\partial I}{\partial t} + \vec{C}.\nabla I \qquad \text{where } \vec{C} = iC_x + jC_y$$

using eq. 22.19 we have

$$\therefore \quad \frac{dI}{dt} = \frac{\partial}{\partial t}\left(\frac{\partial^2 p}{\partial x^2} + \frac{\partial^2 p}{\partial y^2}\right) + C_x \frac{\partial}{\partial x}\left(\frac{\partial^2 p}{\partial x^2} + \frac{\partial^2 p}{\partial y^2}\right) + C_y \frac{\partial}{\partial y}\left(\frac{\partial^2 p}{\partial x^2} + \frac{\partial^2 p}{\partial y^2}\right)$$

$$= \frac{\partial}{\partial t}\left(\frac{\partial^2 p}{\partial x^2} + \frac{\partial^2 p}{\partial y^2}\right) + C_x \left[\frac{\partial^3 p}{\partial x^3} + \frac{\partial^3 p}{\partial y^2 \partial x}\right] + C_y \left[\frac{\partial^3 p}{\partial x^2 \partial y} + \frac{\partial^3 p}{\partial y^3}\right]$$

By choosing the coordinate axes coinciding with the axes of symmetry, we have

$$\frac{\partial^3 p}{\partial x^3} = \frac{\partial^3 p}{\partial x^2 \partial y} = \frac{\partial^3 p}{\partial x \partial y^2} = \frac{\partial^3 p}{\partial y^3} = 0$$

$\therefore$ The above equation reduces to

$$\frac{dI}{dt} = \frac{\partial}{\partial t}\left(\frac{\partial^2 p}{\partial x^2} \frac{\partial^2 p}{\partial y^2}\right)$$

$$= \frac{\partial}{\partial x^2}\left(\frac{\partial p}{\partial t}\right) + \frac{\partial^2}{\partial y^2}\left(\frac{\partial p}{\partial t}\right)$$

$$= \frac{\partial^2 b}{\partial x^2} + \frac{\partial^2 b}{\partial y^2} \qquad \text{where } b = \text{Barometric tendency} = \frac{\partial p}{\partial t}$$

$$\frac{dI}{dt} = \nabla^2 b \qquad .....(22.20)$$

The last equation shows that the rate of intensification $\left(\frac{dI}{dt}\right)$ of the pressure field around a centre is determined by the instantaneous distribution of barometric tendency.

22.10 The Rate of Deepening (or filling) of a Centre of a Pressure System

This is precisely obtained from the eq. 22.17, namely

$$\frac{dp}{dt} = \frac{\partial p}{\partial t} + \vec{C} \cdot \nabla p \qquad \text{.....(i)}$$

As defined above, $\frac{dp}{dt} < 0$ stands for rate of deepening and $\frac{dp}{dt} > 0$ stands for rate of filling.

The local variation of pressure $\left(\frac{\partial p}{\partial t}\right)$ has two components.

(i) rate of change of the pressure system itself, which is called development part of barometric tendency.

(ii) Change due to the movement of the system, which is called movement part of the barometric tendency.

At the centre of the pressure system $\nabla p = 0$. Therefore the eq. (i) reduces to $\frac{dp}{dt} = \frac{\partial p}{\partial t}$ i.e., at the centre, the development tendency = the barometric tendency at a given station.

To extend the above concept of deepening and filling over an area, consider an elemental area δA enclosed by an isobar. Multiplying eq. (i) by δA, we have

$$\frac{dp}{dt}(\delta A) = \frac{\partial p}{\partial t}(\delta A) + \vec{C} . \nabla p(\delta A)$$

Integrating over a closed area A, we get

$$\oint_A \frac{dp}{dt}\delta A = \oint_A \frac{\partial p}{\partial t}\delta A + \vec{C} . \oint_A \nabla p(\delta A)$$

But $\oint_A \nabla p \delta A = 0$ on a closed curve, the above equation reduces to

$$\oint_A \frac{dp}{dt}\delta A = \oint_A \frac{\partial p}{\partial t}\delta A \qquad \text{.....(ii)}$$

It follows from eq. (ii), deepening or filling of a pressure system as a whole is equal to the sum of the barometric tendencies $\left(\frac{\partial p}{\partial t}\right)$ enclosed in the isobars is negative or positive.

This means that $\oint_A \frac{dp}{dt}\delta A < 0$ indicates deepening of the pressure system

and $\oint_A \frac{dp}{dt}\delta A > 0$ implies filling of the pressure system.

22.11 Deepening or Filling of a Trough (or Ridge)

Similarly as above, consider the eq. 22.17

$$\frac{dp}{dt} = \frac{\partial p}{\partial t} + \vec{C}.\nabla p$$

$$= \frac{\partial p}{\partial t} + C\frac{\partial p}{\partial x}$$

$$\left[\text{on the axis of trough line x (or ridge line x) } \nabla p = \frac{\partial p}{\partial x}\right]$$

On the axis of the trough line (or ridge line) $\frac{\partial p}{\partial x} = 0$, and hence the above equation reduces to $\frac{dp}{dt} = \frac{\partial p}{\partial t}$. This reveals that a trough deepens or fills according as $\frac{\partial p}{\partial t} < 0$, or $\frac{\partial p}{\partial t} > 0$.

Note : Deepening of a trough (equivalent or) asymptotically equal to (~) filling of a ridge and filling a trough ~ deepening of a ridge.

22.12 Pressure Tendency Equation

We know that

$$\delta p = g\rho dz \qquad \text{(Hydrostatic equation)} \qquad .....(22.23)$$

Integrating this between z = z to z = ∞, corresponds to p = p to p = 0, we have

$$\int_p^0 \delta p = \int_z^\infty g\rho dz$$

$$p = g\int_z^\infty \rho dz \qquad (\because \text{ g=constant at a place})$$

Differentiating this equation partially w.r.t t = (time), we have

$$\frac{\partial p}{\partial t} = g \int_z^\infty \frac{\partial \rho}{\partial t}\, dz \qquad .....(22.24)$$

we know that

$$\frac{\partial \rho}{\partial t} = -\nabla.\left(\rho\vec{V}\right) \qquad \text{(Equation of continuity)}$$

$$= -\left[\frac{\partial \rho u}{\partial x} + \frac{\partial \rho v}{\partial y} + \frac{\partial \rho w}{\partial z}\right]$$

using this, eq. (22.24) becomes

$$\frac{\partial p}{\partial t} = -g \int_z^\infty \left(\frac{\partial \rho u}{\partial x} + \frac{\partial \rho v}{\partial y} + \frac{\partial \rho w}{\partial z}\right) dz$$

$$= -g \int_z^\infty \left(\frac{\partial \rho u}{\partial x} + \frac{\partial \rho v}{\partial y}\right) dz - g \int_z^\infty \frac{\partial \rho w}{\partial z}\, dz$$

$$\frac{\partial p}{\partial t} = -g \int_z^\infty \left(\frac{\partial \rho u}{\partial x} + \frac{\partial \rho v}{\partial y}\right) dz + g\rho w \qquad .....(22.25)$$

$\left(\text{Since } P = 0 \text{ at } z = \infty\right)$

eq. (22.25) is called tendency equation.

eq. (22.25) can be written as

$$\frac{\partial p}{\partial t} = -g \int_z^\infty \left[\rho\left(\frac{\partial u}{\partial x} + \frac{\partial v}{\partial y}\right) + \left(u\frac{\partial \rho}{\partial x} + v\frac{\partial \rho}{\partial y}\right)\right] dz + g\rho w$$

$$= -g \int_z^\infty \rho\left(\frac{\partial u}{\partial x} + \frac{\partial v}{\partial y}\right) dz - g \int_z^\infty \left(u\frac{\partial \rho}{\partial x} + v\frac{\partial \rho}{\partial y}\right) dz + g\rho w$$

$$\frac{\partial p}{\partial t} = -g\rho\nabla.\vec{V} - g\vec{V}.\nabla\rho + g\rho w \qquad .....(22.26)$$

where $\vec{V} = iu + jv$

$\frac{\partial p}{\partial t}$ = Horizontal velocity divergence + Horizontal advection of mass + Vertical motion of mass.

Questions

1. Derive an equation for the motion of a circular vortex.
2. Define spherical coordinate system. Evaluate $\frac{di}{dt}, \frac{dj}{dt}, \frac{dk}{dt}$ in spherical coordinates.
3. Derive the equations of motion in spherical coordinates.
4. Derive formula for divergence of horizontal flow in spherical coordinates.
5. Let p(x,y,t) be a pressure field.

   show that $\frac{dp}{dt} = \frac{\partial p}{\partial t} + \vec{C}.\nabla p$

   where $\vec{C}$ = velocity of the movement. If $\frac{\partial p}{\partial n}$ denotes the horizontal pressure gradient, then show that $c_n = \frac{-\delta p / \partial t}{\partial p / \partial n}$.
6. Treating axis of trough as x-axis derive an equation for the (i) movement of a trough, (ii) movement of a front, (iii) movement of a col, (iv) movement of a cyclonic and anticyclonic centre.
7. Define intensification and weakening of a pressure system. Derive an equation for the rate of intensification of a centre of a pressure system.
8. Define pressure tendency. Derive pressure tendency equation and hence show that pressure tendency is the sum of horizontal velocity divergence, horizontal advection of mass and vertical motion of mass

CHAPTER 23

Atmospheric Waves

Sinusoidal Waves : Like water waves, sound waves we observe wave phenomena in atmosphere, which governs large scale flow. For simplicity we consider sinusoidal waves (travelling in x - direction).

Let h = h(x, y, z, t) be a dependent variable of the form

$$h = A\,(y, z) \cos k(x - ct) \qquad .....(23.1)$$

where A(y, z) is called amplitude. A(y, z) is normally a function of y and z, but it is also used as a function of single variable or simply as a constant.

k is called wave number and is given by the formula $k = \dfrac{2\pi}{L}$, where L = wave length. k is related to frequency υ by the formula $\upsilon = kc$ where c = speed of wave propagation. υ, is also given by $\upsilon = \dfrac{2\pi}{T}$, where T = period.

eq. (23.1) represents a simple sinusoidal wave which propagates along x-axis with a speed of c.

We know that $e^{i\theta} = \cos\theta + i\sin\theta, \qquad i = \sqrt{-1}$

$$e^{P+iQ\theta} = e^{P} e^{iQ\theta} = e^{P}(\cos Q\theta + i \sin Q\theta)$$

$$\text{Re } e^{p+iQ\theta} = e^{p}\cos Q\theta. \qquad \text{Re = Real part of}$$

$$\text{Im } e^{p+iQ\theta} = e^{p}\sin Q\theta \qquad \text{Im = Imaginary part of}$$

Therefore it is convenient to express eq. (23.1) in the exponential form.

Consider $B(y, z)\, e^{ik[x-ct]}$, where $c = \alpha + i\beta$

$$= Be^{ik(x-\alpha t) + k\beta t}$$

$$= B\, e^{ik\,(x-\alpha t)}\, e^{k\beta t}$$

$$= (\lambda + i\mu)\, e^{k\beta t}\,[\cos k(x-\alpha t) + i \sin k(x-\alpha t)]$$

where $B = \lambda + i\mu$

$$= e^{k\beta t}\,[\lambda \cos k(x-\alpha t) - \mu \sin k(x-\alpha t)] + ie^{k\beta t}\,[\mu \cos k(x-\alpha t) + \lambda \sin k(x-\alpha t)]$$

$$\therefore \quad \text{Re } B\, e^{ik[x-ct]} = e^{k\beta t}\,[\lambda \cos k(x-\alpha t) - \mu \sin k(x-\alpha t)$$

$$= B_A\, e^{k\beta t} \cos[k(x-\alpha t) + \delta]$$

$$= B_A\, e^{k\beta t}\,[\cos k(x-\alpha t)\, .\, \cos \delta - \sin k(x-\alpha t) \sin \delta] \quad .....(23.2)$$

This implies

$$\lambda = B_A \cos \delta$$

or $$\mu = B_A \sin \delta \quad \text{and} \quad \frac{\mu}{\lambda} = \tan \delta$$

$$\lambda^2 + \mu^2 = B_A^{\ 2}$$

Eq. (23.2) is of the form eq. (23.1) except that for the position of maximum. eq. (23.2) is a generalization of eq. (23.1) with C as complex number $C = \lambda + i\mu$

eq. (23.2) $\text{Re}\, .\, Be^{ik(x-ct)} = B_A e^{k\beta t} \cos[k(x-\alpha t) + \delta]$.

Let $B_A^{\ *} = B_A e^{k\beta t}$

It follows from eq. (23.2) that

B^*_A increases with time if $\beta > 0$

B^*_A is constant with time if $\beta = 0$

and B^*_A decreases with time if $\beta < 0$

If we can find the solution approximately to the above equation, then we can classify it according to the rules

$\beta > 0$ unstable or amplified waves

$\beta = 0$ neutral or stable wave

$\beta < 0$ stable or damped waves.

23.1 The Perturbation Method

This method is extensively used in meteorological research. This method mainly used to get solutions to a set of equations which can be handled mathematically.

The aim of this method is to reduce the non-linear differential equations into a set of linear equations.

The general aspects of the perturbation method is as follows :

1. The field variables are classified as basic state and perturbed state.
2. The state of atmosphere that corresponds to the solutions of governing equations is called basic state. The basic state variables are denoted by

   $\bar{u}, \bar{v}, \bar{w}, \bar{p}, \bar{\rho}, \bar{T}$.

3. The perturbed portion is the local deviation of the field from the basic field. Perturbed field variables are denoted by (primes) u', v', w', p', ρ', T'.
4. Total field variables $(\bar{u} + u', \bar{v} + v', ...)$ etc., shall be solutions to the governing equations of the problem. It may be noted that $\bar{u}$, $\bar{v}$ etc. are time and longitude averaged and are independent of time.

   Thus if u(x, t)= $\bar{u}$ + u' (x, t) then, the inertial acceleration $u\frac{\partial u}{\partial x}$ is given by

   $$u\frac{\partial u}{\partial x} = (\bar{u} + u')\frac{\partial}{\partial x}(\bar{u} + u') = (\bar{u} + u')\frac{\partial u'}{\partial x} \qquad \left(\because \frac{\partial \bar{u}}{\partial x} = 0\right)$$

5. While using total field variables ($\bar{u}$ +u' etc) for substitution into the governing equations of the problem a crucial assumption is made, that the perturbation is small, second and higher order perturbed variables are neglected (eg., u'^2, u'v', $u' v'^2$ etc.). This basic assumption reduces the non-linear equations of the problem into linear partial differential equations. This linearisation in some cases provides mathematical solutions to the differential equations of the problem.
6. Linearised equations provide solutions to the sinusoidal type wave equations described above.
7. From the solutions thus obtained conclusions are drawn that the basic state is unstable (β > 0), neutral (β = 0) or stable (β < 0). These provide the conditions of growth and decay of the waves. Further the speed of the wave (c) and the vertical structure of perturbation of wave can be found.

Questions

1. Assuming atmospheric waves as sinusoidal find the condition for the wave to be amplified (or unstable), stable (or neutral) and dampled.
2. Outline the general theory of a perturbation method.

CHAPTER 24

Sound Waves

Perturbation theory eliminates the influence of certain wave disturbances such as sound waves, gravity waves, frontal waves, tidal waves, waves in the upper westerlies. It reduces the non-linear atmospheric motion equations into linear equations which are superimposed on steady field of motion.

Fluid waves result due to the action of restoring forces on fluid parcels which are displaced from their equilibrium positions. The restoring forces on the fluids may be due to gravity, compressibility etc. The simplest examples of linear waves in fluids are acoustic waves and shallow gravity waves. Sound waves are longitudinal waves (waves in which particles move to and fro in the direction of propagation). Sound waves solely result due to adiabatic compression and expansion (rarefaction) of the medium alternately (caused by divergence in the fluid).

1. The basic state consists of an air mass which we assume to be homogeneous atmosphere, so that $\bar{\alpha} = \frac{1}{\rho}$ = constant, and $\bar{u} = \bar{v} = \bar{w} = 0$

   (since there is no motion = static state)

   $$\therefore \quad \bar{p} = \text{constant} \qquad .....(24.1)$$

   Average pressure = const, when we overlook the effect of gravity.

2. The total field thus consists of

   $$u = u', v = v', w = w', p = \bar{p} + p' \text{ and } \alpha = \bar{\alpha} + \alpha' \qquad .....(24.2)$$

3. Consider the governing equations, viz. equations of motion or momentum equations, equations of continuity, thermodynamic energy equations for adiabatic motion. For sound waves we disregard rotation (f = 0), heating (H = 0), and friction F = 0, g = gravity.

$$\therefore \quad \frac{du}{dt} = -\alpha\frac{\partial p}{\partial x} + fv + \alpha F_x \quad \text{reduces to} \quad \frac{du}{dt} = -\alpha\frac{\partial p}{\partial x}$$

$$\frac{dv}{dt} = -\alpha\frac{\partial p}{\partial y} - fu + \alpha F_y \quad \text{reduces to} \quad \frac{dv}{dt} = -\alpha\frac{\partial p}{\partial y}$$

$$\frac{dw}{dt} = -\alpha\frac{\partial p}{\partial z} + eu - g + \alpha F_z \quad \text{reduces to} \quad \frac{dw}{dt} = -\alpha\frac{\partial p}{\partial z} - g$$

$$\frac{dp}{dt} = -\gamma p \nabla . \vec{V}$$

where $\gamma = \frac{C_p}{C_v}$ (a form of thermodynamic relation).

or $$\frac{dp}{dt} + \gamma p \nabla . \vec{V} = 0$$

Also we know $\frac{d}{dt} = \frac{\partial}{\partial t} + \vec{V}.\nabla$

Substituting the total field variables from eq. (24.2), in the above reduced equations of motion we have

(i) $$\frac{\partial u'}{\partial t} = -\bar{\alpha}\frac{\partial p'}{\partial x} \quad \text{since} \left[\frac{du'}{dt} = \frac{\partial u'}{\partial t} + \vec{V}.\nabla u'\right]$$

(ii) $$\frac{\partial v'}{\partial t} = -\bar{\alpha}\frac{\partial \bar{p}}{\partial y}$$

$$\left[\frac{\partial u'}{\partial t} + u'.\nabla u' \simeq \frac{\partial u'}{\partial t} \text{ neglecting higher order terms}\right]$$

$$\left[\vec{V} = iu' + jv' + kw'\right]$$

(iii) $$\frac{\partial w'}{\partial t} = -\bar{\alpha}\frac{\partial \bar{p}}{\partial z} - g, \text{ and}$$

(iv) $$\frac{\partial p'}{\partial t} = -\gamma \bar{p} \nabla . \vec{V} = -\gamma \bar{p}\left(\frac{\partial u'}{\partial x} + \frac{\partial v'}{\partial y} + \frac{\partial w'}{\partial z}\right) \quad .....(24.3)$$

Differentiating (iv) of eq. (24.3) w.r.t to t, we have

$$\frac{\partial^2 p'}{\partial t^2} = -\gamma\bar{p}\left(\frac{\partial^2 u'}{\partial x \partial t} + \frac{\partial^2 v'}{\partial y \partial t} + \frac{\partial^2 w'}{\partial z \partial t}\right) \qquad (\because \bar{p} = \text{const.})$$

$$\frac{\partial^2 p'}{\partial t^2} = -\gamma\bar{p}\left(-\bar{\alpha}\frac{\partial^2 p'}{\partial x^2} - \bar{\alpha}\frac{\partial^2 p'}{\partial y^2} - \bar{\alpha}\frac{\partial^2 p'}{\partial z^2}\right)$$

[From (i) of eq. (24.3), $\frac{\partial u'}{\partial t} = -\bar{\alpha}\frac{\partial p'}{\partial x}$]

$$\therefore \quad \frac{\partial^2 u'}{\partial x \partial t} = -\frac{\partial}{\partial x}\left(\bar{\alpha}\frac{\partial p'}{\partial x}\right)$$

$$= -\bar{\alpha}\frac{\partial^2 p'}{\partial x^2}$$

similarly $\frac{\partial^2 v'}{\partial y \partial t} = -\bar{\alpha}\frac{\partial^2 p'}{\partial y^2}$, $\quad \frac{\partial^2 w'}{\partial z \partial t} = -\bar{\alpha}\frac{\partial^2 p'}{\partial z^2}$]

$$\frac{\partial^2 p'}{\partial t^2} = \gamma\,\bar{\alpha}\,\bar{p}\left[\frac{\partial^2 p'}{\partial x^2} + \frac{\partial^2 p'}{\partial y^2} + \frac{\partial^2 p'}{\partial z^2}\right]$$

$$\frac{\partial^2 p'}{\partial t^2} = \gamma\bar{\alpha}\bar{p}\nabla^2 p' \qquad .....(24.4)$$

If the pressure decreases at some point 'o' (see Fig. 24.1)and does not change elsewhere then the particles accelerate towards O. It will be seen that $\frac{\partial u'}{\partial t} < 0$ on the positive side of the x-axis and $\frac{\partial u'}{\partial t} > 0$ on the negative side of

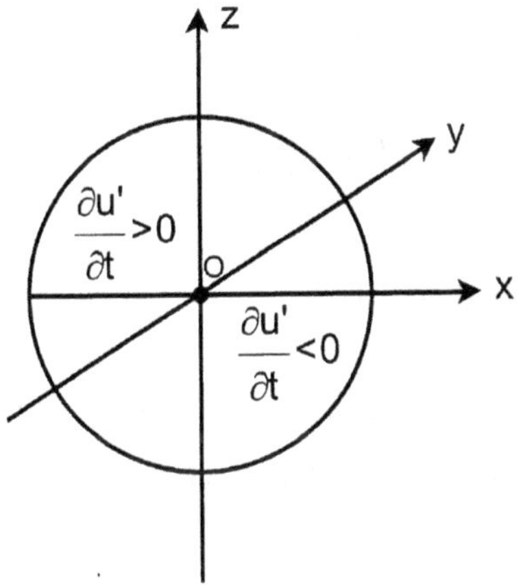

Fig. 24.1

the x-axis. Similar behaviour will be observed along y- and z-axes. After a short time at '0' it will be noticed convergence ($\nabla.\vec{V} < 0$) and $\frac{\partial p'}{\partial t} > 0$, this makes the pressure move back towards $\bar{p}$. This process will continue untill the pressure at 'o' is larger than $\bar{p}$ i.e., $p' > 0$. From the equations of motion we now get divergence $\nabla.\vec{V} > 0$ which results in $\frac{\partial p'}{\partial t} < 0$, and subsequently the pressure again starts decreasing at o.

The oscillation of pressure at o results in alternate fields of divergence and convergence. During this process, the velocity components propagate through the medium. The speed of this propagation will be found as follows.

For simplicity we shall consider the distance r from the center (x–, y– and z–directions are symmetrical about 0). Consequently eq. 24.4 becomes

$$\frac{\partial^2 p'}{\partial t^2} = \gamma\,\bar{\alpha}\,\bar{p}\,\frac{\partial^2 p'}{\partial r^2} \quad \left(\text{Replacing x, by r, } \frac{\partial^2 p'}{\partial x^2} = \frac{\partial^2 p'}{\partial r^2}\right) \qquad .....(24.5)$$

Let $\quad p' = \bar{p} \cos k(r-ct)$ be the wave form.

then $$\frac{\partial p'}{\partial t} = \bar{p} \times -\sin k(r-ct)\ (-ck)$$

$$\frac{\partial^2 p'}{\partial t^2} = -\bar{p} \cos k(r-ct)\ (c^2 k^2) \qquad .....(24.6)$$

$$\frac{\partial p'}{\partial r} = -\bar{p} \sin k(r-ct)\ (k)$$

$$\frac{\partial^2 p'}{\partial r^2} = -\bar{p} \cos k(r-ct)\ (k^2) \qquad .....(24.7)$$

Substituting eq. (24.6) and (24.7) in (24.5), we have

$$-\bar{p} \cos k(r-ct) \,.\, c^2 k^2 = \gamma\bar{\alpha}\bar{p}\ \times - \bar{p} \cos k(r-ct) \times k^2$$

$$c^2 = \gamma\bar{\alpha}\bar{p}$$

$$\therefore \quad c = \pm \sqrt{\gamma\bar{\alpha}\bar{p}} \qquad .....(24.8)$$

This eq. (24.8) gives the speed of propagation of sound wave

The above eq. (24.8) may be written as

$$c = \pm \sqrt{\gamma R \bar{T}} \quad (\text{since } \bar{\alpha}\bar{p} = R\bar{T} \text{ gas equation}) \qquad .....(24.9)$$

If we put

$$C_p = 1004 \text{ m}^2 \text{s}^{-2} \text{deg}^{-1}, \quad C_v = 717 \text{ m}^2 \text{s}^{-2} \text{deg}^{-1},$$

$$\gamma = \frac{C_p}{C_v},$$

$$R = 287 \text{ m}^2 \text{s}^{-2} \text{deg}^{-1} \quad \text{and } \bar{T} = 273 \text{ K}$$

we have

$$c = \pm \sqrt{\frac{1004}{717} \times 287 \times 273} = 331 \text{ ms}^{-1}$$

This value may be considered as the speed of the sound wave in an isothermal atmosphere.

Note : The simplest form of sound waves are senusoidal waves. These cause vibrations of air particles which are described by pressure variations at that point (place). The air pressure rises above atmospheric pressure and then drops below atmospheric pressure. This causes simple harmonic motion of air particles at that point (place).

Question

1. Assuming sound waves to be sinusoidal, disregarding rotation (f = 0), heating (H = 0) and friction (F = 0), applying perturbation method, equation of motion, thermodynamic relation show that the speed of propagation of sound wave (c) is given by

$$c = \pm \sqrt{\gamma \bar{\alpha} \bar{p}}$$

CHAPTER 25

Gravity Waves

There are three types of wave motions in the atmosphere

1. Acoustic waves or sound waves (Longitudinal waves)
2. Gravity waves (transverse oscillations) and
3. Rossby waves or rotational waves.

These equations are primarily governed by the equations of motion.

In sound waves the balance is between $\frac{d\vec{V}}{dt}$ and $-\alpha\nabla p$ and the motion is assumed to be horizontal. The upper air (Radiosonde, pilot balloon) observations of the atmosphere are based on the quasi-static approximation. The main features of the gravity waves is the cross-isobaric flow. The main features of the rotational (Rossby) waves is, the flow is along the isobars and the approximate balance between pressure gradient and coriolis force. Gravity waves move (speed about 300 ms^{-1} for external gravity waves and 50 ms^{-1} for internal gravity waves) faster than the rotational waves (speed about 10 ms^{-1}). Mixed waves are those which are on the border line of gravity waves and rotational waves. Mixed waves are also called mixed Rossby-gravity waves or Inertia-gravity waves. There are deviations from the balanced relation due to local and transit departures. These deviations are observed both in space (all places) and time (all times) with varying intensities. The unbalanced perturbations tend to be balanced and static in due course of time.

$$\left[\text{Note}: \Psi = \int_0^z g\,dz \text{ or } \Psi = gz, \frac{\partial \Psi}{\partial p} = -\alpha, \delta p = \rho\delta\Psi\right]$$

Gravity waves are based on first and third equations of motion and two dimensional continuity equaion for incompressible fluid flow. The mechanism of propagation of gravity waves is the change of pressure due to changing weight of air column above a given horizontal surface. Billow clouds develop due to internal gravity waves.

Let the basic state be at rest, that is $\bar{u} = \bar{v} = \bar{w} = 0$ and $\bar{p}$ = const. Consequently the geo-potential isobaric surfaces coincide with the constant isobaric surfaces. $\bar{\Psi} = \bar{\Psi}(p)$, Total field variables reduce to ($u = \bar{u} + u'$ etc. becomes) $u = u'$, $v = v'$, $w = w'$, $p = \bar{p} + p'$, $\alpha = \bar{\alpha} + \alpha'$

Assume that there are no variations in the meridional direction i.e., $\frac{\partial}{\partial y} = 0$ and $v' = 0$. Further assume that the motion be adiabatic and free from friction.

The thermodynamic equation under adiabatic condition is

$$H = c_p \frac{dT}{dt} - \alpha \frac{d\alpha}{dt} \quad \text{or} \quad \frac{1}{\theta}\frac{d\theta}{dt} = \frac{H}{c_p T}$$

where θ = potential temperture

or $$\frac{d}{dt}\ln\theta = \frac{H}{c_p T} \quad \text{(For adiabatic motion } H = 0)$$

The thermodynamic equation for adiabatic conditions is given by $\frac{d}{dt}(\ln\theta) = 0$

or $$\frac{\partial \ln\theta}{\partial t} + u\frac{\partial \ln\theta}{\partial x} + v\frac{\partial \ln\theta}{\partial y} + \omega\frac{\partial \ln\theta}{\partial p} = 0 \quad \text{in pressure coordinates}$$

or $$\frac{\partial \ln\theta}{\partial t} + \vec{V} \cdot \nabla_H \ln\theta + \omega\frac{\partial \ln\theta}{\partial p} = 0 \qquad .....(25.1)$$

$$\left[\frac{d}{dt} = \frac{\partial}{\partial t} + \vec{V}.\nabla, t = t(x, y, z)\right]$$

where $\nabla_H = i\frac{\partial}{\partial x} + j\frac{\partial}{\partial y}$, Horizontal velocity = $\vec{V}$ = iu + jv

$$\frac{\partial \Psi}{\partial p} = -\alpha \text{ (Hydrostatic equation in pressure coordinates.)} \quad .....(25.2)$$

$$\theta = T\left(\frac{p}{p_0}\right)^{\frac{-R}{C_p}} \quad \text{Poisson's equation}$$

$$\theta = \frac{p\alpha}{R}\left(\frac{p}{p_0}\right)^{-\frac{R}{C_p}} = \frac{\alpha}{R}\left(\frac{p^{1-\frac{R}{C_p}}}{p_0^{-\frac{R}{C_p}}}\right)$$

$$\theta = \frac{\alpha}{R}\,\frac{p^{\frac{C_v}{C_p}}}{p_0^{\frac{-R}{C_p}}} \qquad .....(25.3)$$

Taking logarithms of eq. (25.3) we have

$$\ln\theta = \ln\alpha + \frac{C_v}{C_p}\ln p - \ln R + \frac{R}{C_p}\ln p_0$$

or
$$\ln\theta = \ln\alpha + \frac{C_v}{C_p}\ln p + \text{const.} \qquad .....(25.4)$$

Differentiating partially w.r.t to t; eq. (25.4), becomes

$$\frac{\partial \ln\theta}{\partial t} = \frac{\partial}{\partial t}\ln\alpha \qquad (\because p = \text{const. on a pressure surface})$$

or
$$\nabla\ln\theta = \nabla\ln\alpha \qquad ......(25.5)$$

Substituting this eq. (25.5) in eq. (25.1) we have

$$\frac{\partial}{\partial t}\ln\alpha + \vec{V}.\nabla_H\ln\alpha + \omega\frac{\partial\ln\theta}{\partial p} = 0$$

or
$$\frac{1}{\alpha}\frac{\partial\alpha}{\partial t} + \frac{\vec{V}}{\alpha}.\nabla_H\alpha + \omega\frac{\partial\ln\theta}{\partial p} = 0$$

or
$$\frac{\partial\alpha}{\partial t} + \vec{V}.\nabla_H\alpha + \alpha\omega\frac{\partial\ln\theta}{\partial p} = 0 \qquad .....(25.6)$$

Using hydrostatic equation in pressure coordinates $\frac{\partial\Psi}{\partial p} = -\alpha$, the above equation becomes

$$-\frac{\partial}{\partial t}\left(\frac{\partial\Psi}{\partial p}\right) - \vec{V}.\nabla_H\left(\frac{\partial\Psi}{\partial p}\right) + \omega\left(\alpha\frac{\partial\ln\theta}{\partial p}\right) = 0$$

$$\text{or}\quad \frac{\partial}{\partial t}\left(\frac{\partial \Psi}{\partial p}\right) + \vec{V}.\nabla_H\left(\frac{\partial \Psi}{\partial p}\right) + \sigma\omega = 0, \quad \text{where } \sigma = -\alpha\left(\frac{\partial \ln\theta}{\partial p}\right) \qquad .....(25.7)$$

From eq. (25.4) we have $\ln\theta = \ln\alpha + \frac{C_v}{C_p}\ln p + \text{const.}$

Differentiating this equation w.r.t p we have

$$\frac{\partial \ln\theta}{\partial p} = \frac{1}{\alpha}\frac{\partial \alpha}{\partial p} + \frac{C_v}{C_p}\frac{1}{p}$$

and $$\sigma = -\alpha\left[\frac{1}{\alpha}\frac{\partial \alpha}{\partial p} + \frac{C_v}{C_p}\frac{1}{p}\right]$$

or $$\sigma = -\left[\frac{\partial \alpha}{\partial p} + \frac{C_v}{C_p}\frac{\alpha}{p}\right]$$

$$\sigma = \frac{\partial^2 \Psi}{\partial p^2} + \frac{C_v}{C_p}\cdot\frac{1}{p}\frac{\partial \Psi}{\partial p} \qquad \left(\because \alpha = -\frac{\partial \Psi}{\partial p}\right) \qquad .....(25.8)$$

It follows that $\bar{\Psi} = \bar{\Psi}_{(P)}$ in the basic state and from eq. (25.8) $\bar{\sigma} = \bar{\sigma}(p)$. For simplicity consider two level model of the atmosphere of equal masses.

Let the surface pressure be ≈ 1000 hPa and let 500 hPa–surface divides the upper and lower parts of the atmosphere. Let the representative level of the upper atmosphere be 250 hPa level, and the representative level of the lower atmosphere be 750 hPa level as shown in Fig. 25.1 (where ω is vertical velocity in pressure coordinates).

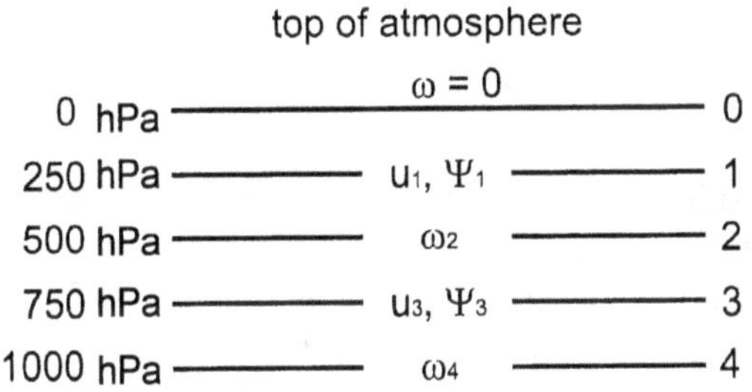

Fig. 25.1 Two level model of atmosphere of equal mass.

The equation of motion $\left(\frac{du}{dt} = -\alpha\frac{\partial p}{\partial x}\right)$ for u-component in the upper and lower layers are :

$$\frac{\partial u_1'}{\partial t} = -\alpha\frac{\partial p}{\partial x} = \frac{-\partial \Psi_1'}{\partial x} \qquad .....(25.9)$$

$$\frac{\partial u_3'}{\partial t} = -\alpha\frac{\partial p}{\partial x} = \frac{-\partial \Psi_3'}{\partial x} \qquad .....(25.10)$$

The continuity equation $\left(\frac{\partial u}{\partial x}+\frac{\partial v}{\partial y}+\frac{\partial \omega}{\partial p}=0\right)$ applied to the upper and lower layers in finite difference form is given by $\left(\text{Note } \frac{\partial v}{\partial y}=0,\ \frac{\partial u}{\partial x}=-\frac{\partial \omega}{\partial p}\right)$

$\frac{\omega_2'-0}{p} = -\frac{\partial u_1'}{\partial x}$, where p = 500 h P_a, the average $\left(\frac{0+1000}{2}\right)$ atmospheric pressure

$$\frac{\omega_4'-\omega_2'}{p} = -\frac{\partial u_3'}{\partial x}$$

or
$$\omega'_2 = -p\frac{\partial u_1'}{\partial x} \qquad .....(25.11)$$

$$\omega'_4 - \omega'_2 = -p\frac{\partial u_3'}{\partial x} \qquad ......(25.12)$$

The thermodynamic eq. (25.7) at middle level -2 becomes

$$-\frac{\partial}{\partial t}\left(\frac{\Psi_1'-\Psi_3'}{p}\right)+\bar{\sigma}_2\,\omega_2' = 0 \qquad \left(\because\ \vec{v}.\nabla_H\frac{\partial \psi}{\partial p}=0\right)$$

or
$$-\frac{1}{p}\frac{\partial}{\partial t}\left(\Psi_1'-\Psi_3'\right)+\bar{\sigma}_2\,\omega_2' = 0 \qquad .....(25.13)$$

where the finite difference is taken at middle level-2 (that is across half the atmosphere between level 1 and level 3). Now we have five eq. (25.9), (25.10), (25.11), (25.12), and (25.13) but we have six variables u'_1, u'_2, Ψ'_1, Ψ'_2, ω'_2, ω'_4. This system is not amenable for solution. Consequently we take lower boundary condition $\omega = \omega_4 = 0$ when $p = p_4$ (= 1000 hPa) and suppose Ψ_4 be the geo-potential at p_4

$$\therefore \qquad g\omega_4 = \frac{d\Psi_4}{dt} \simeq \frac{\partial \Psi_4'}{\partial t} + \omega_4'\frac{\partial \Psi_4'}{\partial p} = 0$$

$$\frac{\partial \Psi_4'}{\partial t} = -\omega_4'\frac{\partial \Psi_4'}{\partial p} = -\bar{\alpha}\,\omega_4' \qquad .....(25.14)$$

The finite difference linear extrapolation is taken as

$$\Psi'_4 = \frac{3}{2}\Psi'_3 - \frac{1}{2}\Psi'_1 \quad .....(25.15)$$

Thus the linear perturbation equations are given by eq. (25.9) to (25.15).

Let each variable is represented by a sinusoidal wave of the form

$$(h)' = (\hat{A})\, e^{ik(x-ct)}$$

where h' is any variable, $(\hat{A})$ =amplitude.

From eq. (25.9) to (25.15) we get with above wave form

(a) $-ikc\,\hat{u}_1 = -ik\hat{\Psi}_1$ From eq. (25.9)

(b) $-ikc\,\hat{u}_3 = -ik\hat{\Psi}_3$ From eq. (25.10)

(c) $\hat{\omega} = -pik\,\hat{u}_1$ From eq. (25.11)

(d) $\hat{\omega}_4 - \hat{\omega}_2 = -pik\,\hat{u}_3$ From eq. (25.12)

(e) $-\frac{ikc}{p}\left(\hat{\Psi}_1 - \hat{\Psi}_3\right) + \bar{\sigma}_2\,\hat{\omega}_2 = 0$ From eq. (25.13)

(f) $\hat{\omega}_4 \simeq \bar{\rho}_4\, ikc\left(\frac{\hat{\Psi}_1}{2} - \frac{3\hat{\Psi}_3}{2}\right)$ From eq. (25.14) .....(25.16)

From eq. (25.16) we can eliminate all variables.

If we retain the variables $\hat{u}_1$, $\hat{u}_3$, we have from eq. (25.16) that [substituting (a), (b) (c) in (e)]

$$(c^2 - \bar{\sigma}_2 p^2)\,\hat{u}_1 - c^2\,\hat{u}_3 = 0 \quad .....(25.17)\text{ (i)}$$

[and using (c), (d) on LHS of (f) and (a), (b) on RHS of (f)]

$$\left(p + \frac{1}{2}\bar{\rho}_4 c^2\right)\hat{u}_1 + \left(p - \frac{3}{2}\bar{\rho}_4 c^2\right)\hat{u}_3 = 0 \quad .....(25.17)\text{ (ii)}$$

The system (25.17) shows two homogeneous linear equations in $\hat{u}_1$, $\hat{u}_3$, and that the condition for non-trivial solution is that the determinant is zero. i.e.,

$$\begin{vmatrix} c^2 - \bar{\sigma}_2 p^2 & -c^2 \\ p + \frac{1}{2}\bar{\rho}_4 c^2 & p - \frac{3}{2}\bar{\rho}_4 c^2 \end{vmatrix} = 0$$

We shall consider two special cases of eq. (25.17).

Case (i) Let $\bar{\sigma}_2 = 0$, implies $\bar{\theta}$ = const. from eq. (25.17) $\because \sigma = -\alpha\left(\frac{\partial \ln\theta}{\partial p}\right)$

and from (25.17) (i), $(c^2 - \bar{\sigma}_2 p^2)\ \hat{u}_1 - c^2\ \hat{u}_3 = 0$. we have $\hat{u}_1 - \hat{u}_3 = 0$

i.e., $\hat{u}_1 = \hat{u}_3$. This implies that there is great deal of divergence in the upper part as in the lower part of the atmosphere.

From (25.17) (ii) $\left(p + \frac{1}{2}\bar{\rho}_4 c^2\right)\hat{u}_1 + \left(p - \frac{3}{2}\bar{\rho}_4 c^2\right)\hat{u}_3 = 0$

we have $\left(2p - \bar{\rho}_4 c^2\right)\hat{u}_1 = 0$, $\left(\text{putting } \hat{u}_1 = \hat{u}_3\right)$

i.e., $2p - \bar{\rho}_4 c^2 = 0$ .....(25.18)

$$c = \pm\sqrt{\frac{2p}{\bar{\rho}_4}} = \pm\sqrt{\frac{p_4}{\bar{\rho}_4}} \quad \text{[since p = 500 hPa and } P_4 = \text{1000 hPa]}$$

$$c = \pm\sqrt{R\bar{T}_4} \quad [\text{since } p_4\,\bar{\alpha}_4 = R\bar{T}_4 \text{ gas equation}] \qquad ....(25.19)$$

eq. (25.19) describes that wave speed at the lower boundary. These are called *external gravity waves.*

Put $R = 287\ m^2 s^{-2}\ {}^0K^{-1}$; $\bar{T}_4 = 273$ (lower boundary conditions)

then we got

$$c = \pm\sqrt{287 \times 273} = \pm\ 280 \text{ m/s.}$$

This shows that the external gravity waves travel as fast as sound waves (speed (c) = 331 ms^{-1}) in the atmosphere.

Case (ii) Let $p = p_4$ = 1000 hPa and $\omega_4 = \frac{dp_4}{dt} = 0$

$$\omega = \frac{dp}{dt} = \frac{\partial p}{\partial t} + \vec{V}.\nabla p + w\frac{\partial p}{\partial z} \qquad .....(25.20)$$

$\vec{V}.\nabla p \approx 0$ (since observed wind is approximately geostrophic) and from observed tendencies we know that

$$\left|\frac{\partial p}{\partial t}\right| << g\rho\,|w| \quad \text{(See pressure tendency)}$$

∴ From eq. (25.20) we have

$$\omega = g\rho w \qquad \left(\because \frac{\delta p}{\delta z} = g\bar{\rho}\right) \qquad .....(25.21)$$

At the lowest boundary we have $w = 0$ and thus $\omega \simeq 0, \bar{\rho}_4 = 0$.

Substituting this in (25.17) (ii)

$$\left(p + \frac{1}{2}\bar{\rho}_4 c^2\right)\hat{u}_1 + \left(p - \frac{3}{2}\bar{\rho}_4 c^2\right)\hat{u}_3 = 0$$

we get $p\hat{u}_1 + p\hat{u}_3 = 0 \qquad (\because \bar{\rho}_4 = 0)$

or $\hat{u}_1 + \hat{u}_3 = 0 \qquad$ (since $p \neq 0$) $\qquad .....(25.22)$

eq. (25.22) implies that the total divergence in the entire atmospheric column is zero. Substituting this $\hat{u}_3 = -\hat{u}_1$ in (25.17) (i) we get $(c^2 - \bar{\sigma}_2 p^2)\,\hat{u}_1 + c^2\,\hat{u}_1 = 0$

or $(2c^2 - \bar{\sigma}_2 p^2)\,\hat{u}_1 = 0 \qquad .....(25.23)$

This equation gives the speed of wave as

$$c = \pm\sqrt{\frac{\bar{\sigma}_2}{2}p^2} \qquad .....(25.24)$$

The above case (ii) special case of waves determined entirely by internal stratification $\bar{\sigma}_2$, and hence called *internal gravity waves.* It is found by observations that

$$\bar{\sigma}_2 \simeq 2 \times 10^{-6} \text{ mks units and } |c| \simeq 50 \text{ ms}^{-1} \qquad .....(25.25)$$

It may be noted here that the speed of external gravity waves is about 280 ms^{-1} while that of internal gravity waves is about 50 ms^{-1}, that is internal gravity waves travel slower than the external gravity waves. Again in case of external gravity waves, in the entire column of the atmosphere the divergence exists $\omega_4 \neq 0$, while in the case of internal gravity waves the divergence of the entire atmospheric column is zero.

In general case we have to solve the determinant

$$\begin{vmatrix} c^2 - \bar{\sigma}_2 p^2 & -c^2 \\ p + \frac{1}{2}\bar{\rho}_4 c^2 & p - \frac{3}{2}\bar{\rho}_4 c^2 \end{vmatrix} = 0$$

From eq. (25.24) and (25.19) we have

$$c_{IN}^2 = \frac{\bar{\sigma}_2 p^2}{2} \text{ and } c_{Ex}^2 = \frac{2p}{\bar{\rho}_4},$$

[c_{IN} = speed of internal gravity waves, c_{Ex} = speed of external gravity waves]

Expanding the determinant

$$pc^2 - \frac{3}{2}\bar{\rho}_4 c^4 - \bar{\sigma}_2 p^3 + \frac{3}{2}\bar{\sigma}_2 p^2 \bar{\rho}_4 c^2 + c^2 p + \frac{1}{2}\bar{\rho}_4 c^4 = 0$$

$$-\bar{\rho}_4 c^4 - \bar{\sigma}_2 p^3 + \frac{3}{2}\bar{\sigma}_2 p^2 \bar{\rho}_4 c^2 + 2pc^2 = 0$$

or $$c^4 + \bar{\sigma}_2 \frac{p^3}{\bar{\rho}_4} - \frac{3}{2}\bar{\sigma}_2 p^2 c^2 - \frac{2pc^2}{\bar{\rho}_4} = 0 \qquad \text{.....(25.26)}$$

From above

$$c_{Ex}^2 + 3c_{IN}^2 = \frac{2p}{\bar{\rho}_4} + \frac{3}{2}\bar{\sigma}_2 p^2$$

$$c_{IN}^2 \times 3c_{Ex}^2 = \frac{\bar{\sigma}_2 p^2}{2} \times \frac{2p}{\bar{\rho}_4} = \frac{p^3 \bar{\sigma}_2}{\bar{\rho}_4}$$

Substituting these values in eq. (25.26) we have

$$c^4 - \left(C_{Ex}^2 + 3C_{IN}^2\right)c^2 + C_{IN}^2 \times C_{Ex}^2 = 0 \qquad \text{.....(25.27)}$$

eq. (25.27) is a quadratic equation in c^2, therefore its solution is

[Note : if $ax^2 + bx + c = 0$, then $x = \dfrac{-b \pm \sqrt{b^2 - 4ac}}{2a}$]

given by

$$c^2 = \frac{c_{Ex}^2 + 3c_{IN}^2 \pm \sqrt{\left(c_{Ex}^2 + 3c_{IN}^2\right)^2 - 4 \times 1 \times c_{IN}^2 c_{Ex}^2}}{2}$$

or $$c^2 = \frac{1}{2}\left\{\left[c_{Ex}^2 + 3c_{IN}^2\right] \pm \sqrt{c_{Ex}^2 + 9c_{IN}^4 + 2c_{Ex}^2 c_{IN}^2}\right\} \qquad \text{.....(25.28)}$$

eq. (25.28) gives four values of c and there is no possibility of instability. Of these four values two values corresponding to the plus sign, gives speeds of external gravity waves while the other two gives the speeds of internal gravity waves. For the same numerical values given above we find

$$c_{12} = \pm 290 \text{ ms}^{-1} \text{ and } c_{34} = \pm 59 \text{ ms}^{-1}.$$

Questions

1. Show that the thermodynamic equation for adiabatic conditions is given by $\frac{d}{dt}(\ln \theta) = 0$

   where θ = potential temperature

2. Using Poisson's equation for potential temperature (θ) show that $\nabla \ln \theta = \nabla \ln \alpha$

   Where $\alpha = \frac{1}{\rho}\left(\frac{1}{\text{density}}\right)$

   Also show that the thermodynamic equation for adiabatic conditions $\frac{d}{dt}(\ln \theta) = 0$ can be put in the form $\frac{\partial \alpha}{\partial t} + \vec{V} \cdot \nabla_H \alpha + \alpha \frac{\omega \partial \ln \theta}{\partial p} = 0$

3. Assuming thermodynamic equation for adiabatic condition to obey

   $$\frac{\partial \alpha}{\partial t} + \vec{V} \cdot \nabla_H \alpha + \alpha\, \omega \frac{\partial \ln \theta}{\partial p} = 0$$

   and hydrostatic equation in pressure coordinate as $\frac{\partial \Psi}{\partial p} = -\alpha$

   (where ψ = geopotential) show that $\sigma = -\alpha \frac{\partial \ln \theta}{\partial p}$ can be put in the form $\sigma = \frac{\partial^2 \Psi}{\partial p^2} + \frac{C_\upsilon}{C_p} \frac{1}{p} \frac{\partial \Psi}{\partial p}$ and the above adiabatic condition equation reduces to

$$\frac{\partial}{\partial t}\left(\frac{\partial \Psi}{\partial p}\right) + \vec{V}.\nabla_H\left(\frac{\partial \Psi}{\partial p}\right) + \sigma\, w = 0$$

4. Discuss the velocity propagation of external gravity waves and internal gravity waves.

CHAPTER 26

Inertia Waves

We shall consider a special case of wave motion, in which earth's rotation is in a pure form.

Consider a possible wave motion which may exist in horizontal motion where pressure force is negligible. Let the basic state be free of motion. Under these conditions the equations of motion reduces to

$$\frac{\partial u'}{\partial t} = fv' \qquad .....(26.1)$$

$$\frac{\partial v'}{\partial t} = -fu' \qquad .....(26.2)$$

eq. (26.1) and (26.2) form a closed system.

Assume the wave disturbance be of the form

$$(h') = (\hat{A})e^{ik(x-ct)} \qquad .....(26.3)$$

where $(\hat{h})$ any variable, $(\hat{A})$ = Amplitude, $i = \sqrt{-1}$

Under this assumption the solution of eq. (26.1) and (26.2) can be obtained directly. If $f = f_0$ we get

$$ikc\,\hat{u} + f_0\hat{v} = 0$$

$$f_0\hat{u} - ikc\,\hat{v} = 0 \qquad .....(26.4)$$

This gives

$$\begin{vmatrix} ikc & f_0 \\ f_0 & -ikc \end{vmatrix} = 0$$

or $$k^2 c^2 - f^2 = 0 \text{ or } \quad c^2 = \frac{f_0^2}{k^2}$$

$$\therefore \quad c = \pm \frac{f_0}{k} \qquad .....(26.5)$$

The values of speed of the wave given by eq. (26.5) are called inertia wave speeds. These waves are different from sound waves and internal and external gravity waves. These depend on wave number k

If $$k = \frac{2\pi}{L}$$

From eq. (26.5) we have

$$c = \pm \frac{f_0 L}{2\pi} \qquad .....(26.6)$$

Put L = 1000 km = 10^6 m, f = 10^{-4}/s

we get $$c = \pm \frac{10^{-4}\, s^{-1} \times 10^6\, m}{2\pi}$$

$$c = \pm\, 15.9\ ms^{-1}.$$

Questions

1. Derive an expression for the speed of inertia waves.

CHAPTER 27

Inertia-Gravity Waves

We have considered gravity waves and inertia waves in pure form. However, generally they exist together like internal & external gravity waves. We shall now consider a model of simultaneous existence of both internal gravity waves and inertia waves.

Consider a two level model with basic state at rest with coriolis force. Atmosphere is divided into two equal parts with respect to mass. The surface level pressure assumed to be 1000 hPa. Let the 500 hPa pressure surface divide the atmosphere into lower and upper parts. The upper part of the atmosphere is represented by 250 hPa level and the lower part of the atmosphere is represented by 750 hPa level as shown in Fig. 27.1.

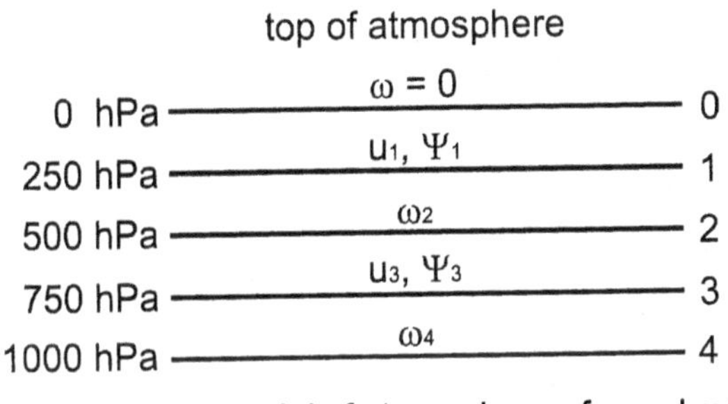

Fig. 27.1 Two level model of atmosphere of equal masses.

The basic state of rest gives in that $\bar{u} = \bar{v} = \bar{w} = 0$. The geo-potential isobaric surface is considered as constant for each isobaric surface. i.e., $\bar{\Psi} = \bar{\Psi}(p)$. In this case the coriolis force $f = f_0$ is taken into account while the pressure force is disregarded. Perturbation dependent variables are x and t. With these assumptions the perturbation equations of motions in the upper and lower layers are now given by.

For upper layer

$$\frac{\partial u_1'}{\partial t} = -\frac{\partial \Psi_1'}{\partial x} + f_0 v_1' \qquad .....(i)$$

[since $\frac{du}{dt} = -\alpha \frac{\partial p}{\partial x} + fv - ew + \alpha F_x, \quad \alpha \delta p = \delta \psi, F_x = 0,$

and $e = 0, \frac{du}{dt} \simeq \frac{\partial u}{\partial t}$]

$$\frac{\partial v_1'}{\partial t} = -f_0 u_1' \qquad .....(ii)$$

[since $\frac{dv}{dt} = -\alpha \frac{\partial p}{\partial y} - fu + \alpha F_y, \ \bar{\Psi} = \bar{\Psi}(p), \ u = \bar{u} + u'$ etc.,

$\frac{\partial}{\partial y} = 0, \frac{dv}{dt} \simeq \frac{\partial v}{\partial t}$]

For lower layer

$$\frac{\partial u_3'}{\partial t} = -\frac{\partial \Psi_3'}{\partial x} + f_0 v_3' \qquad .....(iii)$$

$$\left[u = u', v = v', \omega = \omega', \alpha' = \bar{\alpha} + \alpha', \Psi = gz \text{ geopotential}, \frac{\partial \Psi}{\partial p} = \alpha \right]$$

$$\frac{\partial v_3'}{\partial t} = -f_0 u_3' \qquad .....(iv)(27.1)$$

The equations of continuity applied to the upper and lower layers in the finite difference form is given by

$$\left[\frac{\partial u}{\partial x} + \frac{\partial v}{\partial y} + \frac{\partial w}{\partial p} = 0 \text{ or } \frac{\partial u}{\partial x} = \frac{-\partial w}{\partial p} \right]$$

$$\frac{\omega_2' - 0}{p} = \frac{-\partial u_1'}{\partial x} \qquad \text{or } \omega'_2 = -P \frac{\partial u_1'}{\partial x} \qquad .....(a)$$

$$\frac{\omega_4 - \omega_2}{p} = \frac{\partial u_3'}{\partial x} \qquad \text{or } \omega'_2 = -P \frac{\partial u_3'}{\partial x} \qquad .....(b)$$

Thermodynamic equation in the middle level is given by (Ref) eq. (25.7)

$$-\frac{1}{p} \frac{\partial \left(\Psi_1' - \Psi_3' \right)}{\partial t} + \bar{\sigma}_2 \omega_2' = 0 \qquad ...(c)..(27.2)$$

In eq. (27.2) the boundary condition $\omega_4 = 0$ has been taken in to account. Let the wave disturbance is of the standard sinusoidal form

$$(h') = (\hat{A})e^{ik(x-ct)} \qquad .....(27.3)$$

where (h)' = any variable, $(\hat{A})$ = amplitude. Applying eq. (27.3) to first two equations of eq. (27.1) (27.1) (i), (27.1) (ii) we get

$$-ikC\,\hat{u}_1 - f_0\,\hat{v}_1 + ik\bar{\Psi}_1 = 0 \qquad .....(27.4)$$

$$-ikC\,\hat{v}_1 + f_0\,\hat{u}_1 = 0 \qquad .....(27.5)$$

From eq. (27.5) we get $\hat{v}_1 = \dfrac{f_0\,\hat{u}_1}{ikC}$

Substituting this value in eq. (27.4), we get

$$-ikC\;\hat{u}_1 - f_0 \times \frac{f_0\,\hat{u}_1}{ikC} + ik\hat{\Psi}_1 = 0$$

or $$-iC\,\hat{u}_1 - \frac{f_0^2\,\hat{u}_1}{ik^2C} + i\hat{\Psi}_1 = 0$$

or $$\Psi_1 = C\hat{u}_1 - \frac{f_0^2\,\hat{u}_1}{k^2C} \qquad (\because i^2 = -1)$$

$$\hat{\Psi}_1 = \hat{u}_1\left(C - \frac{C_I^2}{C}\right) \text{ where } C_I = \frac{f_0}{k} \text{ inertia wave speed.} \qquad .....(27.6)$$

eq. (27.1) (iii), (27.1) (iv) equations are analogous to the (27.1)(i), (27.1) (ii) and hence we have similar result.

$$\hat{\Psi}_3 = \hat{u}_3\left(C - \frac{C_I^2}{C}\right) \qquad .....(27.7)$$

Applying eq. (27.3), the first two eq. of (27.2) (a), (b) we find that

$$\hat{\omega}_2 = -pik\hat{u}_1 \qquad .....(27.8)$$

$$-\hat{\omega}_2 = -pik\hat{u}_3 \qquad .....(27.9)$$

Adding $(\hat{u}_1 + \hat{u}_3)(-pik) = 0$

$$\therefore \quad \hat{u}_1 + \hat{u}_3 = 0 \qquad .....(27.10)$$

Subtracting eq. (27.9) from eq. (27.8), we have

$$2\hat{\omega}_2 = -pik\,(\hat{u}_1 - \hat{u}_3)$$

$$\hat{\omega}_2 = -pik\,\hat{u}_1 \quad (\because \hat{u}_3 = -\hat{u}_1 \text{ from eq. (27.10)}) \qquad .....(27.11)$$

From eq. (27.6) and (27.7)

$$\hat{\Psi}_1 = \left(C - \frac{C_I^2}{C}\right)\hat{u}_1,\ \hat{\Psi}_3 = \left(C - \frac{C_I^2}{C}\right)\hat{u}_3$$

$$\therefore \quad \hat{\Psi}_1 + \hat{\Psi}_3 = C\,(\hat{u}_1 + \hat{u}_3) - \frac{C_I^2}{C}(\hat{u}_1 + \hat{u}_3)$$

i.e., $$\hat{\Psi}_1 + \hat{\Psi}_3 = 0 \qquad \text{.....(27.12)}$$

$[\because\ \hat{u}_1 + \hat{u}_3 = 0$ from (27.10)$]$

Substituting eq. (27.6) for $\hat{\Psi}_1$, (27.11) for $\hat{\omega}_2$ and (27.12) for $\hat{\Psi}_1 + \hat{\Psi}_3 = 0$ in the last equation of (27.2) (c) and applying (27.3) wave form we have.

$$\frac{1}{p} ikC \times 2\hat{\Psi}_1 + \bar{\sigma}_2 \hat{\omega}_2 = 0$$

or $$\left[\frac{ikC}{p} \times 2\left(C - \frac{C_I^2}{C}\right) - \bar{\sigma}_2\, pik\right]\hat{u}_1 = 0$$

or $$\frac{2ik}{p}\left[C^2 - C_I^2 - \frac{\bar{\sigma}_2 p^2}{2}\right]\hat{u}_1 = 0$$

which will be satisfied if $C^2 - C_I^2 - C_{IN}^2 = 0$, where $C_{IN}^2 = \dfrac{\bar{\sigma}_2 p^2}{2}$.

It follows from the condition that

$$C = \pm\sqrt{C_I^2 + C_{IN}^2} \qquad \text{.....(27.13)}$$

eq. (27.13) gives the speed of inertia-gravity waves.

Questions

1. Derive an expression for the speed of inertia-gravity waves, and show that it satisfies of $C^2 - C_I^2 - C_{IN}^2 = 0$

   where C = speed of inertia gravity waves

   CI = speed of an inertia waves

   C_{IN} = speed of internal gravity waves.

CHAPTER 28

Rossby Waves (Barotropic Waves)

Rossby waves or long waves move from west to east in middle and high latitudes whose speed is less than the air currents. These waves are smooth and sinusoidal type, cover 4 to 5 waves over a hemisphere. They mainly travel in middle, and upper troposphere. Other waves are super-imposed on these waves and travel much faster than these waves.

An atmosphere in which isobaric surfaces coincide with constant density surfaces is called barotropic atmosphere.

The rotation of the earth has significant influence on the atmospheric wave motion and in particular the north-south (meridional) variation of coriolis parameter (f). This was first studied by C.G. Rossby in 1939. Observational studies indicate that the general order of magnitude of divergence in the atmosphere is less than the order of magnitude of the vorticity i.e. $|\nabla_h \cdot \vec{V}| < |\zeta|$. This has bearing on the fact that the vertical motion is one or two order smaller than horizontal motion i.e., $|w| << |\vec{V}_h|$. Major part of horizontal flow may be considered non-divergent and it can be described by a stream function. Thus in Rossby waves (impure form) we disregard the horizontal divergence and vertical velocity.

Let the flow be horizontal and non-divergent. The vorticity equation is given by

$$\frac{d(\zeta+f)}{dt} = 0 \qquad .....(28.1)$$

or

$$\frac{\partial \zeta}{\partial t} + u\frac{\partial \zeta}{\partial x} + v\frac{\partial \zeta}{\partial y} + \beta v = 0 \qquad .....(28.2)$$

$$\left[\because \frac{\partial f}{\partial t} = \frac{\partial f}{\partial y}\cdot\frac{\partial y}{\partial t} = \beta v\right]$$

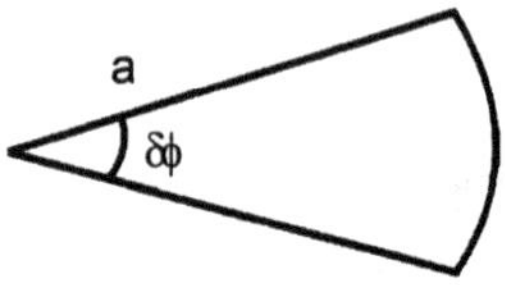

Fig. 28.1

where $\beta = \frac{\partial f}{\partial y} = \frac{\partial f}{\partial \phi}\frac{\partial \phi}{\partial y}$

$$= \frac{1}{a}\cdot\frac{\partial}{\partial \phi}(2\Omega \sin\phi)$$

$$\beta = \frac{2\Omega\cos\theta}{a} \qquad \beta = \text{Rossby parameter} \qquad .....(28.3)$$

a = R + h, where R = radius of the earth; h = height above msl

Consider local cartesian coordinate system. The real variation of f is replaced by

$$f = f_0 + \beta y \qquad .....(28.4)$$

where $f = f_0$ when $y = 0$ and β = constant.

A horizontal plane in which $f = f_0 + \beta y$ is valid is called beta-plane (β-plane).

(i) Now assume the basic state in the β– plane is $\bar{u} = U$ in x-direction and $\bar{v} = 0$.

This is valid in basic state, since $\bar{\zeta} = 0$ (U is zonal current speed).

∴ All terms in eq. 28.2 are identically zero.

(ii) In perturbation, for simplicity, we assume that it depends only on x and t.

Under these assumption, we have

$$\zeta = \frac{\partial v'}{\partial x} - \frac{\partial u'}{\partial y}$$

$$\zeta' = \frac{\partial v'}{\partial x} \qquad .....(28.5)$$

Substituting this value of eq. 28.5 in eq. 28.2 we have

$$\frac{\partial}{\partial t}\left(\frac{\partial v'}{\partial x}\right) + U\frac{\partial}{\partial x}\left(\frac{\partial v'}{\partial x}\right) + \beta v' = 0 \qquad \left(\frac{\partial \zeta}{\partial y} \text{ is taken zero}\right) \qquad .....(28.6)$$

Assume $v' = ve^{ik(x-ct)}$ (usual sinusoidal form)

Then $\dfrac{\partial v'}{\partial x} = \dfrac{\partial v\left(e^{ik(x-ct)}\right)}{\partial x}$

$$\frac{\partial v'}{\partial x} = vike^{ik(x-ct)} \qquad .....(28.7)$$

Substituting 28.7 in 28.6 we have

$$\left[\text{since } \frac{\partial}{\partial t} vike^{ik(x-ct)} = vik(-ikC)e^{ik(x-ct)}\right.$$

and $\left.\dfrac{\partial}{\partial x} vike^{ik(x-ct)} = vi^2k^2e^{ik(x-ct)}\right]$

$$e^{ik(x-ct)}\left[-\hat{v}i^2k^2C + U\hat{v}i^2k^2 + \beta\hat{v}\right] = 0$$

$\therefore \quad k^2C - k^2U + \beta = 0$

or $\quad C = \dfrac{k^2U - \beta}{k^2} = \left(U - \dfrac{\beta}{k^2}\right) \qquad .....(28.8)$

$\therefore$ The speed of waves is given by $C = U - \dfrac{\beta}{k^2}$.

This is Rossby wave formula. This formula shows a definite direction as it is not having any double ($\pm$) sign. The wave moves west ward relative to the basic current U.

(a) If $U > 0$, the wave moves west to east. if $(C > 0)$ $U - \dfrac{\beta}{k^2} > 0$ .....(28.9)

(b) If $U > 0$, but if $(C < 0)$

$U - \dfrac{\beta}{k^2} < 0$ the wave moves from east to west. .....(28.10)

(c) If $U - \dfrac{\beta}{k_s^2} = 0$, stationary Rossby waves, k_s^2 is called stationary wave number. .....(28.11)

Let L_s denote the stationary wavelength of Rossby waves, then

$$L_S = \frac{2\pi}{k_s} \qquad .....(28.12)$$

From eq. (28.11) and (28.12) we have

$$L_S = 2\pi \sqrt{\frac{U}{\beta}} \qquad .....(13)$$

At latitude 45^0, $\beta \simeq 16 \times 10^{-12}\ m^{-1}\ s^{-1}$. If $U = 10\ ms^{-1}$ (which is approximately zonal wind velocity at 500 hPa level in winter), then

$$L_S = 2 \times \frac{22}{7} \sqrt{10/16 \times 10^{-12}} = 4.967 \times 10^6\ m$$

$$L_S = 4967\ km.$$

Rossby waves with length greater than 4967 km move westward (Retrograde waves) while waves with length less than 4967 km move eastward (prograde). The real waves in the atmosphere at 500 hPa level is found to be qualitatively in agreement with the Rossby formula $C = U - \frac{\beta}{k^2}$, when wavelength is not too large. It seems the real atmosphere around middle of the troposphere behaves like horizontal non-divergent motion. With all these assumptions the model with the dynamics given by $\frac{d(\zeta + f)}{dt} = 0$ satisfies in the middle tropospheric flow to the first approximation only.

$\frac{\beta}{k^2}$ some times denoted by $C_R = \left(\frac{\beta}{k^2}\right)$ which is called Rossby speed.

At Lat 45° N, we have

$$C_R = \frac{16 \times 10^{-12}}{4\Pi^2} \times \overset{2}{L} \simeq 0.4 \overset{2}{L}$$

where L = wavelength measured in units of 10^6 m = 1000 km.

Note : Rossby observed wave patterns on routine daily synoptic weather charts. These waves related to the large scale motions in the atmosphere. He found useful (of these waves) for meteorological forecasts. He gave lot of importance to relative vorticity and coriolis parameter in the large scale atmospheric motion. He found (with vorticity equation) in daily synoptic charts and time-averaged charts of middle latitude, the movement of migratory cyclonic storms and quasi-stationary planetary waves.

Length and Time scales are :

	L : length scale	T : Time scale
Planetary-scale	Horizontal length scale ~ 5000 km	5 days or more
Synoptic scale	Horizontal length scale ~ 1000 km	1 day
Meso-scale	Horizontal length scale ~ 100 km	5 hours
Micro-scale	(Less than or equal to 1 km)	10 minutes or less

Meso-scale sub-divisions

Meso - α	200 km < L < 2000 km
Meso - β	20 km < L < 200 km
Meso - γ	2 km < L < 20 km

Questions

1. Describe Rossby waves. Derive an expression for the speed of Rossby waves. Show that if $U > 0$, $C > 0$ the waves move from west to east, if $U > 0$, $C < 0$ the waves move from east to west, if $C = 0$, they are stationary. Find the formula for the wavelength of stationary Rossby waves.

2. Calculate the Rossby parameter β at Lat 45°, given $\Omega = 7.29 \times 10^{-5}$ R = Radius of the earth 6370 km.

   Hint : $\beta = \dfrac{\partial f}{\partial y}$ where $f = 2\,\Omega\,\sin\theta$, $\dfrac{\partial \varphi}{\partial y} = \dfrac{1}{R}$

   $$= \frac{\partial f}{\partial \phi}\,\frac{\partial \varphi}{\partial y}$$

CHAPTER 29

Atmospheric Turbulence (A)

The atmospheric motions vary both in time and space. On time scale we have Biennial Oscillations, Annual variations, seasonal (months) variation, a few days in Rossby and cyclone waves, diurnal pressure wave oscillations and minutes and seconds in smaller scale motions. In space scale large scale motions, Rossby waves are 10^4 km, cyclone waves $10^3 - 5 \times 10^3$ km, meso scale motions $10^2 - 10^3$ km, Convection $1 - 10^2$ km, small scale/local systems in meters or cm or mm. Thus the atmospheric motion spectrum is very broad and ranges over several orders of magnitude. It is impossible to observe variations of pressure, temperature, humidity, wind etc. in all scales of atmosphere in time and space. With all these limitations we cannot ignore the small scale motions (though unobservable) but we can study these small scale motions by relating it to the large scale motions. For this purpose the statistical methods are best suited. We also know that the atmospheric interactions are on different scales as the governing equations are of non-linear form.

Let $u = u(x)$ has many scales, but for simplicity we consider two scales with wave numbers K_1 & K_2.

Let
$$u_1 = A_1 \cos K_1 x + B_1 \sin K_1 x$$
$$u_2 = A_2 \cos K_2 x + B_2 \sin K_2 x \quad .....(29.1)$$

The interaction between the two components may be of the form $u_1\left(\frac{\partial u_2}{\partial x}\right)$. Then from eq. 29.1, we have

$$u_1 \frac{\partial u_2}{\partial x} = (A_1 \cos K_1x + B_1 \sin K_1x)\ (-A_2K_2 \sin K_2x + B_2K_2 \cos K_2x)$$

$$= -A_1A_2K_2 \cos K_1x \sin K_2x + A_1B_2K_2 \cos K_1x \cos K_2x - A_2B_1K_2 \sin K_1x \sin K_2x + B_1B_2K_2 \sin K_1x \cos K_2x$$

$$= -\frac{A_1A_2K_2}{2}\left[\sin(K_1+K_2)x - \sin(K_1-K_2)x\right]$$

$$+ \frac{A_1B_2K_2}{2}\left[\cos(K_1+K_2)x + \cos(K_1-K_2)x\right]$$

$$- \frac{A_2B_1K_2}{2}\left[\cos(K_1-K_2)x - \cos(K_1+K_2)x\right]$$

$$+ \frac{B_1B_2K_2}{2}\left[\sin(K_1+K_2)x + \sin(K_1-K_2)x\right]$$

$$u_1 \frac{\partial u_2}{\partial x} = \frac{1}{2}K_2\ [(A_1B_2 + A_2B_1)\cos(K_1+K_2)x + (B_1B_2 - A_1A_2)\sin(K_1+K_2)x + (A_1B_2 - A_2B_1)\cos(K_1-K_2)x + (B_1B_2 + A_1A_2)\sin(K_1-K_2)x] \qquad .....(29.2)$$

Eq. (29.2) shows new components with wave numbers $(K_1 + K_2)$ and $(K_1 - K_2)$ when $K_1 > K_2$. The component with wave number $(K_1 + K_2)$ may have smaller scale than the original components while the component with wave number $(K_1 - K_2)$ will have scale larger than K_1 (since $K_1 - K_2 < K_1$). The scale of $K_1 - K_2 < K_2$ i.e. $K_1 < 2K_2$ may be larger than K_2. This shows a process of cascade process that is "increase (up) the wave numbers" (in small scale motion). or "decrease (down) the wave numbers" (in large-scale motion).

Note :

$$2 \sin A \cos B = \sin(A + B) + \sin(A - B)$$
$$2 \cos A \sin B = \sin(A + B) - \sin(A - B)$$
$$2 \cos A \cos B = \cos(A + B) + \cos(A - B)$$
$$2 \sin A \sin B = \cos(A - B) - \cos(A + B)$$

29.1 Response Function

Time Average : Let $b = b(x, y, z, t)$ be an arbitrary scalar function (or quantity).

A running time average $\bar{b}$ is defined as

$$\bar{b} = \frac{1}{T}\int_{t-\frac{T}{2}}^{t+\frac{T}{2}} b\,dt \qquad .....(29.3)$$

where $\bar{b} = \bar{b}(x, y, z, t)$ is a function of x, y, z, t in which small scale functions have been averaged.

Let 'b' represents a time series (Fourier type) of the type

$$b = B_0 + a_1 \cos V_1t + a_2 \cos V_2t + a_3 \cos V_3t + \ldots + a_n \cos V_nt$$
$$+ b_1 \sin V_1t + b_2 \sin V_2t + b_3 \sin V_3t + \ldots + b_n \sin V_nt$$

$$b = B_0 + \sum_n (a_n \cos V_nt + b_n \sin V_nt) \qquad .....(29.4)$$

where B_0 = constant

Consider the time average of a general term in (eq. 29.4)

$$\frac{1}{T}\int_{t-\frac{T}{2}}^{t+\frac{T}{2}} (a_n \cos V_nt + b_n \sin V_nt)\, dt$$

$$= \frac{1}{T}\left[\frac{a_n}{V_n}\sin V_nt - \frac{b_n}{V_n}\cos V_nt\right]_{t-\frac{T}{2}}^{t+\frac{T}{2}}$$

$$= \frac{1}{T}\frac{a_n}{V_n}\left[\sin V_n\left(t+\frac{T}{2}\right) - \sin V_n\left(t-\frac{T}{2}\right)\right]$$

$$- \frac{1}{T}\frac{b_n}{V_n}\left[\cos V_n\left(t+\frac{T}{2}\right) - \cos V_n\left(t-\frac{T}{2}\right)\right]$$

$$= \frac{a_n}{T}\frac{2}{V_n}\left[\cos V_nt.\sin\frac{T}{2}V_n\right]$$

$$+ \frac{1}{T}\frac{b_n}{V_n}2\left[\sin tV_n.\sin\frac{T}{2}V_n\right] \qquad .....(29.5)$$

Using (eq. 29.5) averaging (eq. 29.2) we have

$$\bar{b} = \frac{1}{T}\int_{t-\frac{T}{2}}^{t+\frac{T}{2}} b\,dt$$

$$= B_0 + \frac{1}{T}\sum_n 2\left[\frac{a_n}{V_n}\cos V_nt + \frac{b_n}{V_n}\sin V_nt\right]\sin V_n\frac{T}{2}$$

$$= B_0 + \sum_n \frac{2}{V_n T} [a_n \cos V_n t + b_n \sin V_n t] \sin V_n \frac{T}{2}$$

$$\bar{b} = B_0 + \sum_n R_n [a_n \cos V_n t + b_n \sin V_n t] \qquad .....(29.6)$$

where $$R_n = \frac{\sin V_n \frac{T}{2}}{V_n \frac{T}{2}} = \frac{\sin\left(\pi \frac{T}{T_n}\right)}{\pi \frac{T}{T_n}}; \quad \text{put} \quad V_n = \frac{2\pi}{T_n} \qquad .....(29.7)$$

R_n defined in eq. (29.7) above is called the response function. It follows from response function that $R_n \to 1$ as $T_n \to \infty$. Further $R_n \to 0$ when $T_n = T, 0 < R < 1$ for $T < T_n < \infty$. This means that the long period components are not effected by time average defined by eq. (29.3), but it smooths by reducing the amplitude components. Thus the average field $\bar{b}$ represents long period motion, while the short period (high frequency) components are smoothed out.

29.2 Reynolds Stresses

Let $\overline{\rho\vec{V}}$ define the average mass transport in the atmosphere and let

$$\bar{\rho}\,\hat{\vec{V}} = \overline{\rho\vec{V}} \qquad .....(29.8)$$

where $\bar{\rho}$ = average density

$\hat{\vec{V}}$ = Average velocity = $\dfrac{\overline{\rho\vec{V}}}{\bar{\rho}}$

$\vec{V}$ = instantaneous velocity

or $$\hat{\vec{V}} = \frac{\int_{t-\frac{T}{2}}^{t+\frac{T}{2}} \rho\vec{V}\,dt}{\int_{t-\frac{T}{2}}^{t+\frac{T}{2}} \rho\,dt} \qquad .....(29.9)$$

The vector $\vec{\hat{V}}$ as defined by eq. (29.8) expresses that the average mass transport in atmosphere is the product of average density and average velocity.

Let $\vec{V}'$ denote the instantaneous fluctuation in velocity, then

$$\vec{V} = \vec{\hat{V}} + \vec{V}' \quad .....(29.10)$$

[i.e., $u = \hat{u} + u'$, $v = \hat{v} + v'$, $w = \hat{w} + w'$]

Multiplying eq. (29.10) by ρ and averaging we have

$$\overline{\rho\vec{V}} = \overline{\rho\left(\vec{\hat{V}} + \vec{V}'\right)} = \bar{\rho}\vec{\hat{V}} + \overline{\rho\vec{V}'}$$

using eq. (29.8) the above equation becomes

$$\bar{\rho}\vec{\hat{V}} = \bar{\rho}\vec{\hat{V}} + \overline{\rho\vec{V}'}, \text{ thus mean that } \rho\vec{V}' = 0 \quad .....(29.11)$$

we know

$$\frac{\partial\rho}{\partial t} + \nabla.\left(\rho\vec{V}\right) = 0$$

or
$$\frac{\partial\rho}{\partial t} + \frac{\partial\rho u}{\partial x} + \frac{\partial\rho v}{\partial y} + \frac{\partial\rho w}{\partial z} = 0 \quad \text{(Equation of continuity)} \;....(29.12)$$

Using the averaging operator this equation becomes

$$\frac{\partial\bar{\rho}}{\partial t} + \frac{\partial\bar{\rho}\hat{u}}{\partial x} + \frac{\partial\bar{\rho}\hat{v}}{\partial y} + \frac{\partial\bar{\rho}\hat{w}}{\partial z} = 0 \quad .....(29.13)$$

The first equation of motion in local cartesian coordinates is given by

$$\frac{du}{dt} = -\frac{1}{\rho}\frac{\partial p}{\partial x} + fv - ew + \frac{1}{\rho}F_x$$

or
$$\frac{\partial u}{\partial t} + u\frac{\partial u}{\partial x} + v\frac{\partial u}{\partial y} + w\frac{\partial u}{\partial z} = -\frac{1}{\rho}\frac{\partial p}{\partial x} + fv - ew + \frac{F_x}{\rho} \quad .....(29.14)$$

Multiplying eq. (29.12) by u and eq. (29.14) by ρ we have

$$u\frac{\partial\rho}{\partial t} + u\frac{\partial\rho u}{\partial x} + u\frac{\partial\rho v}{\partial y} + u\frac{\partial\rho w}{\partial z} = 0$$

$$\rho\frac{\partial u}{\partial t}+\rho u\frac{\partial u}{\partial x}+\rho v\frac{\partial u}{\partial y}+\rho w\frac{\partial u}{\partial z}=-\frac{\partial p}{\partial x}+\rho fv-\rho ew+F_x$$

Adding these two equations we got

$$\frac{\partial \rho u}{\partial t}+\frac{\partial \rho uu}{\partial x}+\frac{\partial \rho uv}{\partial y}+\frac{\partial \rho uw}{\partial z}=-\frac{\partial p}{\partial x}+\rho fv-\rho ew+F_x \quad .....(29.15)$$

eq. (29.15) is called the flux form of x-momentum equation.

Similarly from the second and third equations of motion we get.

$$\frac{\partial \rho v}{\partial t}+\frac{\partial \rho vu}{\partial x}+\frac{\partial \rho vv}{\partial y}+\frac{\partial \rho vw}{\partial z}=-\frac{\partial p}{\partial y}-\rho fu+F_y$$

(y-momentum equation) ....(29.16)

$$\frac{\partial \rho w}{\partial t}+\frac{\partial \rho wu}{\partial x}+\frac{\partial \rho wv}{\partial y}+\frac{\partial \rho ww}{\partial z}=-\frac{\partial p}{\partial w}-g\rho+\rho eu+F_z \quad .....(29.17)$$

(z-momentum equation)

Now $\overline{\rho uv}=\bar{\rho}\ \overline{uv}=\bar{\rho}\overline{(\hat{u}+u')(\hat{v}+v')}=\bar{\rho}\ \hat{u}\hat{v}+\overline{\rho\ u'v'}$ (since $\overline{u'}=\overline{v'}=0$)

Similarly $\overline{\rho vw}=\bar{\rho}\,\hat{v}\hat{w}+\overline{\rho v'w'}$, $\overline{\rho uu}=\bar{r}\,\hat{u}\hat{u}+\overline{\rho u'u'}$ etc.

using these and averaging eq. (29.15, 29.16, and 29.17), we have

$$\frac{\partial \bar{\rho}\hat{u}}{\partial t}+\frac{\partial \bar{\rho}\,\hat{u}\hat{u}}{\partial x}+\frac{\partial \bar{\rho}\,\hat{v}\hat{u}}{\partial y}+\frac{\partial \bar{\rho}\,\hat{w}\hat{u}}{\partial z}=-\frac{\partial \bar{p}}{\partial x}+\bar{\rho}f\hat{v}-\bar{\rho}e\hat{w}+$$

$$\left[F_x-\frac{\partial \overline{\rho u'u'}}{\partial x}-\frac{\partial \overline{\rho v'u'}}{\partial y}-\frac{\partial \overline{\rho w'u'}}{\partial z}\right] \quad .....(29.18)$$

$$\frac{\partial \bar{\rho}\,\hat{v}}{\partial t}+\frac{\partial \bar{\rho}\,\hat{u}\hat{v}}{\partial x}+\frac{\partial \bar{\rho}\,\hat{v}\hat{v}}{\partial y}+\frac{\partial \bar{\rho}\,\hat{w}\hat{v}}{\partial z}=-\frac{\partial \bar{p}}{\partial y}-\bar{\rho}\,f\hat{u}+$$

$$\left[F_y-\frac{\partial \overline{\rho u'v'}}{\partial x}-\frac{\partial \overline{\rho v'v'}}{\partial y}-\frac{\partial \overline{\rho w'v'}}{\partial z}\right] \quad .....(29.19)$$

$$\frac{\partial \bar{\rho}\,\hat{w}}{\partial t}+\frac{\partial \bar{\rho}\,\hat{u}\hat{w}}{\partial x}+\frac{\partial \bar{\rho}\,\hat{v}\hat{w}}{\partial y}+\frac{\partial \bar{\rho}\,\hat{w}\hat{w}}{\partial z}=-\frac{\partial \bar{p}}{\partial z}-g\bar{\rho}+\bar{\rho}e\hat{u}+$$

$$\left[F_z-\frac{\partial \overline{\rho u'w'}}{\partial x}-\frac{\partial \overline{\rho v'w'}}{\partial y}-\frac{\partial \overline{\rho w'w'}}{\partial z}\right] \quad .....(29.20)$$

Eq. (29.18, 29.19 and 29.20) are time averaged equations of motion for a turbulent fluid, which are equivalent to eq. (29.15, 29.16 and 29.17) respectively (which are non time averaged motions) except for the last three terms indicated in the brackets. These three terms

$$\left[F_x - \frac{\partial \overline{\rho u'u'}}{\partial x} - \frac{\partial \overline{\rho v'u'}}{\partial y} - \frac{\partial \overline{\rho w'u'}}{\partial z}\right]$$

$$\left[F_y - \frac{\partial \overline{\rho u'v'}}{\partial x} - \frac{\partial \overline{\rho v'v'}}{\partial y} - \frac{\partial \overline{\rho w'v'}}{\partial z}\right]$$

$$\left[F_z - \frac{\partial \overline{\rho u'w'}}{\partial x} - \frac{\partial \overline{\rho v'w'}}{\partial y} - \frac{\partial \overline{\rho w'w'}}{\partial z}\right] \quad .....(29.21)$$

are called Reynolds stresses or eddy stresses.

Reynolds stresses given by eq. (29.21) represent the influence of fluctuating velocity of $\vec{V}'$ on the field $\hat{\vec{V}}$ and add to the frictional terms. Now we shall express Reyñold stresses in terms of average fields.

Consider the thermodynamic equation,

$$H = C_v \frac{dT}{dt} + p\frac{d\alpha}{dt}$$

$$= C_v \frac{dT}{dt} + \rho RT \frac{d\alpha}{dt} \quad \text{(using gas equation } p = \rho RT\text{)}$$

$$= C_v \frac{dT}{dt} - \rho \frac{RT}{\rho^2}\frac{d\rho}{dt} \quad \left[\because \alpha = \frac{1}{\rho}\right]$$

$$H = C_v \frac{dT}{dt} + RT\nabla.\vec{V}$$

$$\left[\because \frac{1}{\rho}\frac{d\rho}{dt} + \nabla.\vec{V} = 0 \quad \text{Equation of continuity}\right]$$

$$\therefore \quad C_v \frac{dT}{dt} = -RT\nabla.\vec{V} + H$$

or

$$\frac{\partial T}{\partial t} + u\frac{\partial T}{\partial x} + v\frac{\partial T}{\partial y} + w\frac{\partial T}{\partial z} = -\frac{R}{C_v}T\nabla.\vec{V} + \frac{H}{C_v} \quad .....(29.22)$$

$$\therefore \qquad \rho\left[\frac{\partial T}{\partial t}+u\frac{\partial T}{\partial x}+v\frac{\partial T}{\partial y}+w\frac{\partial T}{\partial z}=-\frac{R}{C_v}T\nabla.\vec{V}+\frac{H}{C_v}\right]$$

and $$T\left[\frac{\partial \rho}{\partial t}+\frac{\partial \rho u}{\partial x}+\frac{\partial \rho v}{\partial y}+\frac{\partial \rho w}{\partial z}=0\right]$$

Adding these we get

$$\frac{\partial T\rho}{\partial t}+\frac{\partial}{\partial x}(\rho uT)+\frac{\partial(\rho vT)}{\partial y}+\frac{\partial(\rho wT)}{\partial z}$$

$$=-\frac{p}{C_v}\nabla.\vec{V}+\frac{\rho H}{C_v} \qquad .....(29.23)$$

(since $\rho RT = p$)

Using $\bar{\rho}\hat{b}=\overline{\rho b}$, defining $b=\hat{b}+b'$ .....(29.24)

and averaging gas equation

$$\bar{p}=R\bar{\rho}\hat{T} \qquad .....(29.25)$$

Averaging eq. (29.23) with the help of eq. (29.24 and 29.25),we have

$$\frac{\partial\bar{\rho}\hat{T}}{\partial t}+\frac{\partial\bar{\rho}\hat{u}\hat{T}}{\partial x}+\frac{\partial\bar{\rho}\hat{v}\hat{T}}{\partial y}+\frac{\partial\bar{\rho}\hat{w}\hat{T}}{\partial z}=-\frac{\bar{p}}{C}\nabla.\hat{\vec{V}}+\frac{1}{C_v}\bar{\rho}\hat{H}$$

$$-\left(\frac{\partial\overline{\rho u'T'}}{\partial x}+\frac{\partial\overline{\rho v'T'}}{\partial y}+\frac{\partial\overline{\rho w'T'}}{\partial z}\right)-\frac{1}{C_v}\overline{p\nabla.\vec{V}} \qquad .....(29.26)$$

The variables in eq. (29.26) are $\hat{V}$, $\hat{T}$, $\bar{\rho}$ and $\bar{p}$ and all fluctuations appearing in Reynolds stresses given by eq. (29.21).

Note : In order to predict average quantities we must have relation between mean quantities ($\hat{u},\hat{v}$...) and the fluctuations (u', v', ...). This relation has not been found satisfactorily. The relation between mean quantities ($\hat{u},\hat{v},\hat{w}$...) and fluctuations (u', v', w' ...) depend on the period of time (T) selected for averaging. If T = 10 minutes, say, fluctuations can be treated as turbulence.

In such cases Reynolds stresses completely dominate over friction force ($\vec{F}$) which is of molecular. Here *we may completely disregard molecular friction in atmospheric motion.* Similarly the heat flux by fluctuating quantities is much larger than the molecular heat conduction. As a consequence the molecular heat conduction may be neglected in H of the thermodynamic equation of atmosphere.

The chief criteria of similitude in fluid dynamics are the Reynolds number Re, Froude number Fr, Strouhal number St and Mach number M which are given below.

1. The Reynolds number Re $= \dfrac{vl}{\upsilon}$,

where v = Characteristic velocity of the fluid for the given problem

l = Characteristic linear dimension

υ = Kinematic viscosity of the fluid.

Characteristic velocity and length can be chosen depending upon the problem. For example for transverse flow of a fluid around a circular cylinder of diameter d, the value of l = d, v is the velocity of the undisturbed fluid. The Reynolds number is the ratio of the inertial forces to the viscous forces in the stream of fluid.

2. The Froude number Fr $= \dfrac{V^2}{gl}$

where V = Velocity of the fluid at a point far from the body it is flowing around

l = linear dimension of the body

g = acceleration due to gravity.

The Froude number is the ratio of internal forces to gravity forces.

3. The Strouhal number St $= \dfrac{VT}{l}$,

where V = Velocity

T = Characteristic time interval

l = Linear dimension

St = the criterion of similitude for unsteady motion of fluids.

4. The Mach number M $= \dfrac{V}{C}$

where V = Velocity of the fluid

C = Velocity of the sound in the fluid.

The Mach number is a measure of the influence of the compressibility of the fluid on its motion or ratio of elastic forces to inertial forces. If $M <<< 1$, the fluid is regarded as incompressible. For compressible fluid flow, if $M < 1$, is called subsonic and if $M > 1$ it is called supersonic.

The principal criteria of similitude for steady-state free convection of heat transfer is an incompressible fluid are given by the Nusselt number Nu, the Prandtl number Pr and Grashof number Gr which are given below.

5. The Nusselt number Nu $= \frac{\alpha l}{K}$,

where α = Coefficient of heat transfer
l = Characteristic Dimension
K = Heat conductivity of the fluid

6. The Prandtl number Pr $= \frac{\upsilon}{a} = \frac{\eta c}{K}$,

where υ = Kinematic viscosity of the fluid.
a = Coefficient thermometric conductivity
c = Specific heat
K = heat conductivity of the fluid

7. Grashof number Gr $= \frac{\alpha_t g l^3}{v^3} \nabla T$,

where α_t = Coefficient of thermal expansion of the fluid
υ = Kinematic viscosity
g = Acceleration due to gravity
l = Characteristic dimensoin
ΔT = Absolute temperature difference of fluid and the wall

29.3 Types of Viscous Fluid Flow

1. Laminar, 2. Turbulent. Laminar flow is an ordered fluid flow in which the paths of neighbouring particles differ slightly from one another, but the fluid flow still be considered as separate layers or laminae, flowing at different velocities without being mixed. In turbulent fluid flow, particles have unsteady random motion along complex irregular paths and the various layers of the fluid are necessarily (completely) mixed together. Laminar flow may be steady or unsteady but the turbulent flow is always unsteady. The turbulent stream is characterized by the two components, namely $u = \hat{u}+u', v=\hat{v}+v', w=\hat{w}+w'$ where u = instantaneous velocity, $\hat{u}$ = time averaged velocity in space defined in response function (temporal mean velocity$\left(\vec{V} = iu + jv + kw\right)$), u' = fluctuations in velocity.

A turbulent flow may be considered as steady if the average velocity $\hat{V}$ is independent of time i.e. $\frac{\partial \hat{V}}{\partial t} = 0$.

At any instant all the molecules in a liquid flow will not have same velocity. The (molecules or) molecular layers can move relative to each other in different velocities. Because of this relative motion there will be friction between different molecular layers which offers resistance to the liquid motion. This property is termed viscosity. In nonviscous fluids the velocity of all the particles at any (cross) section of a pipe are equal and the liquid moves a unit along a pipe. When the fluid is viscous and the velocity is not large the flow is laminar. The lamina of liquid slide over each other. Consider a fluid flow through a pipe of diameter D. Let υ be the kinematic viscosity of the fluid. The velocity of the fluid varies from zero, at the walls, to a maximum at center (the axis of the pipe) Fig. 29.1(a), Fig. 29.1(b). Let $\bar{U}$ = characteristic velocity.

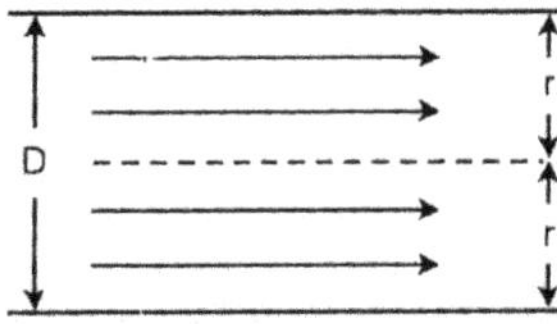

Fig. 29.1(a) Non-Viscous fluid flow

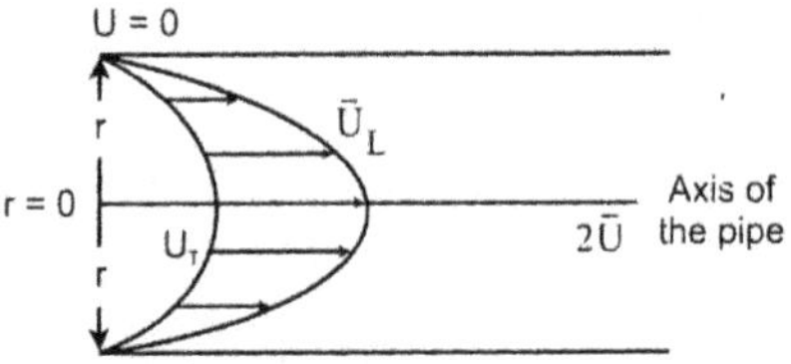

Fig. 29.1(b) Viscous fluid flow

$$\bar{U} = \frac{\text{Total volume transported through the pipe}}{\text{area of cross} - \text{section of the pipe}}$$

If $\bar{U}$ is not-large, the flow is laminar (streamlines and trajectories are parallel to the diameter of the pipe). When $\bar{U}$ exceeds the critical value Re, the flow is turbulent (that is unorganized, irregular and highly variable both in time and space). The critical value Re is called Reynolds number and is given by

$$Re = \frac{\bar{U}D}{\upsilon} \qquad .....(29.27)$$

This is a dimensionless quantity, its value is 2300 for pipe flow.

The laminar flow of Poiseuille's Law profile is given by

$$U_L = U_L(r) = 2\bar{U}\left[1 - \left(\frac{2r}{D}\right)^2\right] \qquad .....(29.28)$$

where r = distance from the center (axis of the pipe). The wind profile is a parabola as $U_L = U_L(r)$ as shown in the Fig. 29.1(b). When turbulent sets in, the wind profile takes the curve from $U_T = U_T(r)$. The turbulent flow is well described by statistical theory like the kinetic theory of gases.

Turbulence is found in connective clouds and in the vicinity of jet stream-core due to large wind shears. Convection is mainly caused by the difference between the adiabatic lapse rate and the actual lapse rate of environment. The Richardson number (Ri) gives the combined effects of temperature lapse rate and wind shear. Ri is defined as

$$Ri = \frac{\frac{g}{\theta}\frac{\partial\theta}{\partial z}}{\left(\frac{\partial\vec{V}}{\partial z}\right)^2} \qquad .....(29.29)$$

where θ = Potential temperature

$\vec{V}$ = Wind velocity

z = height.

Richardson number Ri is a dimensionless quantity.

From the definition of potential temperature θ we have,

$$\theta = T\left(\frac{p_0}{p}\right)^{\frac{R}{C_p}}$$

$$\therefore \qquad \ln\theta = \ln T + \frac{R}{C_p}[\ln p_0 - \ln p]$$

Differentiating w.r.t height z, we have

$$\frac{1}{\theta}\frac{\partial\theta}{\partial z} = \frac{1}{T}\frac{\partial T}{\partial z} - \frac{R}{C_p}\frac{1}{p}\frac{\partial p}{\partial z},$$

or

$$\frac{T}{\theta}\frac{\partial\theta}{\partial z} = \frac{\partial T}{\partial z} - \frac{RT}{p}\cdot\frac{1}{C_p}\cdot(-g\rho), \quad [dp = -g\rho dz \text{ Hydrostatic equation}]$$

$$= \frac{\partial T}{\partial z} + \frac{\rho RT}{p}\cdot\frac{g}{C_p}$$

$$\frac{T}{\theta}\frac{\partial\theta}{\partial z} = \frac{\partial T}{\partial z} + \frac{g}{C_p} \qquad .....(29.30)$$

In adiabatic process $\frac{\partial \theta}{\partial z} = 0$ (potential temperature is conserved with height)

$$0 = \frac{\partial T}{\partial z} + \frac{g}{C_p} \qquad \text{i.e.} -\frac{\partial T}{\partial z} = \frac{g}{C_p}$$

$$\therefore \quad 0 = -\gamma_d + \frac{g}{C_p} \quad \text{or} \quad \gamma_d = -\frac{\partial T}{\partial z} = \frac{g}{C_p} \qquad .....(29.31)$$

where γ_d = dry adiabatic lapse rate.

From (eq. 29.30), we have

$$\frac{T}{\theta} \cdot \frac{\partial \theta}{\partial z} = -\gamma + \frac{g}{C_p} \qquad \left(\because \; -\frac{\partial T}{\partial z} = \gamma \text{ lapse rate}\right)$$

$$\frac{T}{\theta} \cdot \frac{\partial \theta}{\partial z} = -\gamma + \gamma_d \qquad \text{using (eq. 29.31)}$$

$$\text{or} \quad \frac{g}{\theta}\frac{\partial \theta}{\partial z} = \frac{g}{T}(\gamma_d - \gamma) \qquad .....(29.32)$$

From (eq. 29.29 and 29.32), we have

$$R_i = \frac{\frac{g}{\theta}\frac{\partial \theta}{\partial z}}{\left(\frac{\partial \vec{v}}{\partial z}\right)^2} = \frac{\frac{g}{T}(\gamma_d - \gamma)}{\left(\frac{\partial \vec{v}}{\partial z}\right)^2} \qquad .....(29.33)$$

It follows from (eq. 29.33) that $R_i < 0$ if $\gamma > \gamma_d$. Therefore we require small or negative values of R_i to maintain convection. We have noted that Turbulence sets in when $\left(\frac{\partial \vec{v}}{\partial z}\right)$ is large. Hence small values of Ri favours turbulence.

Richardson number (Ri) may be expressed, pressure as vertical coordinate.

$$Ri = \frac{\frac{g}{\theta}\frac{\partial \theta}{\partial z}}{\left(\frac{\partial \vec{v}}{\partial z}\right)^2} = \frac{\frac{g}{\theta} \cdot \frac{\partial \theta}{\partial p} \cdot \frac{\partial p}{\partial z}}{\left(\frac{\partial \vec{v}}{\partial z}\frac{\partial p}{\partial p}\right)^2} = \frac{\frac{g}{\theta}(-g\rho)\frac{\partial \theta}{\partial p}}{\left(\frac{\partial p}{\partial z}\right)^2\left(\frac{\partial \vec{v}}{\partial p}\right)^2} \qquad \left(\because \; \frac{\partial p}{\partial z} = -g\rho\right)$$

$$= \frac{-\frac{1}{\rho\theta}\frac{\partial\theta}{\partial p}}{\left(\frac{\partial\vec{v}}{\partial p}\right)^2} = \frac{-\frac{\alpha}{\theta}\frac{\partial\theta}{\partial p}}{\left(\frac{\partial\vec{v}}{\partial p}\right)^2}$$

i.e., $$\text{Ri} = \frac{\sigma}{\left(\frac{\partial\vec{v}}{\partial p}\right)^2} \quad \text{where } \sigma = -\frac{\alpha}{\theta}\frac{\partial\theta}{\partial p} \qquad \text{.....(29.34)}$$

Ri is small in the layer near to the ground when γ and $\frac{\partial\vec{v}}{\partial z}$ are large, which is favorable for turbulence in this layer.

Let $\gamma = 0.67 \times 10^{-2}$ deg/m, T = 300 ^{0}K and g $\simeq$ 10 m/s^2

Then $$\frac{g}{T}(\gamma_d - \gamma) = \frac{10}{300}(1-0.67)\,10^{-2} \approx 10^{-4}\ s^{-1}$$

and $$\left|\frac{\partial\vec{V}}{\partial z}\right| \simeq 1\ ms^{-1}\ m^{-1} = 1\ s^{-1}$$

$\therefore$ $$\text{Ri} = \frac{\frac{g}{T}(\gamma_d - \gamma)}{\left(\frac{\partial\vec{v}}{\partial z}\right)^2} = \frac{10^{-4}\ s^{-1}}{1 s^{-1}} \simeq 10^{-4}$$

The Reynold number Re $= \frac{\bar{U}D}{\upsilon}$, near the ground $\bar{U}$ = 10 ms^{-1}, D $\simeq$ 10 m, $\upsilon \approx 10^{-5}\ m^2\ s^{-1}$.

$\therefore$ $$\text{Re} = \frac{10m \times 10ms^{-1}}{10^{-5}\ m^2\ s^{-1}} = 10^7$$

In the free atmosphere Ri is much larger than 10^{-4}, (calculated above), since $\left(\frac{\partial\vec{v}}{\partial z}\right)^2$ is much smaller. In the free atmosphere Ri generally varies between 10^1 to 10^2.

Questions

1. Briefly describe the atmospheric motions in time and space. Consider two sinusoidal waves with wave numbers k_1, k_2.

   Discuss the interaction of the two components of the form $u.\left(\frac{\partial u_2}{\partial x}\right)$

2. Let $b = b(x, y, z, t)$ be any arbitrary scalar function of Fourier type. Let

   $\bar{b}$ = running time average be defined as $\bar{b} = \int_{t-\frac{T}{2}}^{t+\frac{T}{2}} b dt$.

   Derive response function R_n.

3. Derive the expressions for Reynolds stresses or eddy stresses.

4. Express Reynold stresses in terms of average fields.

5. Define Reynolds number (Re) and Richadsons number Ri.

   Show that $Ri = \frac{\sigma}{\left(\frac{\partial \bar{V}}{\partial p}\right)^2}$

   where $\sigma = \frac{\alpha}{\theta} \frac{\partial \theta}{\partial p}$.

CHAPTER 30

Atmospheric Turbulence (B)

The factors that operate on atmospheric air flow near the ground (say below 200m agl) are listed below.

1. Turbulence due to irregular and random motions.
2. Degrees of roughness of surface (underlying) (e.g., rough urban areas, smooth sea surface).
3. Heating/cooling of the underlying surface during day and night.
4. Topographic features, local small scale obstacles (e.g., trees, buildings, ridges, valleys)
5. Outside disturbance both natural (convection, down draughts) and manmade explosions.

The characteristic airflow close to the ground contains turbulent component. The principal properties of turbulent flow are given below.

1. Flow is irregular. Both wind speed and direction changes randomly
2. Flow is diffusive. It mixes different properties such as heat, momentum etc.
3. The flow has large Reynolds number $Re = \frac{UL}{\upsilon}$,

   where υ = Kinetic viscosity of air $\simeq 10^{-4}$ m s^{-1}

   U = horizontal wind speed order $\simeq$ 1 m s^{-1}

   L = Order of length scale $\simeq$ 10 m, $Re = \left(\frac{1 \times 10}{10^{-4}}\right) = 10^{6}\ s^{-1}$

4. The flow has three dimensional vorticity fluctuations, namely rotational with eddies of different sizes.
5. The flow is dissipative (K E is extracted from the mean motion and dissipated by viscous shear stresses)
6. The flow is a continuum, that is, it possess all scales of motion and obeys the equations of motion (fluid mechanics)
7. Turbulence is not the property of fluid but a necessary feature of fluid flow. All turbulent flows are similar when Reynolds number is large.

The turbulent motion at a point is described by u, v, w wind components, defined by the time average

$$\hat{u}\,(x, y, z, t) = \frac{1}{T}\int_{t-\frac{T}{2}}^{t+\frac{T}{2}} u(x,y,z,t)\,dt \qquad .....(30.1)$$

Instantaneous velocity components u, v, w, are given by

$$u = \hat{u} + u'$$

$$v = \hat{v} + v'$$

$$w = \hat{w} + w'$$

where u', v', w' are turbulent fluctuations or deviations from mean values [$u = \hat{u}$, $v = \hat{v}$, $w = \hat{w}$] defined by eq. (30.1).

In synoptic weather observations generally T = 10 minutes, $\hat{w} = 0$ close to the ground over flat homogeneous terrain.

$$\text{KE (per unit mass)} = \frac{1}{2}\left(\hat{u}^2 + \hat{v}^2 + \hat{w}^2\right), = \frac{1}{2}\left(\hat{u}^2 + \hat{v}^2 + \hat{w}^2 + \overline{\hat{u}^2} + \overline{\hat{v}^2} + \overline{\hat{w}^2}\right),$$

where $b = \frac{1}{2}\left(\overline{u'^2} + \overline{v'^2} + \overline{w'^2}\right)$ = KE of turbulent motion or turbulent energy.

30.1 Wind Spectrum Near the Ground

The power spectrum of atmospheric wind motion shows two large peaks and one small peak Fig. 30.1(a). The large peak has a period of about 4 to 5 days (about 100 hours) which correspond to the large scale motions of westerlies (Lows and High). The second large peak has period of about 10 to 15 hours. This correspond to the diurnal variations of wind speed. The third peak has a period of about 2 to 3 minutes, this corresponds to the wind motion close to the surface which is almost always turbulent or gusty. There is spectral gap between periods 0.1 to 3 – 4 hours with little energy. This corresponds to the

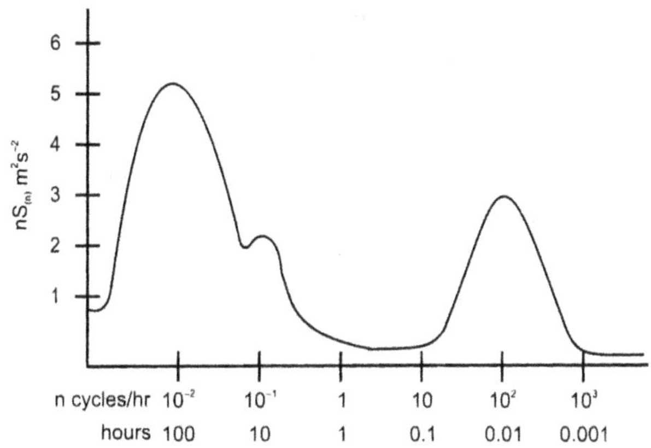

Fig. 30.1(a) Schematic spectrum of wind speed near the ground S(n) is power 'spectral density'. This power spectrum shows how a particular place is affected by various types of atmospheric motions.

absence of persistent motions. This gap is called "mesoscale gap" (10–100 km). An idealized picture of space and time scale atmospheric motion is given in the Fig. 30.1(b) (not to scale).

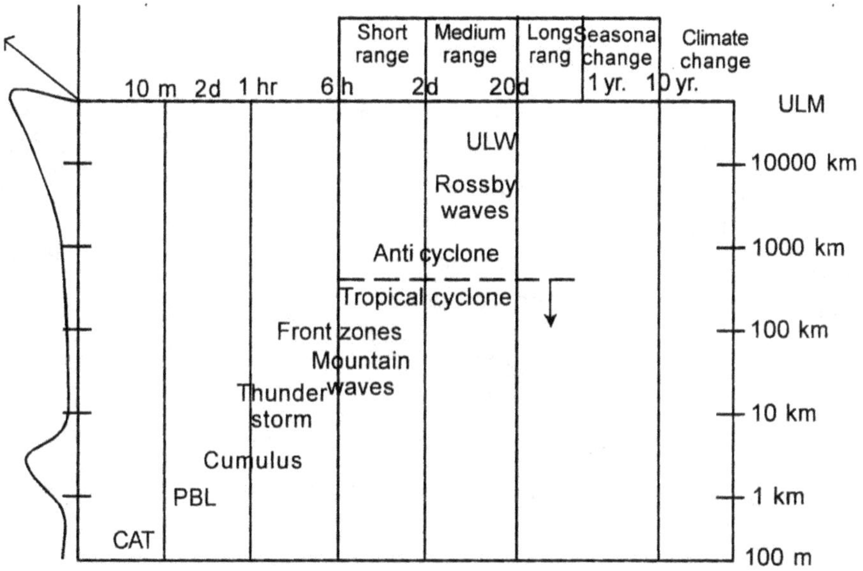

Fig. 30.1(b) Space and Time scales of Atmospheric motions.

PBL = Planetary Boundary layer

ULW = Ultra long waves

CAT = Clear Air Turbulence.

At the surface of the earth, the roughness of the ground makes the wind speed to be zero. This amounts to the fact that near the ground there is a zone of strong vertical variation of wind speed or wind shear. In a dry neutral atmosphere (lapse rate 10 °C/km) the mean wind assumes logarithmic wind profile Fig. 30.2. The height at which the wind speed becomes zero is called roughness height z_0, is measure of the average height of the surface roughness or irregularities.

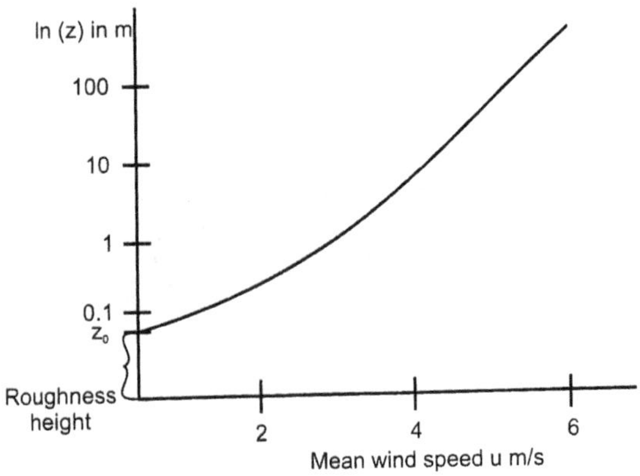

Fig. 30.2 Logarithmic wind profile.

Near the ground turbulent fluctuations or eddies of different magnitudes may be seen. In the lowest (say) 100 m, the largest eddies are a few. Eddy sizes are proportional to the height above ground. Eddies of different magnitudes transport different (amounts) magnitudes of momentum from one level to another. At any height large eddies (carry) transfer more momentum compared to small eddies. The momentum transport may be expressed as

$$\tau_{zx} = -\rho\overline{u'w'} = \rho k \frac{\partial u}{\partial z} \qquad .....(30.2)$$

where τ_{xz} = Transport (or flux) of x-momentum in z-direction and

τ_{xz} = is proportional to the correlation between turbulent fluctuations of u and w components of wind ($\rho\overline{u'w'}$). According to Ekman

$$\tau_{xz} = K \frac{\partial \hat{u}}{\partial z} \qquad .....(30.3)$$

where K = eddy viscosity or eddy diffusion. eq. (30.3) expresses momentum transport as diffusion.

Prandtl introduced the mixing length eddy size 'l' and propounded that

$$K = l^2 \frac{\partial \hat{u}}{\partial z} \qquad .....(30.4)$$

in neutral condition near the ground, height less than 100 m.

The shear stress τ_{xz} is generally represented by friction velocity U_* and given by the relation

$$U_* = \sqrt{\frac{\tau_{xz}}{\rho}} = \sqrt{-\overline{u'w'}} \qquad .....(30.5)$$

The value of U_* near the ground is of the order 10 cm/s, which represents intensity of the turbulent flux of momentum. Atmosphere very rarely assumes neutral state, but it generally varies between stable and unstable condition depending on vertical temperature variation. Γ_d (dry adiabatic lapse rate) expresses the decrease of temperature with height in dry atmosphere ($\simeq 10^0$C/km)

From eq. (30.5), we have

$$U_*^2 = \frac{\tau_{xz}}{\rho} = -\overline{u'w'} = K\frac{\partial \hat{u}}{\partial z} \qquad \text{[from eq. (30.3)]}$$

$$= l^2 \frac{\partial \hat{u}}{\partial z}.\frac{\partial \hat{u}}{\partial z} \qquad \left[\because \ K = l^2 \frac{\partial \hat{u}}{\partial z} \ \text{from eq. (30.4)}\right]$$

$$U_*^2 = (Kz)^2 \left(\frac{\partial \hat{u}}{\partial z}\right)^2$$

where l = Kz, K is called Von Karman's constant = 0.38

$$\text{i.e.} \qquad U_* = Kz\frac{\partial \hat{u}}{\partial z} \qquad \text{or} \qquad \frac{\partial \hat{u}}{\partial z} = \frac{u}{Kz} \qquad .....(30.6)$$

Integrating the last equation w.r.t z (between proper limits $u_* = 0$ to $\hat{u}$ corresponding $z = z_0$ to z).

$$\int_0^{\hat{u}} \frac{\partial \hat{u}}{\partial z} dz = \int_{z_0}^{z} \frac{u_*}{K}\frac{dz}{z}$$

$$\hat{u} = \frac{u_*}{K} \ln\left(\frac{z}{z_0}\right).$$ This is logarithmic wind law.

K_M = Eddy viscosity for momentum exchange

K_H = Eddy viscosity for heat exchange

K_W = Eddy viscosity for water vapour exchange.

Questions

1. (a) State the factors that effect the atmospheric air flow close to the ground (say below 200 m agl).

   (b) State the principal properties of turbulent flow.

2. Describe briefly the wind spectrum close the ground both in time and space.

3. What is roughness height? Derive logarithmic wind law equation.

CHAPTER 31

The Planetary Boundary Layer

1. Laminar Sub Layer

In general molecular viscosity is negligible if $Re = \frac{z_0 u_*}{\upsilon} > 2.5$. In this case aerodynamically the surface is called rough. If $Re = \frac{z_0 u_*}{\upsilon} < 0.13$, the molecular velocity (υ) is not negligible. In this case aerodynamically the surface is called smooth. When $Re < 0.13$, we have a thin layer next to the ground/sea surface which is dominated by molecular viscosity (υ). This thin layer is called laminar sub-layer or skin layer, (Fig. 31.1). The wind profile in this layer is given by

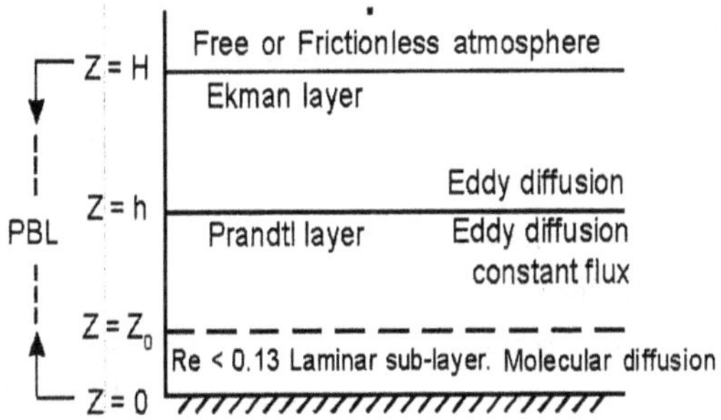

Fig. 31.1 PBL with laminar sub-layer, Prandtl layer and Ekman layer.

$\bar{u} = \frac{u_*^2 z}{\upsilon}$. On many occasions this layer may not exist. Mass and properties exchanges take place through molecular diffusion.

2. The Prandtl-Layer

The lowest layer next to the ground including laminar sub-layer is called Prandtl layer or surface layer or constant flux layer. On some occasions there exists a thin layer next to the ground where molecular viscosity is dominant, which we called it laminar sub-layer. The depth of Prandtl layer ($z = h$)] is 20-60 m and has little mass per unit area. The characteristic feature of this layer is quick variation with height in mean velocity, temperature and mass per unit. Further, it is assumed that turbulent transport of sensible heat (H_d), turbulent stress (τ) and vertical turbulent transport of water vapour (H_e) are all constant. In this layer stress is constant in vertical direction and it (stress) is much larger than pressure force and coriolis force. Wind vector may change in magnitude but not in direction. Thermal wind $\left(\frac{\partial \vec{V}_g}{\partial z}\right)$ is parallel to the wind in this layer.

The PBL : Let $z = H$ be the height above ground level above which ($z > H$) the influence of friction is negligible and which is called free atmosphere. $z = H$ is called the Planetary Boundary Level (PBL). The layer above prandtl layer but [$z > h$ & $z \leq H$] is called Ekman layer. In the Ekman layer there is a balance between pressure force and coriolis, frictional forces

$$\left[0 = -\alpha \nabla p - f\hat{k} \times \vec{V} + \alpha \frac{\partial \vec{\tau}}{\partial z} \qquad \text{where } \hat{k} \text{ is unit vector in } z-\text{direction}\right]$$

Eddy Diffusion : Like molecular viscosity, eddy viscosity (internal friction) is used in fluid mechanics and in atmospheric physics. Like molecular random motion, parcels of fluid are in random motion is considered in fluid mechanics. This random motion of fluid parcels, transfer heat, momentum, moisture and other suspended material from high concentration region to low concentration region. This is called Eddy diffusion.

Significance of PBL : More than 50% of atmospheric KE is lost in PBL due to friction. Though the source of heat for the earth and atmosphere is the sun, yet the atmosphere is driven by the heat from the surface of earth in the form of long wave radiation, sensible heat and latent heat. Atmosphere continuously loses momentum in PBL. In middle latitudes momentum is lost in westerlies while in tropics momentum is lost in easterlies (or momentum gained in westerlies). Momentum loss or gain occurs in PBL. The transport of water vapour (latent heat) into free atmosphere takes place from PBL, of-course depends on the properties of the PBL. Weather at any place depend on the frictional convergence and divergence and air mass modification. All these occur in PBL. Air-sea interaction or momentum exchange between atmosphere and sea take place in PBL, which contributes to the climate of the earth. All aerosols have their source of origin in PBL. Aerosols play vital role in cloud formation, precipitation, radiative balance and atmospheric pollution.

Conceptional basis of various layers of PBL : In skin or laminar sub-layer exchange of momentum, heat and water vapour between surface of the earth and atmosphere takes place through molecular diffusion. In Prandtl and Ekman layers molecular diffusion is insignificant, where exchange of mass and properties mainly take place through eddy diffusion. Molecular conduction of heat occurs when there is temperature gradient (i.e., in laminar sub-layer), while molecular transfer of momentum takes place when there is a velocity gradient or momentum (in Prandtl and Ekman layers).

The depth of laminar sublayer is of the order of few millimeters, denoted by $z = z_0$. Horizontal wind speed in this layer is taken to be zero and lapse rate of temperature is also zero (i.e., isothermal). Prandtl layer extends up to 50-60 m agl above laminar layers. This layer has large vertical gradients of temperature, specific humidity and wind speed. Vertical fluxes of momentum, sensible heat and water vapour are nearly constant with the absence of convergence and divergence. In this layer the coefficient of eddy diffusion is of the order of 10 $m^2 s^{-1}$ which is much larger than the molecular diffusion of sub-layer. In Prandtl layer wind direction is practically constant and wind speed obeys logarithmic law.

$$u = \frac{u_*}{K} \ln\left(\frac{z}{z_0}\right)$$

where z_0, roughness length = constant

K, Von Karman's constant ≃ 0.4.

Ekman layer : It extends above Prandtl layer to about 1 km. On some occasions this layer depth may rise to about 2 km or fall to about 500 m (in stable conditions under large scale subsidence). In this layer wind direction veers (rotates clockwise) in northern hemisphere but backs (rotates anticlockwise) in the southern hemisphere. Wind speed increases with height in accordance with the Ekman spiral.

Prandtl's Mixing length Hypothesis : In molecular diffusion of gases we have molecular free path. In a similar fashion we have mean mixing length in eddy diffusion. According to Prandtl's mixing length hypothesis, a fluid parcel when displaced vertically retains its mean horizontal velocity of its original level up to a certain length and then loses its identity abruptly in the environment. This length is called mixing length similar to the mean free path in molecular viscosity.

Consider an air parcel displaced at a level $z - \frac{l_1}{2}$ with velocity $\hat{u}_z - \frac{l_1}{2}\frac{\partial \hat{u}}{\partial z}$.

Let it rises up to the level $z + \frac{l_1}{2}$ where it abruptly loses its identity in the environment. Let the velocity of the environment or velocity of the parcel after losing its identity be $\hat{u}_z + \frac{l_1}{2}\frac{\partial \hat{u}}{\partial z}$.

Let the turbulent fluctuation

u' = (velocity of eddy parcel $\hat{u} - \frac{l_1}{2}\frac{\partial \hat{u}}{\partial z}$ at $z - \frac{l_1}{2}$) –(velocity of the environment $\hat{u} + \frac{l_1}{2}\frac{\partial \hat{u}}{\partial z}$ at $z + \frac{l_1}{2}$)

$$= \left(\hat{u}_z - \frac{l_1}{2}\frac{\partial \hat{u}}{\partial z}\right) - \left(\hat{u}_z + \frac{l_1}{2}\frac{\partial \hat{u}}{\partial z}\right)$$

$$u' \approx -l_1 \frac{\partial \hat{u}}{\partial z} \qquad \text{.....(31.1)}$$

where $l_1 > 0$ for upward displacement

$l_1 < 0$ for downward displacement.

Similarly the vertical transport of momentum across unit horizontal area in unit time is given by

$$\rho w' u' = -\rho w' l_1 \frac{\partial \bar{u}}{\partial z} \qquad \text{.....(31.2)}$$

Several eddy parcels will be moving up and down which transfer momentum without transfer of total mass. The root mean square of l_1, which we call mixing length, is given by $l = \sqrt{l_1^2}$.

Also we have

$$\tau_{zx} = -\rho\overline{u'w'}, \quad \tau_{zy} = -\rho\overline{v'w'}$$

and $$-\rho\overline{u'w'} = \rho\overline{w'l}\frac{\partial \hat{u}}{\partial z} = A_z \frac{\partial \hat{u}}{\partial z}$$

where A_z eddy exchange coefficient, which depends on the distance from the surface.

$$\rho\overline{v'w'} = \rho\overline{w'l}\frac{\partial \hat{v}}{\partial z} = A_z \frac{\partial \hat{v}}{\partial z}$$

Note : The average diameter of atmospheric molecule is 3.7×10^{-2} μm. At STP the root velocity of air molecules is about 500 m s^{-1}. The number of molecules per cubic centimeters is of the order 3×10^{19}, mean free path is of the order 6 μm. The molecular viscosity of air is $\frac{1}{3}$ Cl, where C = root mean square velocity of molecule and l = mean free path of molecule.

31.1 The Ekman Layer Equation

For geostrophic wind

$$\alpha\frac{\partial p}{\partial x} = fv_g, \qquad \alpha\frac{\partial p}{\partial y} = -fu_g$$

Multiplying these equations by i, j respectively and adding we have

$$\alpha\left(i\frac{\partial p}{\partial x} + j\frac{\partial p}{\partial y}\right) = f\,(iv_g - ju_g)$$

or $$-\alpha\nabla p = f\hat{k}\times[iu_g + jv_g]$$

[$\hat{k}$ = unit vector in z direction]

or $$-\alpha\nabla p = f\hat{k}\times\vec{V}_g$$

[$\vec{V}_g = iu_g + jv_g$ = horizontal wind vector.] .....(31.3)

In Ekman layer, there is a balance of the pressure force, the coriolis force and the frictional force. The equations of motion becomes

$$0 = -\alpha\nabla p - f\,\hat{k}\times\vec{V} + \alpha\frac{\partial\vec{\tau}}{\partial z}. \quad \text{where } \alpha = \frac{1}{\rho}$$

$$0 = f\hat{k} \times \vec{V}_g - f\hat{k} \times \vec{V} + \alpha\frac{\partial\vec{\tau}}{\partial z} \qquad \text{using eq. (31.3)}$$

or
$$f\hat{k} \times (\vec{V}_g - \vec{V}) + \alpha\frac{\partial\vec{\tau}}{\partial z} = 0 \qquad .....(31.4)$$

eq. (31.4) is shown in Fig. 31.2, where double lined vectors shows the balance of forces. OA = $\vec{V}$ (wind vector), OB = $\vec{V}_g$ (geostrophic wind vector)

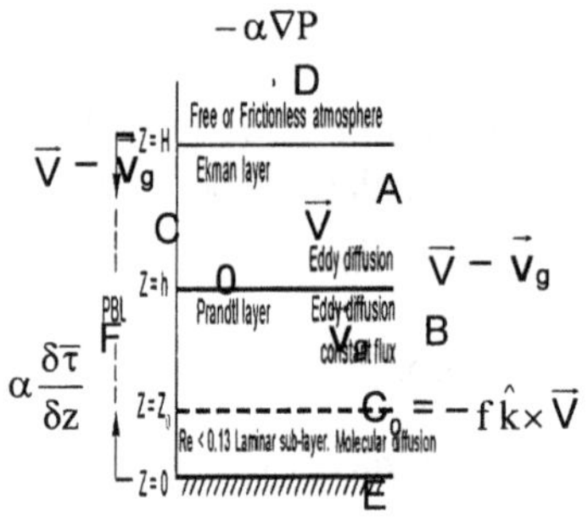

Fig. 31.2

(BA||OC), BA = $\vec{V} - \vec{V}_g$ = OC, OD = $-\alpha\nabla p$ (OD $\perp$ OB)

$$OF = \alpha\frac{\partial\vec{\tau}}{\partial z},$$

OE = C_0 = Coriolis force = $-f\hat{k} \times \vec{V}$, (OE $\perp$ OA)

We know

$$\tau_{zx} = \rho K\frac{\partial u}{\partial z}, \quad \tau_{yz} = \rho K\frac{\partial v}{\partial z} \qquad \text{where K is Von Karmans constant}$$

Assuming K is independent of z we have

$$\frac{\partial\tau_{zx}}{\partial z} = \rho K\frac{\partial^2 u}{\partial z^2}, \qquad \frac{\partial\tau_{yz}}{\partial z} = \rho K\frac{\partial^2 v}{\partial z^2}$$

or
$$\frac{1}{\rho}\frac{\partial\tau}{\partial z} = K\frac{\partial^2\vec{V}}{\partial z^2} \qquad (\text{where } \tau = i\tau_{zx} + j\tau_{yz}, \ \vec{V} = iu + jv) .....(31.5)$$

Substituting eq. (31.5) in (31.4), we have

$$f\hat{k} \times \left(\vec{V}_g - \vec{V}\right) + K\frac{\partial^2 \vec{V}}{\partial z^2} = 0. \quad \text{Ekman layer equation}$$

or $$K\frac{\partial^2 \vec{V}}{\partial z^2} = f\hat{k} \times \left(\vec{V} - \vec{V}_g\right) \qquad .....(31.6)$$

Let x-axis be along the geostrophic wind i.e., $\vec{V}_g = iV_g$. Then the above eq. (31.6) becomes

$$K\frac{\partial^2 (iu + jv)}{\partial z^2} = f\hat{k} \times [iu + jv - iVg]$$

Equating like terms we have

$$K\frac{\partial^2 u}{\partial z^2} = -fv, \qquad K\frac{\partial^2 v}{\partial z^2} = f(u - v_g) \qquad .....(31.7)$$

These are ekman layer equations.

Let pressure force be constant with height, which implies that V_g = constant (i.e. geostrophic thermal wind is zero $\frac{\partial Vg}{\partial z} = 0$)

Define $W = (u - V_g) + iv$ a complex quantity. .....(31.8)

Multiplying the second equation by $\left(\sqrt{-1}\right) = i$ and adding the first equation of (31.7) viz

$$K\frac{\partial^2 u}{\partial z^2} + fv = 0$$

$$iK\frac{\partial^2 v}{\partial z^2} - if(u - V_g) = 0$$

Adding we have

$$K\frac{\partial^2 W}{\partial z^2} - if(u - V_g + iv) = 0. \qquad \left[\text{since } \frac{\partial^2 V_g}{\partial z^2} = 0\right]$$

$$\frac{\partial^2 W}{\partial z^2} - i\frac{f}{K}W = 0$$

i.e., $$\frac{\partial^2 W}{\partial z^2} - \left[\sqrt{\frac{if}{K}}\right]^2 W = 0 \qquad .....(31.9)$$

The solution of differential eq. (31.9) is given by

$$W = C_1 e^{\sqrt{i}\sqrt{\frac{f}{k}}z} + C_2 e^{-\sqrt{i}\sqrt{\frac{f}{k}}z} \quad \text{where } C_1, C_2 \text{ are constants.}$$

or $$W = C_1 \times e^{\sqrt{\frac{f}{2k}}(1+i)z} + C_2 e^{-\sqrt{\frac{f}{2k}}(1+i)z} \quad \left(\text{since } \sqrt{i} = \frac{1+i}{\sqrt{2}}\right)$$

[Note : $2i = 1 - 1 + 2i = (1 + i)^2$

$$\sqrt{2i} = 1 + i \qquad \text{or } \sqrt{i} = \frac{1+i}{\sqrt{2}}$$

$$W = C_1 e^{\alpha(1+i)z} + C_2 e^{-\alpha(1+i)z} \qquad \text{.....(31.10)}$$

where $\alpha = \sqrt{\frac{f}{2k}}$

Applying boundary conditions when $z = \infty$, $u + iv \to V_g$ (wind approaches to geostrophic wind)

i.e,

$$W = u - Vg + iv = 0 = C_1 e^{a(1+i)\infty} + C_2 e^{-a(1+i)\infty}$$

$$\Rightarrow C_1 = 0 \qquad \because e^{-\infty} = 0$$

then eq. (31.10) reduces to $W = C_2 e^{-\alpha(1+i)z}$

or $$W = C_2 e^{-\sqrt{\frac{f}{2k}}(1+i)z} \qquad \text{.....(31.11)}$$

or $$u - V_g + iv = C_2 e^{-\sqrt{\frac{f}{2k}}(1+i)z} = C_2 e^{-\alpha(1+i)z}$$

Applying the other boundary conditions $z = 0$, $u + iv = 0$ this gives [from eq. (31.11)] $C_2 = -V_g$

$\therefore$ eq. (31.11) reduces to

$$u - V_g + iv = W = -V_g e^{-\sqrt{\frac{f}{2k}}(1+i)z} = -V_g e^{-\alpha(1+i)z}$$

or $$u + iv = V_g \left[1 - e^{-\alpha(1+i)z}\right]$$

$$u + iv = V_g \left[1 - e^{-\alpha z}.e^{-i\alpha z}\right]$$

[Note : $e^{-i\theta} = \cos\theta - i\sin\theta$]

$$u + iv = V_g \left[1 - e^{-\alpha z}(\cos\alpha z - i\sin\alpha z)\right]$$

or $$u = V_g\left[1 - e^{-\alpha z}(\cos\alpha z)\right]$$

$$v = V_g\, e^{-\alpha z} \sin \alpha z \qquad .....(31.12)$$

eq.(31.12) is called Ekman spiral solution.

Swedish oceanographer V. W. Ekman derived similar solution for the surface wind drift current in the ocean. The hodograph of the solution is shown in Fig. 31.3.

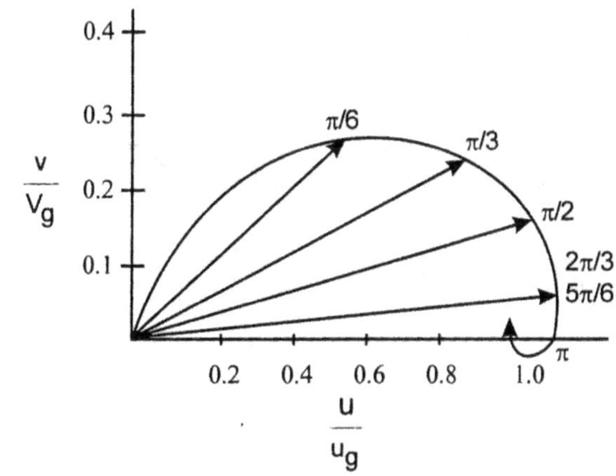

Fig. 31.3

Questions

1. (a) What is laminar sub-layer? State the criteria aerodynamically the surface is rough and smooth.
   (b) What is Prandtl layer, planetary boundary layer (PBL), Ekman layer?
   (c) What is eddy diffusion?
2. (a) Describe the importance of PBL
   (b) Describe the conceptional basis of various layers of PBL.
3. Describe the Prandtl mixing length hypothesis. Derive an expression for the turbulent fluctuation u' in terms of mixing length, vertical momentum transport.
4. Derive Ekman layer equations
5. Using Ekman layer equations derive Ekman spiral solution.

CHAPTER 32

The General Circulation of the Atmosphere - (A)

Atmospheric motion is a complex phenomenon and it is very difficult to put forth a theory using hydrodynamic and thermodynamic equations. Broadly speaking the general circulation may be described as the average atmospheric flow on the earth. It can be described convincingly by statistical terms. The atmospheric flow statistics are generated from the aggregate of daily patterns. Thus the general circulation may be described as the time and space average phenomena of winds, temperatures, pressure, humidity and their variations during seasons, years or months.

To suppress the random variations of the phenomena the average quantities are considered over a long period (more than 30 years or 3 sunspot cycles). Zonal (east-west direction) averages (of winds etc.) and meridional (north-south direction) averages (of temperature, water vapour, heat transport etc.) are considered well suited for describing many phenomena of atmosphere. General circulation also takes into consideration of energy relations in the atmosphere. A few averages and transports will be considered here.

Longitudinal time average components of general circulation consists of stationary and monsoonal circulations, which change little with time. The perturbation method is dealt with mean zonal components and perturbation components. We shall consider zonally averaged flow particularly momentum and energy budgets.

(a) The atmospheric motion is mainly driven by solar radiation and the angular momentum of the earth. We have already studied that the primary force driving the atmosphere is the uneven heating of the globe by solar radiation. The uneven heating results in continuous transfer of energy from lower latitudes to higher latitudes. This unequal heating of the earth and atmosphere generates potential energy, part of which is converted into kinetic energy (lifting warm air and sinking of cold air), and the rest is dissipated by the friction. There must be a balance of rate of generation and dissipation of energy. It is estimated that only 1% of insolation (about 2 watts/sq. meter) is absorbed by the earth atmosphere runs this atmospheric heat engine.

(b) By the law of inertia earth's atmosphere moves with the earth around its axis of rotation. Further we have, Angular momentum is proportional to (i) the rate of spin of the earth [Angular speed of earth = about 465 m /sec at the equator] (or angular velocity) and (ii) square of the distance of the air parcel from the axis of rotation. Since the earth and atmosphere are rotating uniformly, the angular momentum is conserved. The angular momentum is high at the equator and gradually decreases as we move towards pole and it becomes zero at the poles. Air moving poleward acquires gradually higher eastward velocities. The general circulation of the atmosphere results from this poleward transfer of angular momentum.

32.1 Atmospheric Averages and Transports

Let b be an arbitrary scalar quantity of atmosphere is a function of longitude (λ), latitude (ϕ), pressure (p) and temperature (t). i.e. b = (λ, ϕ, p, t).

The space (area) average is defined as

$$b_s = \frac{1}{S} \int_S b\,ds \qquad \text{.....(32.1)}$$

where S = Total area under consideration

ds = Elemental area

Global Average: Consider an elemental area of sphere bounded

(Fig. 32.1) by longitudes $\left(\lambda - \frac{\delta\lambda}{2}, \lambda + \frac{\delta\lambda}{2}\right)$ and, Latitudes $\left[\phi - \frac{\delta\phi}{2}, \phi + \frac{\delta\phi}{2}\right]$ and

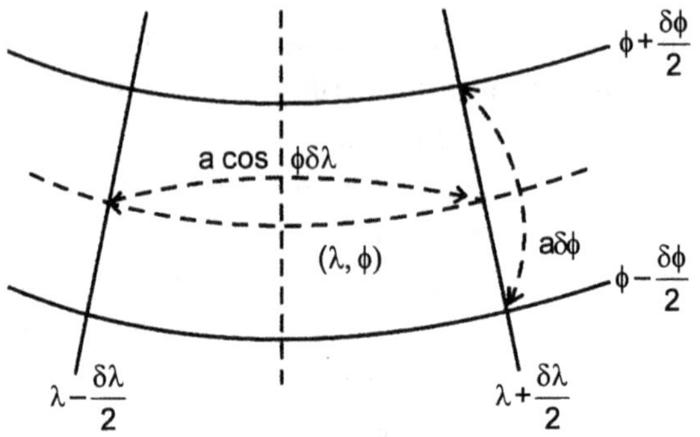

Fig. 32.1

Let a = radius of the sphere. Then

$$ds = a^2 (\cos \phi)\, d\lambda d\phi \qquad \text{.....(32.2)}$$

Area between the two latitudes circles ϕ_1, ϕ_2 ($\phi_1 < \phi_2$) is given by

$$S = 2\pi a^2 (\sin \phi_2 - \sin \phi_1)$$

$$\text{Then } b_s = \frac{1}{2\pi a^2 (\sin\phi_2 - \sin\phi_1)} \int_{\phi_1}^{\phi_2} \int_0^{2\pi} b\, a^2 (\cos\phi)\, d\lambda d\phi \qquad \text{.....(32.3)}$$

If $\phi_1 = -\frac{\pi}{2}$, $\phi_2 = +\frac{\pi}{2}$, the integral of eq. (32.3) gives the global average.

Zonal Average : The average of b along a latitude circle defined as

$$b_z = \frac{1}{2\pi} \int_0^{2\pi} b d\lambda \qquad \text{.....(32.4)}$$

where $b = (\phi, p; t)$ and not λ.

Zonal averages are widely used for general circulation studies. The deviation of b from zonal average b_z is called eddy component and is denoted by b_E. Thus $b_E = b - b_z$. The summation (or integration) of b_E along a latitude circle is zero, that is

$$(b_E)_z = 0 \qquad \text{.....(32.5)}$$

If a zonal average is a product of two scalar fields such $(bC)_z$, then we have the eddy components as

$$(b_E C_E)_z = (bC)_z - b_z C_z$$

which follows from

$(bC)_z = b_z C_z + (b_E C_z)_z + (b_z C_E) + (b_E C_E)_z$ in which $(b_E)_z = 0$ and $(C_E)_z = 0$

Time Average : The time average of b over a time period T is given by

$\bar{b} = \frac{1}{T}\int_0^T b\,dT$, where T is large and $\bar{b}$ is considered as independent of time.

Atmospheric fluxes (or atmospheric transports) : Let $\vec{V} = iu + jv + kw$ a three dimensional velocity vector. Let $\vec{b}_v$ denote the transport vectorial quantity which has three components $\vec{b}_v = ib_u + jb_v + kb_w$.

where b_u = Zonal component

b_v = Meridional component

b_w = Vertical component.

The total transport across a latitude circle is given by

$$\int_0^{2\pi} b_v\, a\cos\phi\, d\lambda = 2\pi a\cos\phi\,(b_v)_z$$

where $$(b_v)_z = \frac{1}{2\pi}\int_0^{2\pi} b_v\, d\lambda$$

32.2 The Momentum Budget

We shall consider the zonal averaged u–component of wind at the ground-(u_z). The long term observed average distribution of u_z at the ground is :

(i) In lower latitudes easterlies (which continue to prevail quite high elevations)

(ii) in middle latitudes westerlies (which increase with height and obey thermal wind equation) and

(iii) In polar regions weak easterlies.

In order to study the process that is responsible for changes in the zonal average wind we shall apply the zonal averaging process to the first equation of motion in spherical coordinates.

The approximate first equation of motion in spherical coordinates is given by

$$\frac{du}{dt} = -\frac{\alpha}{a\cos\phi}\frac{\partial p}{\partial \lambda} + fv + \frac{uv}{a}\tan\phi + F_\lambda$$

and $$\frac{d}{dt} = \frac{\partial}{\partial t} + \frac{u}{a\cos\phi}\frac{\partial}{\partial \lambda} + \frac{v}{a}\frac{\partial}{\partial \phi} + \omega\frac{\partial}{\partial p} \quad .....(32.6)$$

The equation of continuity in spherical coordinates is given by

$$\frac{1}{a\cos\phi}\frac{\partial u}{\partial\lambda}+\frac{1}{a}\frac{\partial v}{\partial\phi}-\frac{v}{a}\tan\phi+\frac{\partial\omega}{\partial p}=0 \qquad \text{.....(32.7)}$$

where $\alpha=\frac{1}{\rho}$, $a=R+h$

Multiplying eq. (32.7) by u and add to eq. (32.6)–viz

$$\frac{\partial u}{\partial t}+\frac{u}{a\cos\phi}\frac{\partial u}{\partial\lambda}+\frac{v}{a}\frac{\partial v}{\partial\phi}+\omega\frac{\partial u}{\partial p}=$$

$$-\frac{\alpha}{\alpha\cos\phi}\frac{\partial p}{\partial\lambda}+fv+\frac{uv}{a}\tan\phi+F_\lambda \qquad \text{.....(i)}$$

$$u\left[\frac{1}{a\cos\phi}\frac{\partial u}{\partial\lambda}+\frac{1}{a}\frac{\partial v}{\partial\phi}-\frac{v}{a}\tan\phi+\frac{\partial\omega}{\partial p}\right]=0 \qquad \text{....(ii)}$$

Adding eq. (i) + (ii) we have

$$\frac{\partial u}{\partial t}+\frac{1}{a\cos\phi}\frac{\partial u^2}{\partial\lambda}+\frac{1}{a\cdot}\frac{\partial uv}{\partial\phi}++\frac{1}{a}\frac{\partial u\omega}{\partial p}-\frac{uv}{a}\tan\phi$$

$$=\frac{-\alpha}{a\cos\phi}\frac{\partial p}{\partial\lambda}+fv+\frac{uv}{a}\tan\phi+F_\lambda$$

or

$$\frac{\partial u}{\partial t}+\frac{1}{a\cos\phi}\frac{\partial u^2}{\partial\lambda}+\frac{1}{a}\frac{\partial uv}{\partial\phi}+\frac{\partial u\omega}{\partial p}-\frac{2uv}{a}\tan\phi$$

$$=-\frac{\alpha}{a\cos\phi}\frac{\partial p}{\partial\lambda}+fv+F_\lambda$$

$$\frac{\partial u}{\partial t}+\frac{1}{a\cos\phi}\frac{\partial u^2}{\partial\lambda}+\frac{1}{a\cos^2\phi}\frac{\partial\left(uv\cos^2\phi\right)}{\partial\phi}+\frac{\partial uw}{\partial p}$$

$$=\frac{-\alpha}{a\cos\phi}\frac{\partial p}{\partial\lambda}+fv+F_\lambda \qquad \text{.....(32.8)}$$

Taking zonal average of eq. (32.8), we have

$$\frac{\partial u_z}{\partial t}=-\frac{1}{a\cos^2\phi}\frac{\partial\{(uv)_z\cos^2\phi\}}{\partial\phi}-\frac{\partial(uw)_z}{\partial p}+fv_z+F_{\lambda z} \qquad \text{.....(32.9)}$$

since $\left(\frac{\partial}{\partial\lambda}\right)_z=0$

Eq. (32.9) indicates that $\left(\frac{\partial u_z}{\partial t} = \right)$ the zonally average u-component changes and is equal to the sum of (a) + (b) + (c) + (d)

where

(a) $-\frac{1}{a\cos^2\phi}\frac{\partial\{(uv)_z\cos^2\phi\}}{\partial\phi}$ = horizonal convergence of meridional transport of moment.

(b) $-\frac{\partial(u\omega)_z}{\partial p}$ = vertical convergence of the vertical transport of u-momentum.

(c) fv_z = effect of coriolis term,

(d) $F_{\lambda,z}$ = Frictional effect.

32.3 Vertical Average together with Zonal Average

Vertical average (b_M) is defined as

$$b_M = \frac{1}{p_s}\int_0^{P_s} b\,dp \qquad .....(32.10)$$

where p_s = pressure at surface and the simplified boundary conditions are $\omega = 0$ at $p = 0$ and $p = p_s$.

We know the approximated equation of continuity (in spherical coordinates) is

$$\frac{1}{a\cos\phi}\left(\frac{\partial u}{\partial\lambda} + \frac{\partial v\cos\phi}{\partial\phi}\right) + \frac{\partial\omega}{\partial p} = 0$$

By taking zonal average this equation reduces to

$$\frac{\partial V_z\cos\phi}{a\cos\phi\,\partial\phi} + \frac{\partial\omega_z}{\partial p} = 0 \qquad \left[\because \frac{\partial u_z}{\partial\lambda} = 0\right] \qquad .....(32.11)$$

By taking vertical average we have

$$\frac{\partial\{V_{zm}\cos\phi\}}{a\cos\phi\,\partial\phi} + \frac{\partial\omega_{zm}}{\partial p} = 0 \qquad \left[\text{since } \frac{\partial\omega_{zm}}{\partial p} = 0\right]$$

or

$$\frac{\partial\{V_{zM}\cos\phi\}}{\partial\phi} = 0 \qquad .....(32.12)$$

Since at north and south poles V_{zM} is zero and from eq. (32.12) $V_{zM}\cos\phi$ = const, the value of V_{zM} must be zero i.e. $V_{zM} = 0$

Let $$F = g\frac{\partial \tau_\lambda}{\partial p}$$ .....(32.13)

we know from zonal average

$$\frac{\partial u_z}{\partial t} = -\frac{1}{a\cos^2\phi}\frac{\partial\{(uv)_z\cos^2\phi\}}{\partial\phi} - \frac{\partial(u\omega)_z}{\partial p} + f\,V_z + F_{\lambda,z}$$ ....(32.14)

Taking vertical average of this, we have

$$\frac{\partial u_{z,M}}{\partial t} = -\frac{1}{a\cos^2\phi}\frac{\partial\{(uv)_{z,M}\cos^2\phi\}}{\partial\phi} + \frac{g}{p_s}\tau_{\lambda z,s}$$ .....(32.15)

$$\left[\because \frac{\partial(u\omega)_{z,M}}{\partial p} = 0 \ \& \ V_{z,M} = 0\right]$$

At the top of the atmosphere the stress vanishes, we can express

$\tau_{\lambda,z,s}$ as $\tau_{\lambda,z,s} \simeq -C_d\rho_s V_s u_{z,s}$ .....(32.16)

where C_d = Drag coefficient

ρ_s = Surface density of air

V_s = Surface wind speed

eq. (32.15) is negative in regions where $u_{z,s} > 0$ and positive in regions where $u_{z,s} < 0$. This means that surface stress decreases u-momentum in westerlies and increases in the easterlies. Consequently the low (and very high) latitudes are sources of westerly momentum and middle latitudes are sinks of westerly momentum. On an average over a long period the westerly momentum is constant. Overall there is divergence of meridional transport of momentum in low latitudes (and to some extent in very high latitudes) while there is convergence in middle latitudes.

32.4 Atmospheric Energy

Atmospheric energy is mainly observed in three forms :

1. Potential energy (PE) 2. Kinetic energy (KE) and
3. Internal energy (IE)

These energies per unit mass are given by

PE = gz where z = is the height above mean sea level;

$$KE = \frac{1}{2}\vec{V}.\vec{V}$$

where $\vec{V}$ = Velocity in three dimensions and

$$IE = C_v T$$

where C_v = Specific heat at constant volume, T = Temperature.

If dv = elemental volume, ρ = density of air/fluid, then for mass of element ρdv

We have

$$PE = gz\rho dv; \quad KE = \frac{1}{2}(\vec{V}.\vec{V})\,\rho dv \text{ and } IE = C_v T\,\rho dv \quad .....(32.17)$$

If ε denotes total energy then

$$\varepsilon = \int_v gz\rho dv + \int_v \frac{1}{2}(\vec{V}.\vec{V})\rho dv + \int_v C_v T\,\rho dv$$

$$= \int_v \left(gz + \frac{1}{2}(\vec{V}.\vec{V}) + C_v T\right)\rho dv \quad .....(32.18)$$

Rate of change of energy per unit mass is given by

$$\frac{dPE}{dt} = g\frac{dz}{dt} = gw \quad .....(32.19)$$

$$\frac{dKE}{dt} = \frac{d}{dt}\left(\frac{1}{2}\vec{V}.\vec{V}\right) = -\alpha\vec{V}.\nabla p - gw + \alpha\vec{V}.\vec{F} \quad .....(32.20)$$

(see equations of motion in local coordinates (12.25)

and $$\frac{dIE}{dt} = C_v\frac{dT}{dt} = H - P\frac{d\alpha}{dt} = H - p\alpha\nabla.\vec{v} \quad .....(32.21)$$

[using first law of thermodynamics and equational of continuity

$$H = C_v\frac{dT}{dt} + p\frac{d\alpha}{dt} \text{ and } \frac{d\rho}{dt} + \rho\nabla.\vec{v} = 0]$$

∴ Rate of change of total energy = eq. (32.19) + eq. (32.20) + eq. (32.21)

$$\frac{d\varepsilon}{dt} = gw - \alpha\vec{V}.\nabla_p - gw + \alpha\vec{V}.\vec{F} + H - p\alpha\nabla.\vec{V}$$

$$\frac{d\varepsilon}{dt} = -\alpha\nabla.(p\vec{V}) + H + \alpha\vec{V}.\vec{F} \quad .....(32.22)$$

[Note : $\frac{d\alpha}{dt} = -\frac{1}{\rho^2}\frac{d\rho}{dt} = \frac{-1}{\rho^2} \times -\rho\nabla.\vec{V}$ (using $\frac{d\rho}{dt} = \rho\nabla.\vec{V}$)

$$= \frac{1}{\rho}\nabla.\vec{V} = \alpha\nabla.\vec{V}]$$

or $$\rho\frac{d\varepsilon}{dt} = -\nabla.\left(p\vec{V}\right) + \rho H + \vec{V}.\vec{F}$$

$$\rho\frac{d\varepsilon}{dt} = -\nabla.\left(RT\rho\vec{V}\right) + \rho H + \vec{V}.\vec{F} \qquad .....(32.23)$$

where ρH = heating per unit volume

$p = RT\rho$ = gas law.

we know $\frac{d}{dt} = \frac{\partial}{\partial t} + \vec{V}.\nabla$

$$\therefore \qquad \rho\frac{d\varepsilon}{dt} = \rho\left[\frac{\partial\varepsilon}{\partial t} + \vec{V}.\nabla\varepsilon\right] = \frac{\partial\varepsilon\rho}{\partial t} - \frac{\varepsilon\partial\rho}{\partial t} + \rho\vec{V}.\nabla\varepsilon \qquad .....(i)$$

from equation of continuity we have

$$\frac{d\rho}{dt} + \rho\nabla.\vec{V} = 0 \text{ or } \frac{\partial\rho}{\partial t} + \vec{V}.\nabla\rho + \rho\nabla.\vec{V} = 0$$

or $$\frac{\partial\rho}{\partial t} + \nabla.\left(\rho\vec{V}\right) = 0$$

$$-\frac{\partial\rho}{\partial t} = \nabla.\left(\rho\vec{V}\right) \qquad .....(ii)$$

Substituting this value of eq. (ii) in the above eq. (i) we have

$$\rho\frac{d\varepsilon}{dt} = \frac{\partial\varepsilon\rho}{\partial t} + \varepsilon\nabla.\rho\vec{V} + \rho\vec{V}.\nabla\varepsilon$$

or $$\rho\frac{d\varepsilon}{dt} = \frac{\partial\varepsilon\rho}{\partial t} + \nabla.\left(\varepsilon\rho\vec{V}\right) \qquad .....(32.24)$$

From eq. (32.23) and eq. (32.24) we have

$$\frac{\partial(\varepsilon\rho)}{\partial t} + \nabla.\left(\varepsilon\rho\vec{V}\right) = -\nabla.\left(RT\rho\vec{V}\right) + \rho H + \vec{V}.\vec{F}$$

$$\text{or} \quad \frac{\partial(\varepsilon\rho)}{\partial t} = -\nabla.\left(\varepsilon\rho\vec{V}\right) - \nabla.\left(RT\rho\vec{V}\right) + \rho H + \vec{V}.\vec{F}$$

$$= -\nabla.\left[\left((\varepsilon + RT)\rho\vec{V}\right)\right] + \rho H + \vec{V}.\vec{F}$$

$$= -\nabla.\left[(C_vT + PE + KE + RT)\rho\vec{V}\right] + \rho H + \vec{V}.\vec{F}$$

$$\text{or} \quad \frac{\partial}{\partial t}\left[\{PE + KE + IE\}\rho\right] = -\nabla\left[\{C_pT + PE + KE\}\rho\vec{V}\right] + \rho H + \vec{V}.\vec{F} \quad .....(32.25)$$

$$(\because C_vT + RT = C_pT)$$

= [Convergence of transports of sensible heat (C_pT), + PE + KE] + ρH + $\vec{V}.\vec{F}$

Note : = $\vec{V}.\vec{F}$ dissipation of KE due to frictional force.

Let $\vec{W} = [C_pT + PE + KE]\,\rho\vec{V}$ = transport vector having PE, KE and transport of sensible heat.

Then eq. (32.25) can be written as

$$\frac{\partial\varepsilon\rho}{\partial t} = -\nabla.\vec{W} + \rho H + \vec{V}.\vec{F} \quad .....(32.26)$$

Let $\rho H_F = -\vec{V}.\vec{F}$; where H_F = Fictional heat. If H_{NF} = Non-frictional heating, then $H = H_F + H_{NF}$

Then from eq. (32.26) we have

$$\frac{\partial\varepsilon\rho}{\partial t} = -\nabla.\vec{W} + \rho H_{NF} \quad .....(32.27)$$

where H_{NF} includes all heating except frictional heating.

32.5 General Circulation of the Atmosphere - (B)

In 18th century Gorge Hadley, a British meteorologist, propounded the general circulation of the atmosphere. According to this model, the heated equatorial air ascends from the surface and moves pole wards in the upper atmosphere (at about 5 km alt) where it cools and sinks down to the earth. The sunken air at the surface again moves equator ward and a small part to pole. This can be explained by a simple experiment as shown in Fig. 32.2(a), Fig. 32.2(b), Fig. 32.2(c), Fig. 32.2(d).

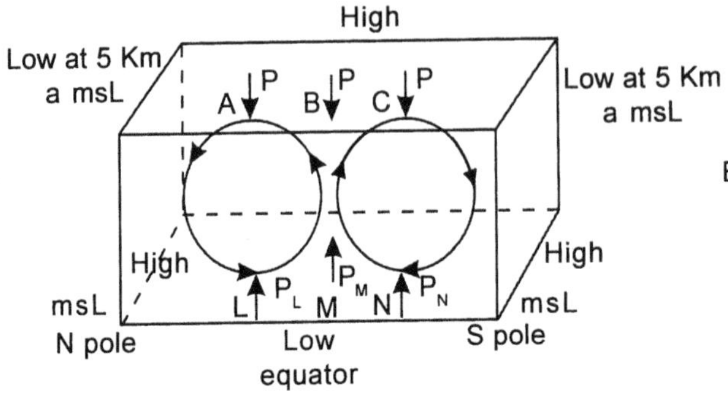

Fig. 32.2(a)

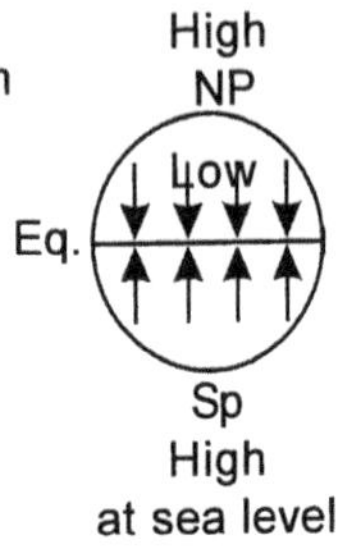

Fig. 32.2(b)

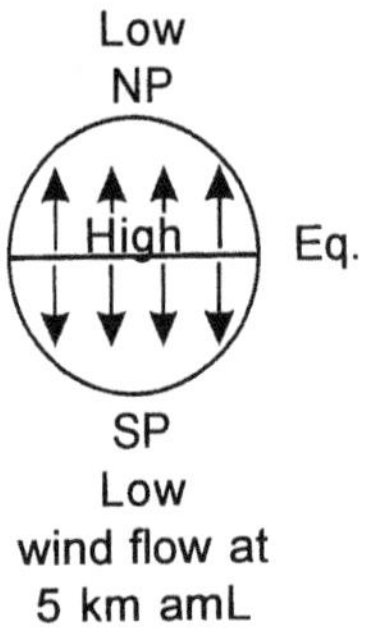

Fig. 32.2(c)

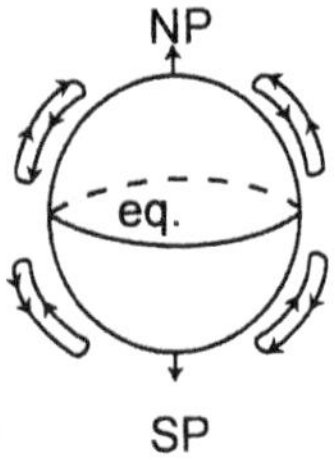

Fig. 32.2(d) Hadley convective circulation model

Consider a rectangular cuboid containing water in a horizontal position. The pressure at the top of the cuboid at A, B, C are equal (say p) and the pressure at bottom points L, M, N just below A, B, C respectively be p_L, p_M, p_N which are equal initially. If we heat at M (The center of the bottom of the cuboid) P_M decreases and convection currents rise to B and then travel towards A and C from where they descend to L and N and then move to M as shown in Fig.(a). In a similar way, heating at the equator by the insolation causes decrease in pressure with wind flow towards equator (as in Fig 32.2 (b)) and rise of convection currents to an altitude of about 5 km where the pressure rises which will be more than at the poles ($B_p > C_p$, $B_p > A_p$) at that level. As a result wind will flow from equator to poles wards at 5 km altitude (as in Fig 32.2 (c)). At poles cooling causes air to sink downwards to earth, where high pressure areas develop while at 5 km altitude low pressures. This is the simplest treatment of meridional circulation of Hadley Cell (Fig. 32.2 (a)). However the atmospheric circulation is not so simple which we shall see presently.

At any place, close to the surface of the earth, wind is never steady neither in direction nor in speed. It is very difficult to describe the varied atmospheric motions in a simple way. As mentioned earlier, the general circulation of the atmosphere is described as average phenomenon in a lucid patterns by statistical averages.

Trade winds : The most dominant feature of the general circulation is reflected in trade winds in both hemispheres. Trade winds nearly blow over half of the globe. They originate near sub-tropical high pressure (STH) cells and blow from North East (NE) direction towards equator in the northern hemisphere and from South East (SE) direction towards equator in the southern hemisphere. In olden days these winds helped the trade vessels constantly blowing in periarticular direction and thus got the name trade winds. Trade winds are strongest in winter. From STH belts wind blows north to south (south to north) direction but deflected by the rotation of the earth (by coriolis force) from NE (or SE) direction.

ITCZ/Doldrums : Between the trade winds there is a region near equator where light variable winds accompanied by heavy rain, thunderstorms and squalls are observed. These are traditionally called doldrums. In this zone sailing ships/vessels were be calmed (stranded)for several days or weeks over the sea and thus much feared. This zone lies in between the trade winds and in the equatorial trough. Modern meteorologists prefer to call this sector as ITCZ (inter tropical convergence zone). Doldrums region is mostly covered with clouds and rains (merky skies) along with nearly calm or variable winds. It may be noted here that the average high temperature zone lies in this equatorial trough. The thermal equator (also called meteorological equator) lies at about lat 5° N. The equatorial trough migrates with the Sun. Equatorial through is centred at about lat 5° S in January and about (12° N) in July. It oscillates about –20° latitudes between seasons. The mean position is lat 5° N which is the Meteorological equator.

Horse Latitudes : With the available present knowledge and data convective air rising at the equatorial zone does not travel up to poles but descendes to a large extent at about lat 30° N (S) where sub-tropical high pressure (STH) belts are located on surface level. The areas of STH belts are pre-dominated by clear skies, light or variable winds. This belt migrates with the sun with a lag of one or two months. Mariners call these areas horse latitudes because in these latitudes sailing ships/vessels were be calmed and traders sometimes use to throw away horses overboard into sea, which were being transported from Europe to America. Thus got the name horse latitudes.

The Equatorial Westerlies : In summer hemisphere there is a zone of westerly winds in equatorial region between trade winds, which migrates with the equatorial through. Over Indian Ocean it is located at about lat 12-15° N during summer (June-July) and at about 2-3 °S during winter (December, January). However over Pacific and Atlantic Oceans this is not observed.

The Middle Latitude Westerlies : Between STH and sub-polar low (SPL) there is a belt of westerlies which are prominent particularly in the southern hemisphere and found around the world. They are called "roaring forties" or "howling fifties" (Located at lat 40° and 50° respectively). They are also called ferrel westerlies (see Fig. 32.3).

Polar Easterlies : Between poler high sub-poler low there are quasi permanent easterlies. These are shown in Fig. 32.4.

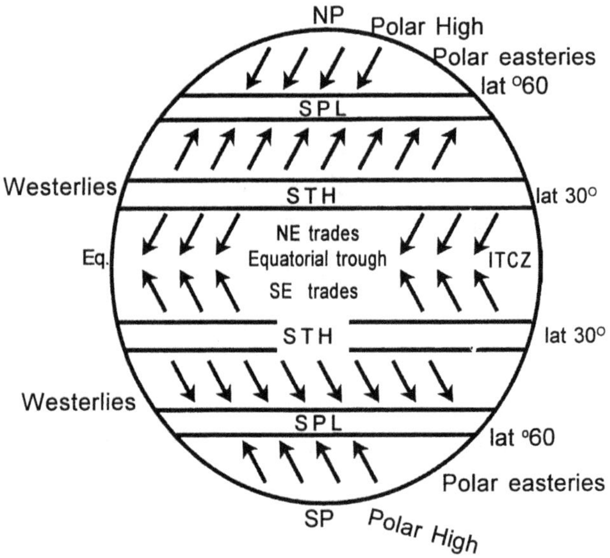

Fig. 32.3 Sea level global general wind pattern.

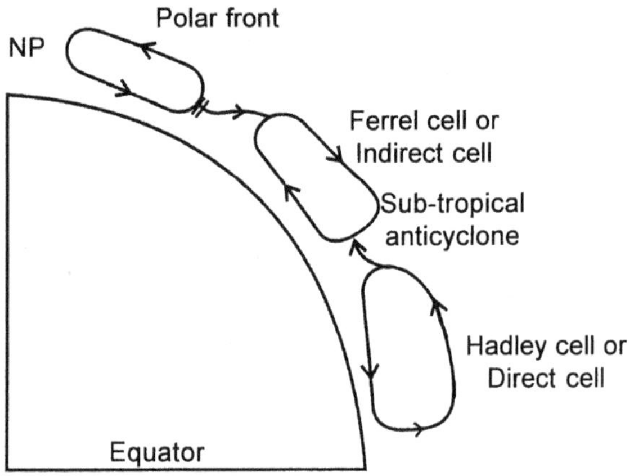

Fig. 32.4 Three cell Meridional circulation of the North Hemisphere.

Ferrel cell : In 1856 a British school teacher modified the Hadley cell model to explain the roaring forties. According to him the Hadley cell lies between the equator and lat 30°, and another cell in opposite direction of wind motion to that of Hedley cell between Lat 30° and 60°. This cell is called Ferrel cell or indirect cell. Another weak cell exists between lat 60° and pole which has similar circulation as the Hadley cell with rising of winds equator ward and sinking of winds at pole. Thus we have three cell model of general circulation of wind Fig. 32.4. It may be noted here that inspite of many excellent laboratory studies there is as yet no complete explanation of the general circulation of the atmosphere.

The above described principal wind systems are modified by Land and Sea surfaces. Land areas heated up quickly during day and cools down quickly during night. In contrast sea surface area heating and cooling does slowly and hence maintains more even temperature throughout the year. In summer, continents behave like heat sources and sea surface areas behaves like heat sinks. In winter the reverse is observed. As a result of this temperature differences, the atmospheric pressure also varies. This causes seasonal winds blowing between oceans and continents. This type of seasonal winds observed prominently in case of monsoons.

As described in the beginning the general circulation is an average phenomena over a long period. In middle and high latitudes the wind variations are prominent. In these latitudes warm subtropical air that moves to high latitudes mixes with the cold winds blowing from the sub-polar regions. This mixing of different air mass winds form the polar front.

Air Mass : An air mass is a large body of air which has horizontally (hundreds of Km) uniform properties of temperature, moisture content, temperature lapse rate. A front is an interface (like the front of any armies)between two air masses of different characteristics particularly of temperature and density. However there does not occur any sharp boundary and hence some meteorologist call it a frontal zone. Similar to the migration of equatorial trough, the polar front also oscillates back and forth from north to south. In this region of oscillation, polar front cyclones develop and generally move from west to east. In contrast to this tropical cyclones move from east to west.

Atmospheric motions thus differ greatly in their size, duration of life ranging from small turbulent eddies to large cyclones. All these motions are fed by the solar energy and cause the circulation of water through the atmosphere (Hydrological cycle).

Note :

1. Fronts may occur where there is a cyclonic wind shift or a cyclonic wind shear. Fronts generally occur along trough line of a low pressure but not along a ridge line or in anticyclonic shear zone. At a front isobars are kinked or pointed away from the low pressure. Isobars across a front make an angle less than 180° towards low pressure where wind discontinuity is observed.
2. A front may be expected in the zone where wind has a component across isotherms that decreases in the direction of downstream.

The general circulation of the atmosphere may thus be classified as 1. Planetary winds 2. Terrestrial winds and 3. Continental winds.

Planetary winds are further classified as (i) Equatorial winds or Doldrums (ii) Trade winds (iii) Mid-LatitudeWesterlies or Horse latitudes (iv) Prevailing Westerlies or Roaring forties or Howling fifties (v) Polar winds or polar Easterlies.

A secondary atmospheric circulation consists of Land and Sea breeze. Valley winds (Katabatic and Anabatic winds), Mountain waves, Tidal and Volcanic winds, eclipse, land slide, Cyclonic storms. These will be discussed separately.

32.6 Zonal Index

The variation in the mid-latitude westerlies (Lat 35°– to 55°) indicate some patterns, particularly in winter months when the zonal westerlies are strong. The variation lasts for about one to two moths when the westerlies develop waves, troughs and ridges become prominent and finally take cellular shape. The strength of the westerlies between Lat 35° – 55° N is called zonal index.

High Zonal Index : When the westerlies are strong (and located a little north of its average position), with marked east-west oriented isobars (or pressure systems) and weak meridional flow is called High zonal Index.(Fig. 32.5(a)). In this state the exchange of mass from north and south is not significant.

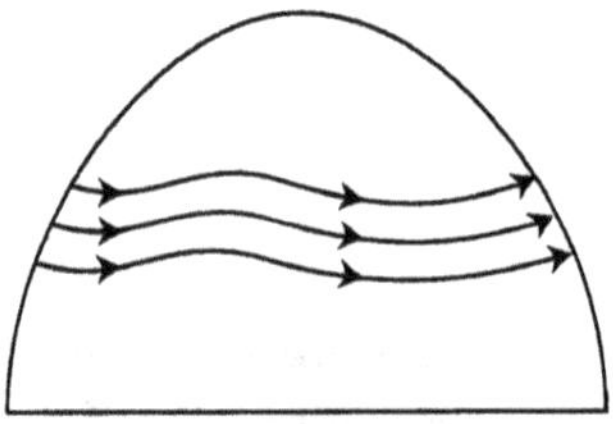

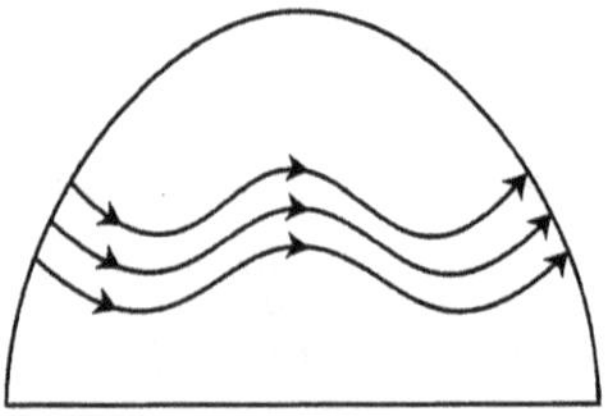

Fig. 32.5(a) High Zonal Index Cycle. **Fig. 32.5(b)** Zonal Index Oscillation.

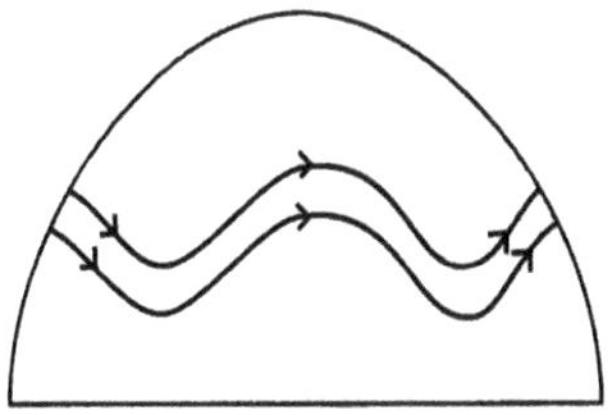

Fig. 32.5(c) Zonal Index Oscillation.

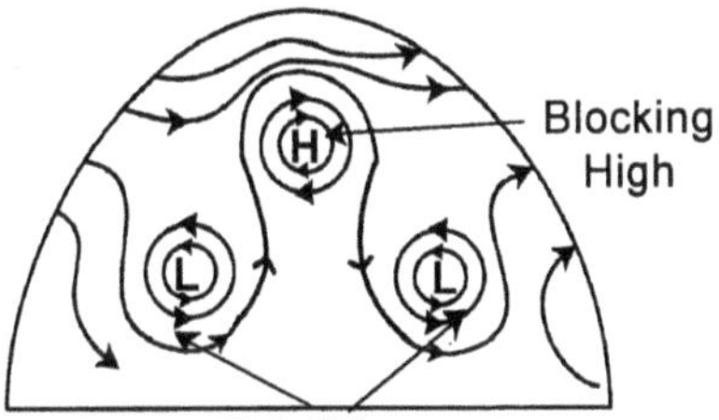

Fig. 32.5(d) Low Zonal Index.

Zonal Index with Oscillations : When the westerlies are strong and located south of its average position with deep troughs and ridges with meridional (north and south) mixing of mass is called zonal index with Oscillations (Fig. 32.5(b) and Fig. 32.5(c)). In this case westerly Jet expands and strengthens with undulation. Meridional heat transfer takes place. Cold air mass from higher latitudes move towards equator with simultaneous warm equatorial air mass moves pole wards.

Low Zonal Index : When the zonal westerlies completely break off with formation of cut-off lows and blocking highs is called low zonal Index. [Fig. 32.5(d)]. In this state westerlies are weak while prominent meridional movement of air is noted. Persistent cloudy weather lies in cut off lows (in low latitudes) which are blocked by anticyclones (called blocking Highs or anticyclones)in higher latitudes having fair weather.

Questions

1. How is atmospheric flow statistics generated and general circulation described with its help ? What is the main source of energy to drive the atmospheric motions.
2. Describe the construction of (a) Global average (b) zonal average (c) Time average

3. Show that the zonal average of u-component of wind (u_z) indicates $\frac{\partial u_z}{\partial t}$ is the sum of horizontal convergence of meridional transport of momentum, vertical convergence of vertical transport of u-momentum and effects of Coriolis term and friction.

4. Derive an expression for vertical average together with zonal average. Hence show that surface stress decreases u-momentum in westerlies and increases in the easterlies.

5. If ε denotes the total energy PE + KE + IE , show that

$$\frac{\partial(\varepsilon\rho)}{\partial t} = -\nabla.\left[(\varepsilon + RT)\rho\vec{V}\right] + \rho H + \vec{V}.F$$

6. Describe briefly:

   (a) Tradewinds, (b) Doldrums, (c) Horse latitudes, (d) Equatorial westerlies, (e) Middle altitude westerlies (f) Polar easterlies, (g) Hadley cell (h) Ferrel cell, (i) Air mass and fronts.

7. Describe briefly:

   (a) Zonal Index, (b) High zonal index, (c) Zonal index with oscillations (d) Low zonal Index with cut-off lows and blocking highs.

APPENDIX

Some Useful Units and Constants

Units of Force

$F = ma$ Force = mass × acceleration

1. The unit of force that gives a mass of standard 1 Kg an acceleration of 1 m/s^2 is called Newton.

   $F(N) = m\ (kg) \times a\ (m.s^{-2})$

2. The unit of force that gives a mass of 1 gm an acceleration of 1cm/s^2 is called dyne.

   $F\ (dyne) = m\ (gm) \times a\ (cm\ s^{-2})$, 1 dyne = 10^{-5} N (1N = 10^5) dynes.

3. The unit of force that gives a mass of 1 slug an acceleration 1 ft/s^2 is called one pound.

   $F\ (lb) = m\ (slug) \times a\ (ft\ s^{-2})$.

Units of Work

Work = $W = F\ .\ S$ = Force × distance.

1 Joule = (1 Newton) (1 meter). 1 J = 1 N. M.

1 erg = (1 dyne) (1 cm) 1 erg = 1 dy. cm.

1 Joule = (1 Newton) (1 meter) = (10^5 dynes) (100 cm) = 10^7 erg

1 J = 10^7 erg 1 erg = 10^{-7} J

Power $P = \frac{W}{T} = \frac{\text{work}}{\text{Time}}$ IHP = 746 watts

$1 \text{ watt} = \frac{1J}{S} = 1Js^{-1}$; $1 \text{ mW} = 10^{-3}$ W mW = milliwatts

$1 \text{ kW} = 10^3$ W, $1 \text{ MW} = 10^6$ W KW = Kilo watt

MW = Mega watt.

1 kWh = one kilowatt-hour = one kilowatt-hour is the work done in one hour (by an agent working) at the constant rate of one kilowatt. 1 J/s = 1 w

$$1 \text{ kWh} = 10^3 \text{W} \times (60 \times 60\text{s}) = 3.6 \times 10^6 \text{Ws}$$

$$= 3.6 \times 10^6 \text{J}$$

1 kWh = 3.6 MJ (Mega Joules)

$$\text{Pressure} = \frac{\text{Force}}{\text{Area}} \quad p = \frac{F}{A}$$

1 mb or hPa = 100 N/m^2

1 Pa = 1 N/m^2

Heat

Calorie : The amount of heat required to raise the temperature of 1 gm of water (14.5 to 15.5 °C) through 1 °C is called Calorie or gram-calorie.

Relative humidity (%)

$$\frac{\text{Actual vapour pressure}}{\text{Saturated vapour pressure at the same temperature}} \times 100$$

$$\frac{\text{Partial pressure of water vapour}}{\text{Vapour pressure at the same temperature}} \times 100\ (\%)$$

1 gm.Calorie = 4.1868 J. (J = Joule)

1 ly/min = 1 $cal/cm^2/min$ = 0.6975 KW/m^2

= 0.06975 W/cm^2

= 69.75 mW/cm^2 [1 KW = 10^3W; 1 mW = 10^{-3}W]

= 1 MW = 10^6 W

$$\frac{\text{Energy}}{\text{area}} : 1 \text{ langely} = \frac{1 \text{Calorie}}{cm^2} = 4.186 \times 10^7 \text{ ergs/cm}^2$$

$$\frac{\text{Power}}{\text{area}} : 1 \text{ langely} = \text{min}^{-1} = 697.6 \text{ watts/cm}^2$$

Solar constant

$$S_0 \simeq 1.95 \text{ Cal cm}^{-2} \text{ min}^{-1} = 1354 \text{ W m}^{-2}.$$

$$\simeq \upsilon_o \text{ ly/minute or } 263 \text{ K ly/year.}$$

$$\text{Specific humidity} = \frac{\text{mass of water vapour}}{\text{mass of moist air containing the vapour}}$$

Average annual depth of evaporation over global land area ≃ 495 mm and precipitation ≃ 760 mm area

Average annual depth of evaporation over global oceanic area ≃ 1251 mm and precipitation ≃ 1140 mm area

Average annual depth of global evaporation ≃ 1030 mm

daily evaporation ≃ 2.8 mm

while precipitation ≃ 1030 mm

Global land surface area ≃ 149×10^6 km^2

Global oceanic surface area ≃ 361×10^6 km^2

Global surface area ≃ 510×10^6 km^2

Speed of light in vacuum 2.998×10^8 m/s ≃ 3.0×10^8 m/s

Velocity of sound waves in air at 20 C° is 344 m/s

Universal gravitational constant

$$G = 6.673 \times 10^{-11} \text{ N m}^2/\text{kg}^2$$

Gas constant R = 8.314 J/mol k

Acceleration due to earths gravity at msL at different latitudes.

Lat	**m/s^2**	**lat**	**m/s^2**
00 (equator)	9.780	60°	.819
20°	9.786	80°	9.831
40°	9.802	90° (poles)	9.832

Variation 'g' with altitude at Lat 45°

Alt	**m/s^2**	**Alt**	**m/s^2**
		16 km	9.76
0 (msL)	9.81	32 km	9.71
1000 m	9.80	100 km	9.60
4000 m	9.79	500 km	8.53
8000 m	9.78	1000 km	7.41

Gravitational attraction of the sun

$= 275 \text{ m/s}^2$ or 28 g

Mass of the seen $\simeq 1.99 \times 10^{30}$ kg

Mass of the earth $\simeq 6. \times 10^{24}$ kg

Radius of the sun $\simeq 6.96 \times 10^{8}$ m

Average density of the sun $\simeq 1.4 \times 10^{3} \text{ kg/m}^3$

Average density sun's core $\simeq 1.6 \times 10^{5} \text{ kg/m}^3$

Average density of the earth $\simeq 5.52 \times 10^{3} \text{ kg/m}^3$ which is roughly 4 times the density of the sun.

Average density of the earth's core $\simeq 9.52 \times 10^{3} \text{ kg/m}^3$

Average density of the earth's crust $= 2.8 \times 10^{3} \text{ kg/m}^3$

Probable density of laboratory vacuum $\simeq 10^{-6} \text{ kg/m}^3$

Density of the ice $\simeq 0.92 \times 10^{3} \text{ kg/m}^3$

Density of water at °C and 1 atm. Pressure $\simeq 1.000 \times 10^{3} \text{ kg/m}^3$

Density of water at 100 °C and 1 atm. Pressure $= 0.95835 \times 10^{3} \text{ kg/m}^3$

at 4 °C = 0.99997

at 20 °C = 0.9982

at 50 °C = 0.98804

Pressure (hpa) and equivalent geopotential height (ψ)

Pressure hp_a	ψ km	Pressure hp_a	ψ km
1000	110 gpm	300	9.16
800	1.46	200	11.78
700	3.01	100	16.18
600	4.21	50	20.58
500	5.57	30	23.85
400	7.19	10	31.05

Angular diameter of the sun as abserved from the earth

$= 9.3 \times 10^{-3}$ radiance

Accederation of gravity at 0 °C : 9.78 m/s²,

at 20 °C : 9.786 m/s²,

at 40 °C : 9.80 m/s²

Equatorial radius of the earth $R_e \simeq 6.3784 \times 10^6$ m

Polar radius of the earth $R_p = 6.3569 \times 10^6$ m

Area of the earth $\simeq 5100 \times 10^{14}$ m^2

Volume of the earth $= 1083 \times 10^{21}$ m^3

Mass of the earth $\simeq 6.0 \times 10^{21}$ tons

Ω angular velocity of the earth $= 7.29 \times$ 10-5 radians

Length of the meridian of the earth $\simeq 4.0009 \times 10^7$ km

Length of the equator of the earth $\simeq 4.0077 \times 10^7$ km

Velocity of a point on the equator of the earth = 1669 kmph or 464 m/s

Orbital speed of the earth 29770 m/s or 29.8 km/s

Perhelion = earth seen distance on 1st January (nearest) $= 147 \times 10^6$ km.

Ap helion = earth seen distance on 1 st July (farthest) $= 152 \times 10^6$ km.

Viscosity η

At Temperature	Water Centipoise	Air Micropoise
0	1.792	171
20	1.005	181
60	0.469	200
100	0.284	218

Mass of the atmosphere $\simeq 5.6 \times 10^{18}$ kg

Mass of the Hydrosphere $\simeq 1.4 \times 10^{21}$ kg

Age of the earth $\simeq$ 4.5 billion years 4.5×10^9 years

The mass of oxygen in the atmosphere 10^{18} kg

The volume of water vapour in the atmosphere $\simeq 14 \times 10^{12}$ m^3

The mass of CO_2 in the atmosphere $\simeq 2.3 \times 10^{15}$ kg

The average temperature of the photosphere of the seen $\simeq$ 6000 °K

The average temperature of the earth $\simeq$ 15 °C

Solar energy intercipted by the earth

$\simeq 3.67 \times 10^{21}$ cal/day

$\simeq 2.55 \times 10^{18}$ cal/minute

N_A = Avogadro number ≃ 6.0225 × 10^{23} molecules/mol

R = Gas constant = 8.314 J/mol/K

R_d = gas constant for dry air = 2875/kg K

Molecular weight of dry air ≃ 28.97

For dry air : C_p = 1004 J/kg.K

C_v = 717 J/kg.K

r = 1.40

Some molecular heat capacities at low temperature

Gas	C_p J/mol.K	C_v J/mol.K	$r = \frac{C_p}{C_v}$
Nitrogen	29.07	20.76	1.40
Oxygen	29.41	21.10	1.40
Hydrogen	28.74	20.42	1.41
Carbondioxide	36.94	28.46	1.30

Γ_d = Dry adiabatic lapse rate ≃ 9.8 °C/km

Γ_s = Saturated adiabatic lapse rate ≃ 5 °C/km

Γ_A = Auto convective (homogeneous atmospheric) lapse rate ≃ 34.1 °C/km

References

1. *Introduction to Dynamic Meteorology;* J.R. Holten.
2. *Dynamic Meteorology;* Haltiner & Martin.
3. Tropical Meteorology; Riehl
4. *Some Applications of Statistics to Meteorology;* H.A. Panofsky & G.W. Barier.
5. *Univeristy Physics;* Francis W. Sears, Mark. W. Zemansky & Hugh.D. Young.
6. *Handbook of Physics*; B. Yavorsky.
7. *Our planet – The Earth*; AV. Byalko
8. *WMO Publications*: Compendium of Meteorology Vol I part 2, Compendium Lecture Notes and Technical Notes.

Index

D

E

F

G

H

I

K

L

M

N

O

P

R

S

T

U

V

W

Z

www.ingramcontent.com/pod-product-compliance
Lightning Source LLC
Chambersburg PA
CBHW050521190326
41458CB00005B/1620
9789352300006